Psychology for Sustainability

Psychology for Sustainability, 4th Edition—known as *Psychology of Environmental Problems: Psychology for Sustainability* in its previous edition—applies psychological theory and research to so-called "environmental" problems, which actually result from human behavior that degrades natural systems. This upbeat, user-friendly edition represents a dramatic reorganization and includes a substantial amount of new content that will be useful to students and faculty in a variety of disciplines—and even to people outside of academia as well.

The literature reviewed throughout the text is up-to-date, and reflects the burgeoning efforts of many in the behavioral sciences who are working to create a more sustainable society.

The 4th Edition is organized in four sections. The first section provides a foundation by familiarizing readers with the current ecological crisis and its historical origins, and by offering a vision for a sustainable future. The next five chapters present psychological research methods, theory, and findings pertinent to understanding, and changing, unsustainable behavior. The third section addresses the reciprocal relationship between planetary and human wellbeing. And the final chapter encourages readers to take what they have learned and apply it to move behavior in a sustainable direction by presenting a variety of theoretically and empirically grounded ideas for how to face this challenging task with positivity, wisdom, and enthusiasm.

This textbook may be used as a primary or secondary textbook on a wide range of courses in Ecological Psychology, Environmental Science, Sustainability Sciences, Environmental Education, and Social Marketing. It also provides a valuable resource for professional audiences of policymakers, legislators, and those working on sustainable communities.

Britain A. Scott is Professor of Psychology at the University of St. Thomas.

Elise L. Amel is Professor of Psychology and Director of Environmental Studies at the University of St. Thomas.

Susan M. Koger is Professor of Psychology at Willamette University in Oregon.

Christie M. Manning is Associate Director of the Educating Sustainability Ambassadors program and a Visiting Assistant Professor of Environmental Studies at Macalester College.

Psychology for Sustainability

4th Edition

Britain A. Scott
Elise L. Amel
Susan M. Koger
Christie M. Manning

Routledge
Taylor & Francis Group

NEW YORK AND LONDON

First published 2016
by Routledge
711 Third Avenue, New York, NY 10017

and by Routledge
2 Park Square, Milton Park, Abingdon, Oxon, OX14 4RN

Routledge is an imprint of the Taylor & Francis Group, an informa business

Third edition published 2010 by Psychology Press

Library of Congress Cataloging-in-Publication Data

Scott, Britain A.
  Psychology for sustainability / authored by Britain A. Scott, Elise L. Amel, Susan M. Koger, Christie M. Manning.
    pages cm
  "4th Edition of The Psychology of Environmental Problems."
  Includes bibliographical references and index.
  1. Environmental responsibility. 2. Environmental psychology. 3. Environmental psychology—History. I. Amel, Elise L. II. Koger, Susan M. III. Manning, Christie M. IV. Koger, Susan M. Psychology of environmental problems. V. Title.
  GE195.7.S36 2015
  363.7—dc23
  2014040334

ISBN: 978-1-84872-579-9 (hbk)
ISBN: 978-1-84872-580-5 (pbk)
ISBN: 978-1-315-72271-9 (ebk)

Typeset in ITC Stone Serif
by Apex CoVantage, LLC

Printed and bound in the United States of America by Publishers Graphics, LLC on sustainably sourced paper.

Contents

Foreword

This wonderful volume both informs and inspires us to address environmental problems with the powerful tools of psychology. I am honored to have been asked by the authors to introduce this edition by writing this Foreword, not only because I get to add my two cents, but mainly because it gives me the opportunity to celebrate the considerable progress which a fourth edition reflects. As the feminists said back in the 1970s: "We've come a long way, baby . . ."

When I conceptualized and wrote the first edition during the years of 1993–1994, there was no psychology of environmental problems, no psychology for sustainability, no conservation psychology, nor green psychology. Less than 1% of the empirical work cited in this edition had been conducted. There were very few psychologists remotely interested in environmental problems. Those who called themselves "environmental psychologists" primarily focused on built environments, such as office design and indoor lighting. And the few emerging "ecopsychologists," while writing beautifully, did not conduct empirical research.

Yet, around the world, environmentalists were hard at work in political realms promoting a sustainable world through improving the quality of water and air, protecting endangered species, recycling waste materials, cleaning up toxic waste sites, defending native forests and coral reefs, reducing greenhouse gasses, and slowing soil erosion and ozone damage. Psychologists might have been concerned about these problems as individuals (as I was), but had little to say about any of those issues as psychologists. As an indoor activity, psychology was split off from the natural world.

Against this cultural backdrop, I experienced a pivotal moment in the winter of 1988, which I described in the Introduction to the first edition:

While on a sabbatical from my teaching position, I was living in Copenhagen . . . and went to visit a friend in Hamburg. We were walking along the shore of the Elbe River one day; it was November and everything felt cold, gray, damp, and dreary. The walk took us past some beautiful Victorian homes and I tried to visualize how pleasant they would be in the summer sun, facing the water. I could see well-dressed little children in white lace, frolicking along the water's edge with their nannies looking on. So this is how the wealthy Germans live. As we continued on, my friend asked what I would like to eat for dinner that evening. I suggested fish, since here we were along the water's edge. My friend answered that fish was very difficult to get and not very good. "Why," I asked.

"Here we are so near the water." "Oh," my friend responded, "the water is dead here. It's been dead for years. Nothing grows in it." . . . Suddenly, I stepped into a new world: a world where the industrial pollutants of a city could actually kill water. Not just water in an isolated lake, but water in a big river. Now I saw the graveyard of an entire ecosystem in which not one living organism existed. No seaweed, no fish, no amoeba, nothing. Just blackness, lapping up against the landscaped grounds of the beautiful estates. What I experienced in those next steps was an important shift in my worldview. Walking along the shore path in Hamburg, I saw that the physical world made human civilization possible: those Victorian estates rested on the industrial wealth from shipping, manufacturing, and merchandising. But human civilization was destroying the physical world in return. Those beautiful estates, financed by the wealth of Hamburg's industries, must now face the deathly result of that civilization, the black liquid that laps up against their shores.

(First edition, p. xii–xiv)

So I asked myself: What does psychology have to say about this predicament in which human beings find themselves at the turn of the century? My answers at that time were almost completely theoretical—both because I am inspired by theory, but also because there were so few empirical studies to draw upon. My hope was that I might be able to show the relevance of psychology for building a sustainable world, and that I might stimulate others to take up the project.[1]

I couldn't then have imagined that within a mere 20 years, there would be such a rich store of empirical research, useful measures, important data-driven questions, insightful operational definitions, and further theoretical contributions as are so lucidly described in this fourth edition. Nor could I have hoped for a group of such talented writers to bring this new empirical work into clearly accessible and well-organized discussion, beautifully crafted for the undergraduate, but also valuable for both the curious layperson and the more experienced scholar.

In addition to informing and instructing, this volume accomplishes the even rarer but more crucial goal for an undergraduate text: inspiring. The problems this text addresses are huge and overwhelming for most people. Without clear tools, clear examples, and clear thinking, we are prone to denial, distortion, or distraction. This volume encourages and supports our sustained attention to these issues, while not underestimating their complexity.

This edition elegantly updates, elaborates, and collates recent empirical work on topics from earlier editions, but also adds new ones that contextualize and extend the psychology of sustainability. To enhance the reader's understanding of the historical context of this work, discussion has been added on William James, the history of American environmentalism, and inspiring contemporary environmental leaders. To better inform the reader of the empirical basis of psychology, a thorough presentation of research methods has been developed. To extend the coverage of psychology for sustainability, many new topics have been addressed, such as the role of positive emotion, habits, and green defaults; Big Five personality theory; self-determination theory; expectancy theory; SMART goals; sleep deprivation; and the psychological impacts of over- and inactivity, among others.

Inevitably, with the expansion of some topics, others have been dropped. Two which I hope will reappear in future editions are insights from psychoanalysis and from ecofeminism. Psychoanalysis because I am convinced that fear and anxiety drive unconscious defense mechanisms that prevent us from deeply confronting terrifying issues like climate change and resource depletion. And ecofeminism because I have been inspired by courageous women who have led us in the serious work of addressing environmental problems, be they Rachel Carson, Lois Gibbs, Joanna Macy, or the four authors of the current edition. That is not to say that there haven't also been courageous men (see Note 1), but the passion and leadership of women in this field is unassailable. Why is that? It is because the impact of gender isn't understood yet. However, to see these two hopes realized would require more empirical work in these domains. Anyone interested?

Regardless, this edition succeeds beautifully in the scholarly challenge of crafting prose that is lucid as well as accurate, informal as well as rigorous. Without dumbing down the discussion, this edition opens doors and warmly welcomes readers of all levels into a serious and inspired discussion of psychology's contributions and potentials for building a sustainable world. That was my hope for the first edition, and the authors have surpassed me many times over in this one.

I praise their informal but serious tone because we will need to invite many more scholars from many more domains into the project in the decades ahead, and let's face it, academic textbooks are not known for their warm friendliness. The respectful, rigorous, but informal voice in this volume demonstrates the authors' talents for teaching bright young scholars in liberal arts institutions (a blessing I also shared). While introducing technical terms, the discussion is remarkably uncluttered by jargon. Consequently, this volume will speak to all those interested in the question of how psychology can be useful in building a sustainable world: an environmental studies student who has never studied psychology before, or a thoughtful community leader who is considering policy issues to address environmental threats, or any curious and concerned citizen, for that matter.

Finally, this volume demonstrates that, in or out of the classroom, human beings are both agents of environmental problems as well as victims of them, we are part of, and players in, a natural world that is both brutal and beautiful, we are simultaneously evolving toward our possible demise AND our possible salvation. This scholarly work will help raise the odds of our survival on our exquisite, small, blue planet, although our survival is not at all certain.

Yet, what else is there to do, but work hard and hope that our efforts align with sustainability? I am deeply grateful to the present authors for their remarkable achievements, for the students they will inspire, and for the growing family of scholars contributing insights into the psychology of sustainability.

Deborah Du Nann Winter, Ph.D.
Professor Emeritus of Psychology
Whitman College
Walla Walla, Washington

NOTE

1. There were some courageous (male) mavericks along my path, of course. Among them, I am deeply grateful to Elliot Aronson for his early work on water conservation, who also wrote intimately and beautifully for undergraduates (and who was kind enough to write the Foreword for the First Edition); Stuart Oskamp's leadership on sustainability issues in psychology (who wrote the Foreword for the Second Edition) and David Myers' prescience in discussing sustainability as a social issue in his popular social psychology text (and who generously wrote the Foreword for the 3rd edition)—to name just a few.

Preface

This book represents the fourth edition of Deborah Du Nann Winter's text, originally entitled, *Ecological Psychology: Healing the Split between Planet and Self* (1996). The book's title and authorship has changed (twice), and the field has changed even more in the decades since its original publication. In this preface, we authors will introduce ourselves, explain the goals of the book, and describe the current edition's new organization and content.

The four of us are psychology professors who have long been interested in the ways in which our field can inform solutions to the pressing planetary concerns described in this book. Deborah's was an early voice in that regard, and her text inspired and emboldened us—and countless others—to teach classes and conduct research on the important interconnections between psychology and environmental issues.

Like many professionals, we enjoy the fruits of industrialized civilization, while struggling to reconcile what we know about the ecological crisis with how we live. Environmental devastation is driven by both greed and need. It is due to the overconsumption of global resources by the world's rich and to the desperate depletion of local resources by the world's poor. We focus more on the former in this book because we fall into that category, and we expect that most of you, our readers, do as well. We believe that people from the wealthiest nations have the most opportunity and responsibility to make crucial changes. As privileged people, we are economically secure enough to have the luxury of considering larger questions of survival than just that of our own. We tell you these things about ourselves because authors are always alive and potent in any intellectual work, no matter how stringent their attempts to be objective.

As teachers, we find that some of our students feel uncertain about the future. They have a sense of foreboding stemming from a recognition that the world is quickly approaching the limits of industrial growth. They realize that their generation, and the generations that follow, will have to pay the costs of the unsustainable practices of their predecessors. Yet, our students are also optimistic, and it is their optimism for which we are most grateful, and to which we speak in this book. In our many years of college teaching, we have seen that most of our students truly want to help the world; they seek careers not only for good money, but for good meaning; their hearts are still as open as their minds; and they ask really good questions about values, choices, and purpose.

Many scholars agree that we are at a pivotal point in human history. The future is uncertain, and forecasted scenarios are downright terrifying. We know the problems, and yet can feel paralyzed or doubtful about how to respond. That's to be expected, given the unprecedented challenges we're facing as a society and as a species. In such times, the best we can do is to identify our most deeply held values and live accordingly and with integrity; consider how we can be of service to our communities and larger world; speak for the voiceless, including future human generations and members of other species; conserve and protect what is beautiful in ourselves and the world around us; be kind and cooperative; all while pausing to appreciate successes and mourn losses, and attending to the gift of the present moment—it's the only moment over which we have any control. Martin Luther is quoted as saying, "If I knew the world would end tomorrow, I would still plant an apple tree today" (Lindberg, 2000, p. 276).

GOAL, ORGANIZATION, AND CONTENT OF THIS BOOK

Like prior editions, this text applies psychological theory and research to "environmental" problems. We think this endeavor is important because there really are no environmental problems. Rather, environmental degradation results from *human behavior*, reflecting a mismatch between how humans meet their needs and wants, and the natural ecological order. However, this edition represents a dramatic reorganization and includes a substantial amount of new content. Our intent was to make the book more "user-friendly" and practical to students and faculty in a variety of disciplines—and perhaps to people outside of academia as well. The literature reviewed throughout the text is up-to-date, and reflects the burgeoning efforts of many in the behavioral sciences who are working to create a more sustainable society. The book is organized in four sections:

Part 1: What on Earth Are We Doing?

The goal of the first three chapters is to provide a foundation for subsequent chapters by familiarizing readers with the current ecological crisis and its origins, and by providing a vision for a sustainable future.

- *There Are No* Environmental *Problems* (Chapter 1) identifies and describes some of the principal ways humans are living unsustainably; it serves as a primer on environmental science.

- *How Did We Get Here? From Western Thought to "Wise Use"* (Chapter 2) contextualizes the current ecological crisis with a historical survey of relevant cultural, technological, economic, and political developments; it represents an introduction to environmental studies.

- *Where Do We Go from Here? Developing an Ecological Worldview* (Chapter 3) proposes basic principles grounded in ecology, followed by examples of behaviors and systems compatible with these principles; this chapter is an introduction to sustainability.

Part 2: Psychology for a Sustainable Future

The second section of the book reviews psychological theory and research findings pertinent to understanding, and changing, unsustainable behavior.

- *Psychology Can Help Save the Planet* (Chapter 4) introduces readers to environmental psychology, ecopsychology, and conservation psychology, and provides an overview of methods used by researchers who study the psychology of sustainability.

- *The Power of the (Unsustainable) Situation* (Chapter 5) describes how situational and social influences shape unsustainable behavior, and how they can be enlisted to promote sustainable behavior.

- *It's Not Easy Thinking Green* (Chapter 6) explains that many ecologically problematic behaviors can be traced back to innate thinking processes that make it challenging for humans to comprehend the scope of environmental degradation and their role in it. The chapter describes how these same processes can be overcome and harnessed to encourage sustainable thinking.

- *Putting the "I" in Environment* (Chapter 7) covers a variety of individual differences among people that predict environmental behavior, including knowledge, beliefs, attitudes and values, personality, and identity.

- *To Be (Green), or Not to Be (Green) . . . It's a Question of Motivation* (Chapter 8) presents theories that describe how external and internal factors combine to influence the motivation people feel to behave sustainably (or not).

Part 3: What's Good for the Planet Is Good for Us

In the third section of the book, we address the reciprocal relationship between planetary and human wellbeing.

- *Making Ourselves Sick: Health Costs of Unsustainable Living* (Chapter 9) documents the detrimental effects of industrialized living and polluted environments on human mental and physical health.

- *Healing the Split between Planet and Self: We All Need to Walk on the Wild Side* (Chapter 10) explores theory and research supporting the idea that living close to nonhuman nature is beneficial for human development and functioning—and may, in fact, be essential for achieving optimal experiences and realizing our full potential.

Part 4: Getting Psyched for Sustainability

The goal of the final chapter is to encourage readers to take what they have learned and apply it to move behavior in a sustainable direction. *Getting Psyched for Sustainability: Being the Change We Want to See* (Chapter 11) presents a variety of theoretically and empirically grounded ideas for how to face this challenging task with positivity, wisdom, and enthusiasm.

Acknowledgments

We are grateful to the people who inspired and supported us through our efforts in developing, writing, and revising this text. We appreciate Paul Dukes at Taylor & Francis who signed this edition, and his assistant, Xian Gu, for his administrative support. Britain and Elise acknowledge the University of St. Thomas Faculty Development Center for release time and funding. Sue appreciates the sabbatical granted by Willamette University.

Thank you to Britain's students, Sabastian Boyle-Mejia, Jenna Erickson, Sean Goossens, Leon Henderson, Angela Kurth, Mark Painter, Maren Starzinski, Meghan Strauss, and Brynn Sytsma for their invaluable feedback on early drafts of the chapters. Thank you to the University of St. Thomas Psychology department chair Greg Robinson-Riegler for supporting the use of a draft manuscript in the Psychology of Sustainability course. And thank you to our academic colleagues and friends Ronald Amel, Catherine Daus, Gayla Lindt, Roxanne Prichard, and Oriel Strickland for their expertise and keen suggestions.

Many thanks to Erin Scott at Wyldehare Creative for her graphics wizardry.

Bless you, Amy Steingas, for your meticulous formatting and checking of references.

Russ, Rich, Kris, and Frank, we so appreciate everything you did to accommodate our mental, emotional, and physical absence during the lengthy and intense writing process. Without your loving support, this book would not have been possible.

Figure Credits

The following images are courtesy of Erin Scott: Figure 1.1 (p. 4), Graphic 1.1 (p. 11), Graphic 1.2 (p. 12), Figure 1.3 (p. 13), Graphic 1.3 (p. 15), Graphic 2.1 (p. 37), Cartoon 3.2 (p. 70), Graphic 3.1 (p. 71), Figure 3.1 (p. 75), Cartoon 3.3 (p. 77), Graphic 4.1 (p. 99), Figure 4.1 (p. 101), Graphic 4.2 (p. 103), Figure 4.2 (p. 104), Figure 4.3 (p. 106), Figure 4.4 (p. 107), Figure 4.5 (p. 108), Figure 4.6 (p. 109), Figure 4.7 (p. 110), Figure 4.8 (p. 112), Figure 4.9 (p. 115), Figure 4.10 (p. 116), Figure 4.11 (p. 116), Figure 4.12 (p. 118), Figure 5.1 (p. 123), Figure 5.2 (p. 124), Figure 5.3 (p. 125), Figure 5.4 (p. 131), Figure 5.5 (p. 132), Figure 5.6 (p. 134), Figure 5.7 (p. 135), Graphic 5.1 (p. 140), Graphic 5.2 (p. 142), Figure 6.1 (p. 148), Figure 6.2 (p. 149), Figure 6.3 (p. 152), Graphic 6.1 (p. 157), Figure 6.4 (p. 160), Figure 6.5 (p. 161), Figure 6.8 (p.169), Graphic 6.2 (p. 171), Figure 7.1 (p. 181), Figure 7.2 (p. 183), Graphic 7.1 (p. 184), Box 7.1 (p. 185), Figure 7.3 (p. 187), Figure 7.4 (p. 189), Figure 7.5 (p. 192), Figure 7.6 (p. 193), Figure 7.7 (p. 194), Table 7.1 (p. 197), Graphic 8.1 (p. 204), Graphic 8.2 (p. 204), Figure 8.1 (p. 206), Figure 8.2 (p. 207), Figure 8.3 (p. 208), Figure 8.4 (p. 210), Figure 8.5 (p. 211), Figure 8.6 (p. 213), Figure 8.7 (p. 213), Figure 8.8 (p. 215), Figure 8.9 (p. 218), Graphic 8.3 (p. 220), Graphic 8.4 (p. 221), Figure 8.10 (p. 222), Graphic 8.5 (p. 224), Figure 8.11 (p. 226), Figure 9.1 (p. 234), Figure 9.2 (p. 235), Figure 9.3 (p. 237), Graphic 9.2 (p. 242), Figure 9.4 (p. 245), Figure 9.5 (p. 246), Graphic 9.4 (p. 250), Figure 9.6 (p. 250), Table 9.1 (p. 252), Box 10.1 (p. 266), Figure 10.6 (p. 274), Figure 10.8 (p. 280), Figure 11.1 (p. 298), Figure 11.2 (p. 301), Figure 11.4 (p. 302), Figure 11.5 (p. 303), Figure 11.6 (p. 306), Figure 11.9 (p. 309), Figure 11.10 (p. 310), Figure 11.11 (p. 312), Figure 11.12 (p. 313), Figure 11.13 (p. 314), Figure 11.14 (p. 316), Figure A.1 (p. 320), Figure A.2 (p. 320)

The following images come from the public domain of Wikipedia/Wikimedia: Figure 1.5 (p. 18), Graphic 2.3 (p. 42), Graphic 2.4 (p. 42), Figure 2.4 (p. 48), Figure 2.6 (p.55), Graphic 2.9 (p.56), Figure 2.9 (p.60), Figure 3.4 (p.85), Figure 10.1 (p.264), Figure 10.2 (p.268), Figure 10.3 (p.269), Figure 10.4 (p.270)

Cartoon 1.1 (p. 8): with permission from Stuart McMillen

Graphic 1.4 (p. 21): courtesy of Stencilease

Cartoon 1.2 (p. 22): with permission from Kirk Anderson

Cartoon 2.1 (p. 32): with permission from Chris Madden

Cartoon 2.2 (p. 33): with permission from Mark Anderson at andertoons

Graphic 2.2 (p. 40): with permission from Retroclipart

Graphic 2.5 (p. 52): with permission from aldoleopold.org, Alamy

Graphic 2.6 (p. 52): with permission from aldoleopold.org, Alamy

Graphic 2.7 (p. 54): with permission from Shutterstock

Cartoon 3.1 (p. 68): with permission from Cartoonstock

Graphic 3.2 (p. 72): courtesy of RandomHouse

Cartoon 3.4 (p. 78): with permission from Kirk Anderson at kirktoons

Cartoon 3.5 (p. 88): with permission from Marian Henley

Cartoon 5.1 (p. 129): with permission from Cartoonstock

Cartoon 5.2 (p. 138): with permission from Cartoonstock

Cartoon 6.1 (p. 154): with permission from Universal Press Syndicate

Cartoon 6.2 (p. 173): with permission from oneworld.org Tiki

Cartoon 7.1 (p. 179): with permission from Shutterstock

Cartoon 7.2 (p. 190): with permission from Cartoonstock

Cartoon 7.3 (p. 195): with permission from Cartoonstock

Graphic 7.2 (p. 200): courtesy of Green Woman Store

Cartoon 9.1 (p. 232): with permission from Cartoonstock

Cartoon 9.2 (p. 239): with permission from Universal Press Syndicate

Graphic 9.1 (p. 240): with permission from Shutterstock

Graphic 9.3 (p. 243): with permission from Shutterstock

Cartoon 9.3 (p. 244): with permission from Cartoonstock

Graphic 9.5 (p. 254): courtesy of Environmental Protection Agency

Cartoon 9.5 (p. 256): with permission from Dave Coverly, speedbump cartoons

Cartoon 9.6 (p. 258): with permission from Cartoonstock

Graphic 10.1 (p. 276): with permission from Shutterstock

Graphic 10.2 (p. 277): with permission from Shutterstock

Graphic 10.3 (p. 284): courtesy of OBH Council

Graphic 10.4 (p. 287): with permission from Shutterstock

Graphic 10.5 (p. 289): with permission from Shutterstock

Graphic 10.6 (p. 290): with permission from Shutterstock

About the Authors

Dr. Britain A. Scott is Professor of Psychology at the University of St. Thomas where she has taught since 1996. She earned her Ph.D. in social psychology from the University of Minnesota. Britain is coauthor of a website for instructors encouraging integration of environmental issues into psychology courses: *Teaching Psychology for Sustainability* at www.teachgreenpsych.com.

Dr. Elise L. Amel is Professor of Psychology and Director of Environmental Studies at the University of St. Thomas where she has taught since 1997. She earned her Ph.D. in industrial-organizational psychology from Purdue University. Elise has successfully led efforts for systemic change at the University of St. Thomas, such as adding sustainability as a university-wide strategic priority, and providing faculty development opportunities to infuse sustainability across the curriculum.

Dr. Susan M. Koger is Professor of Psychology at Willamette University in Oregon, where she has taught for over 20 years. She earned her Ph.D. in physiological psychology at the University of New Hampshire. Sue coauthored the previous two editions of this text with Deborah Du Nann Winter and is the coauthor of *Teaching Psychology for Sustainability*.

Dr. Christie M. Manning is Associate Director of the Educating Sustainability Ambassadors program and a Visiting Assistant Professor of Environmental Studies at Macalester College. She earned her Ph.D. in cognitive and biological psychology from the University of Minnesota. Christie collaborates with nonprofits, government agencies, and local grass-roots groups to encourage household sustainability and build community resilience.

PART 1

What on Earth Are We Doing?

The goal of this first section of the book is to familiarize readers with the current ecological crisis and its origins, as well as provide a vision for a sustainable future. Chapter 1 reviews some of the principal ways humans are living unsustainably; it serves as a primer on environmental science. Chapter 2 contextualizes the current ecological crisis with a historical survey of relevant cultural, technological, economic, and political developments; it represents environmental studies. Chapter 3 presents foundational principles grounded in ecology, followed by examples of behaviors and systems compatible with these principles; this chapter is an introduction to sustainability.

CHAPTER 1

There Are No Environmental Problems

- Biology's Bottom Line: Carrying Capacity
- Overconsumption: Our Ecological Footprint
  - Energy
  - Water
  - Food
  - Material Goods
- Conclusion

The environmental crisis is an outward manifestation of a crisis of mind and spirit. There could be no greater misconception of its meaning than to believe it is concerned only with endangered wildlife, human-made ugliness, and pollution. These are part of it, but more importantly, the crisis is concerned with the kind of creatures we are and what we must become in order to survive.

(Lynton K. Caldwell, quoted by Miller, 2002, p. 1)

What will your future be like? If you are similar to your peers, you have hopes of a happy life with your family and friends.[1] You desire good physical health and your own comfortable space in which to live. You expect to own more—and better—things than you currently do. You plan to travel. And, of course, you assume you will have easy access to basic necessities like electricity, heat, food, and water.

Yet, you might also have a notion, ranging from an inkling to a grave fear, that this scenario is threatened, that your future might not be so rosy. If this has not occurred to you, just skim the local, national, and world news with your eyes peeled for stories about energy debates, toxic pollution, nuclear waste, species extinctions, water shortages, overflowing landfills, plastic gyres in the oceans, topsoil loss, over-population, and a changing climate. As you educate yourself, you will begin to realize that many aspects of our[2] current lifestyles simply cannot be taken for granted or maintained long-term, particularly for those of us who live in the United States. The sobering fact is that because of the way we are living, we are severely compromising planetary resources, and consuming them too quickly and carelessly to keep

Which of the following statements best represents you?

☐ I have not heard about environmental problems.

☐ I have heard about environmental problems, but I don't believe they are true.

☐ I have heard about environmental problems, but I am uncertain whether they are true.

☐ I believe there are environmental problems, but I am not ready to make changes because of them.

☐ I believe there are environmental problems and I would like to make changes because of them.

☐ I believe there are environmental problems and I have made some changes because of them.

☐ I believe there are environmental problems and I have made many changes because of them.

☐ I believe there are environmental problems and I have completely changed how I live because of them.

FIGURE 1.1 *Readiness to change.*
Adapted from Amel, Manning, & Scott (2009).

demand in balance with the supply. If Mother Earth had a Facebook page, her status update would be, "WTF, peeps?"

Large surveys suggest that at least some people are aware of these problems. For example, the most recent Yale survey on *Climate Change in the American Mind* found that two-thirds of Americans believe global warming is happening; about the same number think it will cause harm to future generations of people and to other species; 40% think it is a threat to themselves, their families, or their local communities; and one in three think it is already hurting people in the United States (Leiserowitz, et al., *Climate change in the American mind*, 2014). But just knowing about the problems doesn't mean people are ready to take action (see Figure 1.1).

Compared to a couple of decades ago, more people are doing little things such as recycling their newspapers, bottles, and cans, but when it comes to the big picture, most people generally behave according to established habits. A vague sense of pessimism about the future coexists with a "business as usual" attitude. For example, in spite of the fact that about half of Americans say they are "worried" about global warming (Leiserowitz, et al., *Climate change in the American mind*, 2014; Saad, 2013), most people continue to routinely drive rather than walk or bike, take vacations across the country and around the world, heat their homes to 72 degrees, use leaf blowers instead of rakes and dryers instead of clotheslines, throw usable stuff away, buy new stuff . . . and try not to think about the fact that the planet cannot possibly support all of this for much longer.

Not surprisingly, people have difficulty contemplating planetary collapse. We find it too depressing, too overwhelming, perhaps too terrifying. So, we turn our attention to present concerns such as family obligations, work or school, and paying bills. Such a response is understandable and consistent with an evolutionary perspective. Human perceptual systems evolved in an environment where threats were sudden and immediate; our ancient ancestors had no need to track gradually worsening problems that took many years to manifest. As a result, the human species is short-sighted and has difficulty responding to potentially catastrophic, but slowly developing, harmful conditions. Rather than working to prevent crises, people have a strong tendency to delay action until problems are large scale and readily apparent, at which time they attempt to respond. Unfortunately, by then, it may be too late.

Despite such hardwiring, the human species is capable of dramatic and rapid cultural evolution, as the pace of the agricultural, industrial, and technological revolutions reveals. For example, as undergraduates, the four of us authors used *typewriters* for papers after spending hours searching printed publication indexes to find citations for journal articles that we had to track down in the stacks of bound periodicals. Now, it feels normal to us to use high-speed computers, online databases with full-text pdf files, and electronic networking and file sharing tools. The point is this: Humans are quick to adapt. The human capacity for rapid change could reverse current ecological trends, given sufficient public attention and political will (Ehrlich & Ehrlich, 2008).

Many people reassure themselves that technological fixes will save the earth, but while technological and engineering expertise is certainly needed to reverse ecological damage—just as such knowledge was used to produce it—the problems that threaten the survival of life on this planet are too huge, too complicated, and too urgent to be solved by advances in technology alone. Human beings have always altered their physical environment in order to survive, but the pace and scale of current environmental change knows no precedent. And the longer people wait to take action, the worse the problems will become. Most importantly, pinning hopes on technology misses the primary cause of the current predicament and *the* crucial tool for lasting solutions: human behavior.

The theme of this book is that all so-called environmental problems are actually *behavioral* problems. Ecological systems don't have problems in and of themselves; the problems stem from people's behaviors as consumers, corporate decision makers, city planners, and governmental legislators. *Ecologically incompatible beliefs, values, worldviews, and actions* are ultimately responsible for the rapid deterioration of the natural systems on which every creature depends for survival. Thus, these problems require more than just technological solutions. As *individuals,* we need to make changes in how we satisfy our needs and fulfill our desires, how we express ourselves and our values, how we participate in our communities, how we experience our relationship to nature, and even, perhaps, in how we understand the meaning of our lives.

You probably are somewhat familiar with the contemporary ecological crisis. Still, in order to provide a foundation for the rest of the book, the following sections represent an overview of several of the big issues confronting humanity. The scope is limited in the interest of space, but should give you at least some idea of the challenges that lie ahead.

BIOLOGY'S BOTTOM LINE: CARRYING CAPACITY

It may surprise you to learn that the first scientific calculations of global warming due to human emissions of carbon dioxide were published back in 1896 (Weart, 2013). In 1914, the North American passenger pigeon was declared extinct due to hunting, yet this species had once been so abundant that flocks blackened the entire sky as they took hours to pass overhead (Blockstein, 2002). By the 1930s, negative health effects of new, toxic substances such as polychlorinated biphenyls (PCBs) were being reported in factory workers and confirmed in laboratories (Versar, Inc., 1979). Indeed, for *more than 100 years*, scientists have documented **anthropogenic** (human caused) threats to the survival of the **biosphere**, a term coined in 1875 to mean the entire global ecosystem and all of its inhabitants.

As you will read in Chapter 2, however, widespread public awareness of these problems was not raised until the 1960s. Even some scientists were slow to recognize how industrial activities such as burning fossil fuels and synthesizing chemicals were having serious systemic repercussions. But by 1992, 1,670 prestigious scientists, including over 100 Nobel Laureates, had signed a "World Scientists' Warning to Humanity," urging public attention to the "human activities that inflict harsh and often irreversible damage on the environment and on critical resources" (Union of Concerned Scientists, 1992). Today, it is very clear that the planet and all of us who reside here are in dire straits.

A fundamental concept for understanding what is happening is **carrying capacity**, a term used by biologists to describe the maximum number of any species a habitat can support. If the territory is isolated and the population cannot migrate to a new one, the inhabitants must find a way to live in balance with its resource base. Alternatively, if the population grows too quickly so that it depletes its resources suddenly, the population will crash.

Such crashes have happened in both nonhuman and human populations. Islands, which segregate ecosystems and prevent migration, provide the clearest examples. For instance, in 1944, the U.S. Coast Guard imported 29 yearling reindeer to the isolated St. Matthew Island in the Bering Sea (between Alaska and Russia). The island was ideal for the propagation of reindeer, so by 1963, the population had grown from 29 to over 6,000. However, the terrain became badly overgrazed, food supplies dwindled, and the population crashed in the winter of 1964. The island could have supported about 2,300 reindeer, but after the crash, only 3% of that figure survived (Catton, 1993).

Archaeological evidence from Easter Island, off the coast of Chile, shows that a very complex human population grew there for 16 centuries. To support themselves, the islanders cut more and more of the surrounding forests so that eventually soil, water, and cultivated food supplies were depleted. The population crashed in the seventeenth century, falling from 12,000 in 1680 to less than 4,000 by 1722. In 1877, only 111 people still survived (Catton, 1993). On the mainland, some human population crashes have been hastened by the fact that societies weakened by resource shortages become more vulnerable to being wiped out by other humans. However, the Sumerians of Mesopotamia and the Maya of the Yucatan region provide two clear examples of crashes due simply to exceeded carrying capacity.

The Sumerians were the first literate society in the world, leaving detailed records of their civilization and its decline between 3000 and 2000 B.C. The complicated agricultural system that supported their population also depleted the quality of their soil through salinization and siltation. In the words of environmental historian Clive Ponting (1991):

> The artificial agricultural system that was the foundation of Sumerian civilization was very fragile and in the end brought about its downfall. The later history of the region reinforces the point that all human interventions tend to degrade ecosystems and shows how easy it is to tip the balance towards destruction when the agricultural system is highly artificial, natural conditions are very difficult, and the pressures for increased output are relentless. It also suggests that it is very difficult to redress the balance or reverse the process once it has started.
>
> (p. 72)

Similarly, the Maya, who developed what are now parts of Mexico, Guatemala, Belize, and Honduras, built a complex civilization on the fragile soil of tropical forests. Clearing and planting supported a population from 2000 B.C. to A.D. 800. As the population grew, land was not allowed time to recover between plantings. The deforestation caused significant changes in weather patterns, further reducing crop yields. In about A.D. 800, the population crashed; within a few decades, cities were abandoned, and only a small number of peasants continued to survive in the area.

In the past, population crashes have occurred in one part of the world without seriously affecting people in other regions. Today, however, the threat of a crash on a *planetary level* is looming (Diamond, 2005; Kuntsler, 2005). The earth is essentially a large island, with no way to import resources or immigrate to a less degraded habitat. Like isolated geographical regions, the earth itself has its limits.

Human population growth is accelerating at an exponential rate, which makes the problems even more pressing. **Exponential growth** is deceptive because it starts off slowly, but accelerates quickly. It occurs when a quantity increases by a fixed percentage of the whole, which means that it will double after a certain interval, rather than grow incrementally (i.e., linear growth). The concept of exponential growth is so important to understanding the current predicament that it's worth spending a moment with a conceptual example. Imagine that you have a bottle with one bacterium in it, which will double every minute. Assume that it is now 11:00 p.m. and the bottle will be completely full by midnight. When will the bottle be half full? If you suggest 11:30, you are thinking in terms of linear growth, rather than exponential growth. Actually, the bottle will be half full at 11:59 because the bacteria double every minute. Next question (and this involves a little more imagination): When do you think the bacteria might start to notice that things are getting a little crowded? Probably not even at 11:55, because at this point, the bottle is still only 3% full. Remember, exponential growth begins slowly but accelerates quickly. Final question: Suppose the Royal Bacteria Society sponsored Sir Francis Bacterium to leave the bottle and go exploring for new space, and suppose Sir Francis got really lucky and found three new bottles, quadrupling the space for

With their habitat's resources stretched, the population
had been unable to survive the extremes of winter.

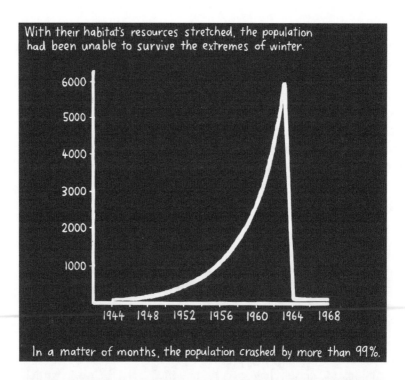

In a matter of months, the population crashed by more than 99%.

The island's untapped natural resources became the reindeer's
source of prosperity, and also the seeds of their demise.

The island was only so big, its resources only so many.

How big is our island?

their society. How much time did he buy? Although it may seem at first that their problems are solved, it would actually only give them two more minutes until all four bottles were completely full.[3]

The human population picture is startling because the past two centuries have seen such quick acceleration. At mid-seventeenth-century rates, it would have taken 250 years to double the world's population, but the most recent doubling has occurred in just the 45 years since 1968, which is when Stanford University professor Paul Ehrlich published his best-seller, *The Population Bomb* (see Figure 1.2). Although industrialized countries have managed to bring birthrates down, population continues to swell in the world's less developed countries, which are home to four-fifths of the planet's human beings. Contrast the rate in the United States, which hovers around 2.1 births per woman, to the average of 5.2 births per woman in the countries of sub-Saharan Africa (with at least ten of these countries averaging six or more) (Haub & Kaneda, 2013). At present rates, the population of the 51 countries in sub-Saharan Africa is expected to *more* than double by 2050, and by that time, in spite of decreasing birth rates in the developed countries, the world's population will have increased from 7.2 billion (at the time of this writing) to 9.7 billion (Haub & Kaneda,

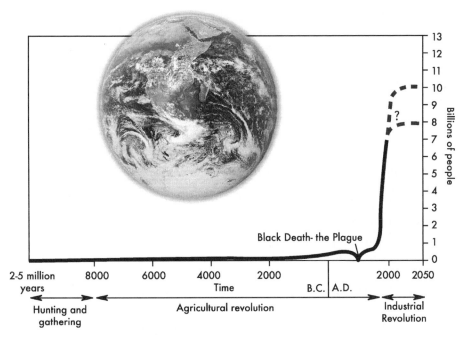

FIGURE 1.2 *World population curve.*

The J-shaped curve of exponential world population growth, with projections to 2050 (not to scale). The current world population of 7.2 billion is projected to reach eight to ten billion this century. Data from World Bank and United Nations; photo courtesy of NASA. (Adapted from Miller, 2007, p. 6).

2013). Like the bacterium example, finding new supplies of food or energy or even a new planet or two would only very temporarily alleviate the strain.

So, has the human population reached the earth's carrying capacity? An important distinction here is the difference between *biological* carrying capacity and *cultural* carrying capacity.

> [T]he question "How many people can the earth support?" cannot be answered using only ecological concepts. Human choices influence the earth's human carrying capacity along with natural constraints. . . . The question is obviously incomplete. Support with what kind of life? With what technology? For how long? Leaving what kind of earth for the future?
>
> (Cohen, 1995, pp. 17, 166)

Estimates of how many people the earth can reasonably support have varied wildly. A review of 69 studies found a range from 500 million to 1×10^{21} billion(!) with a "best-point estimate" at just under eight billion, a level we have nearly reached already.

The highest estimates rely on the hypothetical idea that the billions and billions of people would have a very simple standard of living, probably akin to our Stone Age ancestors. When it comes to human populations, cultural carrying capacity is considerably less than biological carrying capacity, so long as we use more resources than are necessary to keep us alive and reproducing. For example, energy is the most basic resource; its consumption is measured in terms of the daily number of kilocalories (i.e., "calories") that people use to meet their needs. It is estimated that ancestral humans living hunter–gatherer subsistence lifestyles each required a daily average of about 2,000 calories (in the form of food) for all their activities. Today, those of us living in the developed world consume roughly the same amount of dietary calories each day,[4] but require *an additional 600,000 calories of energy per person* for the machines and systems that support our lifestyles (Miller & Spoolman, 2012). Clearly, some of us are using more than our fair share of what the planet has to offer. And by doing so, we are the ones exacerbating the divide between biological and cultural carrying capacity.

Another way of conceptualizing cultural carrying capacity is with the formula for environmental impact used by population scientists (Ehrlich & Ehrlich, 1991; Ehrlich & Holdren, 1971):

$$\mathbf{I} = \mathbf{P} \times \mathbf{A} \times \mathbf{T}$$

where
I is the impact of any group or nation
P is population size
A is per capita affluence, as measured by consumption
T is technology employed in supplying that consumption

With this formula, one can see that doubling a population will double its impact if affluence and technology remain constant; it is likely that this is what will happen in the countries in sub-Saharan Africa over the next three decades. One can also see how it is the case that the 20% of people living in the affluent, high-tech, industrialized world are actually having a much bigger ecological impact than the 80% living in the developing world.

To see population as only a Third World problem is a fallacy. Population in the United States and other industrialized countries must continue to decrease significantly if current levels of resource consumption do not. Or, if population does not decrease, resource consumption *will*, either systematically with planning or suddenly through ecological collapse.

To see a real-time world population ticker, visit
www.worldometers.info

OVERCONSUMPTION: OUR ECOLOGICAL FOOTPRINT

One way to quantify consumption is with the use of the **ecological footprint,** a measure of how fast people (individuals or populations) use resources and generate waste in comparison to how rapidly nature can absorb the waste and replenish the resources (Wackernagel & Rees, 1996). Variables include food, housing, home energy and transportation, goods, and services. The footprint is reported as an estimate of the hypothetical amount of land and ocean needed to support the lifestyle, either in hectares or in terms of the number of planet earths that would be required if everyone lived that way.

Globally, human demand is exceeding the regenerative capacity of the planet by about one half; that is, collectively, humans are consuming resources and generating waste so rapidly that it would take one and a half earths to keep pace over time. This represents a striking increase since the early 1970s, which is when the world's ecological footprint first started exceeding biocapacity.

83% of the world's population now live in countries that use more biocapacity to support production activities than they have available within their boundaries. The deficit is covered through the overexploitation of domestic natural capital stocks (e.g., through overharvesting and overfishing), net import of resources, and the use of the global commons (for instance by emitting CO_2 from fossil fuel into the atmosphere).

(Galli, Wackernagel, Iha, & Lazarus, 2014, p. 125)

As a group, those living in the United States have the largest ecological footprint, consuming more resources and generating more waste than any other country on the planet.[5] If everyone in the world lived like U.S. residents do, it would take *four* planet earths to keep up (Grooten, 2012). Yet, there is only one.

So how are humans overconsuming and overproducing waste? In the following sections, we will review some data on primary resources measured by the ecological footprint: energy, water, food, and material goods.

Energy

When you think of your own energy use, what comes to mind? Stop for a minute or two and try to identify all the ways you consume energy in your daily routine. (We'll wait. . .) Okay, what's on your list? The electricity you use to charge your phone? Gas for your car? You probably thought of the power to run your computer—but did you think of the power consumed by all the servers and routers that fire up every time you access the Cloud? Maybe you thought of the fuel you use to cook food, but did you think of the energy it took to produce that food and get it from the farm to your grocery store?

The energy used in homes and for personal transportation may be easy to identify, but the bulk of what we (indirectly) consume is essentially invisible to us as

Calculate YOUR Ecological Footprint at
www.footprintnetwork.org

individuals; it includes energy used in manufacturing, agriculture, mining, and construction; fuel for trucks, trains, planes, boats, and barges that carry goods; and power used in public buildings such as shopping malls, offices, hospitals, schools, and restaurants.

When you were making your list of ways you use energy, did you happen to ponder the *sources* of energy you tap? Like the amount of energy we use, the origin of our energy is something about which many of us remain in the dark (pun intended). One of us authors once overheard a conversation in which a person was confronting another about wasting electricity. It went something like this: "You really shouldn't waste electricity." "Why not? What does it matter?" "Well, where do you think electricity comes from?" "I don't know—the air?" Hopefully, you are not quite as uninformed as this naive individual was. Still, do you know where your energy comes from?

According to the U.S. Energy Information Administration (2013b), the primary energy sources in the United States are petroleum (oil), natural gas, coal, nuclear, and renewable energy, in that order. Electricity is a secondary source that is generated from these primary sources (see Figure 1.3). Only about 1% of electricity in the United States is generated by oil, but oil is responsible for 93% of the energy consumed by various forms of transportation, and 34.7% of energy overall. Using oil as a case example, it is clear why it would take four planets if everyone lived like those of us in the United States do.

Although it is home to less than 5% of the world's population, the United States uses over 20% of the world supply of oil, 18,490,000 barrels *per day*. The next largest consumer is China, which uses 11.5% of the world's oil, but also is home to nearly 19% of the world's population (that is, the United States has only one-quarter the population of China, but uses nearly twice the fuel). Seventy percent of oil consumed in the United States is used for transportation fuels such as jet fuel and gasoline (which accounts for about half of the oil used for transportation). Per person, U.S. residents guzzle about four-and-a-half times as much gasoline as people living

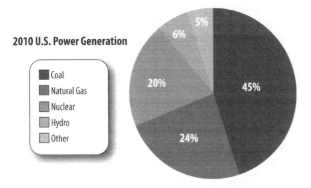

2010 U.S. Power Generation

- Coal
- Natural Gas
- Nuclear
- Hydro
- Other

FIGURE 1.3 *Sources of electricity in the United States.*
Adapted from the U.S. Energy Information Administration.

in the United Kingdom and Germany, 22 times as much as people in China, and 92 times as much as people living in India (The World Bank, 2014).

So far, we have been using the terms "consume" and "use" interchangeably, but consumption can actually be split into two categories: energy that is actually used and energy that is wasted. In fact, *the majority of energy consumed in the United States is wasted*, about 61% according to the latest estimate generated by the Lawrence Livermore National Laboratory (Fischer, 2013). This number is based on data about the efficiency of car engines, heating systems, light bulbs, and the like; but, it is undoubtedly an underestimate, given that it does not take into account **behavior-related inefficiencies** such as idling the car motor, keeping the thermostat set to "comfy" when away from home, and leaving the lights on in an empty room. Nor does it address the preventable waste that occurs when consumers select energy inefficient automobiles, furnaces, and appliances, when more efficient choices are available.

Not only do we waste energy, we *produce* waste in the generation of energy. Recall that the ecological footprint calculation is based both on consumption and on the creation of waste that must be absorbed by the biosphere. Different energy sources produce different by-products. Two of the most concerning are nuclear waste and carbon dioxide (CO_2).

Nuclear power plants provided 12.3% of the world's electricity in 2012 (Nuclear Energy Institute, 2014). The United States leads the world in the amount of nuclear power generated, almost twice as much as the second runner-up, France. The portion of U.S. electricity coming from nuclear reactors has remained at about 20% since 1990 (U.S. EIA, 2013), compared to 75% for France. Thirty-one countries have at least one nuclear reactor, and 13 of these countries currently rely on nuclear energy to supply at least one-quarter of their total electricity (Nuclear Energy Institute, 2014). Around the world, there are 436 nuclear reactors in operation and 70 more are under construction (World Nuclear Association, 2013).

Nuclear power is often touted as "clean" energy because of the fact that it does not produce CO_2 or emissions that contribute to air pollution and acid rain. However, the waste it does produce is particularly and immediately hazardous to all life forms, and cannot be absorbed by the biosphere—at least not at a reasonable pace. The waste created by nuclear fission of uranium is **radioactive**, meaning that it emits subatomic particles that can easily penetrate living tissue, causing cell damage that may manifest as radiation sickness (nausea, weakness, hair loss, skin burns, organ failure) or death in cases of high-level acute exposure, and can lead to cancer or genetic mutation in cases of chronic exposure.

The most significant high-level waste from a nuclear reactor is the used nuclear fuel left after it has spent three years in the reactor generating heat for electricity. Low-level waste is made up of lightly-contaminated items like tools and work clothing from power plant operation and makes up the bulk of radioactive wastes. Items disposed of as intermediate-level wastes might include used filters, steel components from within the reactor and some effluents from reprocessing. High level wastes make just 3% of the total volume of waste arising from nuclear generation, but they contain 95% of the radioactive content.

(World Nuclear Association, 2014)

The radioactivity of high-level wastes takes from 1,000 to 10,000 years to decay to the amount present in the original uranium ore. And some of the radioactive isotopes present in this waste take *hundreds of thousands of years* to decay. How much of a hazard this presents "depends on how concentrated it is" (World Nuclear Association, 2012).

We are exposed to radiation every day from natural and man-made sources. Estimates of the yearly dose to which people are exposed range from around 100 to 600 millirems (1000 millirems = 1 rem). Whether this low-level gradual exposure contributes to health problems has not been established; what is known, however, is that sudden exposure to higher amounts is dangerous. According to the U.S. Nuclear Regulatory Commision (2007),

High-level wastes are hazardous to humans and other life forms because of their high radiation levels that are capable of producing fatal doses during short periods of direct exposure. For example, ten years after removal from a reactor, the surface dose rate for a typical spent fuel assembly exceeds 10,000 rem/hour, whereas a fatal whole-body dose for humans is about 500 rem (if received all at one time). Furthermore, if constituents of these high-level wastes were to get into ground water or rivers, they could enter into food chains. Although the dose produced through this indirect exposure is much smaller than a direct exposure dose, there is a greater potential for a larger population to be exposed.

Clearly, the question of how and where to store spent nuclear fuel for tens of thousands of years is paramount.

Currently, spent fuel is stored on-site at nuclear power plants. Some is stored in pools of water which cool the fuel and act as a radiation shield. The rest is stored outside in steel or concrete casks. Both of these methods are considered temporary, but as of yet, there are no permanent storage facilities for high-level waste. The most popular vision for how to isolate this waste for the long, long term is to dig **deep geologic depositories**, sequestering the waste 1,000 feet or more underground. Concerns about this method include the potential risk of seismic activity and the problem of transporting the waste from the temporary storage facilities to the permanent one. Picture the outcome if a semitruck carrying spent fuel jackknifed on the highway.

Accidents occurring during the transportation of nuclear waste are not the only potential ecological and public health calamity associated with this energy source. The other scenario is a mishap at the nuclear plant itself. In fact, accidents have been happening at nuclear power plants since 1952. The International Atomic Energy Authority, which does not actually keep a comprehensive record of all accidents, uses a 1–7 rating scale, where 1 = anomaly, 2 = incident, 3 = serious incident, 4 = accident with local consequences, 5 = accident with wider consequences, 6 = serious accident, and 7 = major accident. In the past 50 years, there have been at least 33 serious incidents and accidents, the most significant being the level 7 explosion at Chernobyl in 1986 (which was caused by unforeseen complications during a test of the plant's cooling systems), and the most recent being the meltdown of three reactors at Japan's Fukushima plant in 2011 after they were flooded by an earthquake-caused tsunami (Rogers, 2011).

You may be aware that the majority of the world's electricity (about two-thirds) is generated by burning fossil fuels, primarily coal and natural gas. These sources create waste in several ways, beginning with the processes used to extract them. For example, after it is mined, coal is washed in a chemical bath in a **coal impoundment** pond, producing slurry, which contains dangerous heavy metals such as mercury, lead, and arsenic, as well as a variety of carcinogenic compounds. Impoundments cover acres and can contain millions of gallons of sludge. When a coal impoundment fails (which happens with some regularity), the spill surges into nearby streams, rivers, and lakes, threatening wildlife and contaminating drinking water (The National Academies, 2002). At the time of this writing, such a spill has just happened in West Virginia, leaving about 100,000 residents without safe drinking water and prompting the governor to declare a state of emergency (Nicks & Stout, 2014).

The form of fossil fuel waste that gets the most attention, however, is the CO_2 that is produced when these energy sources are *combusted* to provide heat for buildings, and power for vehicles, machinery, and home appliances. Unlike the situation with radioactive nuclear waste, the planet *does* have a system for naturally absorbing CO_2 (i.e., through the respiration of plants). The problem is that humans have released *extra* carbon into the atmosphere that would otherwise have remained trapped underground. Compounding the problem, humans have also diminished the earth's capacity for absorbing carbon by dramatically decreasing the size of the planet's green spaces. Carbon dioxide (along with methane, nitrous oxide, and water vapor) traps heat in the atmosphere. The resulting **greenhouse effect** is necessary to stabilize atmospheric temperatures and maintain a climate suitable for life on the planet. Gas levels vary naturally to some extent, but burning fossil fuels has created an unprecedented increase, which is clearly correlated with planetary warming patterns (see Figures 1.4 and 1.5).

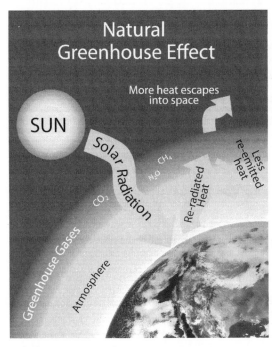
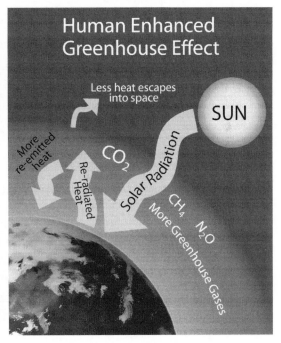

FIGURE 1.4 *The greenhouse effect.*
Naturally occurring greenhouse gases trap some of the sun's heat, keeping the earth warm enough to support life, but human activities, such as burning fossil fuels, are increasing greenhouse gas levels, leading to more heat being trapped in the atmosphere, thus warming the planet.
Courtesy of National Park Service

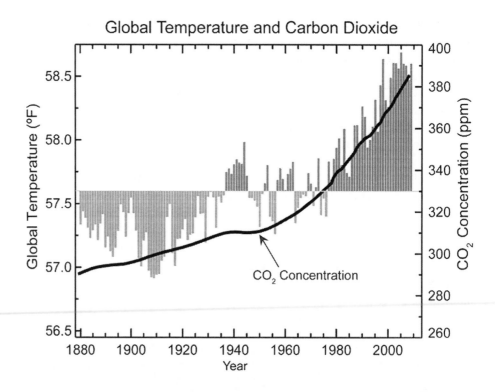

FIGURE 1.5 *Global temperature and CO$_2$ concentrations since 1880.*
Source: National Oceanic and Atmospheric Administration

Some people dismiss the idea of anthropogenic global warming, because they believe the documented increase in temperature is just natural fluctuation. Others are skeptical about the idea that the planet is getting hotter because they can point to examples of record low temperatures. So as to evaluate these positions, it can be helpful to keep a few things in mind. First, in the words of the National Oceanic and Atmospheric Administration (2014),

> The globally-averaged temperature for 2013 tied as the fourth warmest year since record keeping began in 1880. It also marked the 37th consecutive year with a global temperature above the 20th century average. . . . all 13 years of the 21st century (2001–2013) rank among the 15 warmest in the 134-year period of record.

Second, it is more descriptive to use the term **climate change** than "global warming," because among its many effects, the planetary rise in temperature is causing shifts in climate regions and more extreme weather events. Third, the scientific debate regarding the reality of human-caused climate change effectively ended in 1995 with the first report from the Intergovernmental Panel on Climate Change (IPCC), an international coalition of 2,000 eminent climatologists and other scientists. In its most recent report, the IPCC stated,

> The atmospheric concentrations of carbon dioxide, methane, and nitrous oxide have increased to levels unprecedented in at least the last 800,000 years. Carbon dioxide concentrations have increased by 40% since pre-industrial times, primarily from fossil fuel emissions and secondarily from net land use change emissions.
>
> (IPPC, 2013, p. 9)

The report unequivocally documents simultaneous increases in atmospheric and ocean temperatures, decreases in arctic ice and snow cover, and rises in sea levels. The authors conclude: "It is *extremely likely* that human influence has been the dominant cause of the observed warming since the mid-20th century" (IPCC, 2013, p.15; the italics are theirs). Anyone who tells you that there is significant disagreement among scientists about climate change caused by burning fossil fuels is simply ignorant of the facts.

Water

Energy isn't the only resource that is being overconsumed by people living in the industrialized world. Like other animals, humans need water to drink, but these days, we also use water for daily showers, lawn sprinklers, car washes, hot tubs, swimming pools, decorative fountains, and water parks. We use it in industrial processes—intensive agricultural irrigation and to cool spent nuclear fuel rods and wash coal. And we use it to create artificially lush habitats for ourselves, such as the luxurious cities of Las Vegas and Dubai (which, in case you don't know, are located in arid deserts). The thing is that there is only so much water.

Coleridge's (1798) ancient mariner was prescient when he lamented, "Water, water every where, nor any drop to drink." For water to be drinkable, it must be *fresh*water rather than saltwater. Freshwater comes from rivers, lakes, human-built reservoirs, and underground aquifers. Aquifers are refilled by rainfall seeping into the ground. But rain tends to run off rather than seep when it hits overgrazed turf, desiccated cropland, or pavement. Surface runoff causes erosion, siltation, and flooding rather than aquifer replenishment. Runoff also picks up pollutants such as pesticides, motor oil, and trash on its way to lakes and rivers. Coastal aquifers are susceptible to saltwater contamination when their levels drop, making them unsuitable sources of drinking water without expensive and energy-intensive desalination.

Currently, more groundwater is being withdrawn than is replenished—some for drinking, but mostly for nonessential purposes. The combined direct and indirect water consumption of the average U.S. resident is estimated to be the equivalent of about 33,000 glasses a day (waterfootprintnetwork.org). A combination of increasing demand on water supplies, decreasing precipitation due to unusual weather patterns, and disruption of the water replenishment cycle is resulting in widespread and dangerous droughts. At the time of this writing, the governor of California has just declared a statewide drought emergency; 99% of the state is abnormally dry. The year 2013 was the driest on record, and San Francisco had less rain than any year since the first records were kept during the 1849 gold rush (Myers, 2014). News reports are highlighting the heightened risk of wildfires and increasing smog over Los Angeles. Meanwhile, a trend is developing in Texas, which is also experiencing its worst drought in recorded history, where affluent residents are digging private wells, tapping into the region's groundwater supply specifically to keep their sprawling lawns green without paying for city water (Satija & Root, 2013). Also in the news this month was the worsening of the drought that has plagued the Colorado River for the past 14 years. The Colorado supplies most of the water consumed in the Southwestern United States (Wines, 2014).

At current rates of population growth and industrial development, serious worldwide water shortages are a very real probability within the next decade; already more than one billion people in the developing world do not have regular access to safe drinking water. The 2014 World Economic Forum's Global Risk Report identified water crises as the third most troubling concern (after economic crises and high unemployment/underemployment) to the more than 700 government, business, and nonprofit leaders they surveyed. Water shortages threaten food supplies, energy production, and of course human health. Yet those of us in the developed world pollute and waste gallons of drinkable water every time we flush the toilet.

When your water goes down the drain, what happens to it? Just as the original source of your water is a local lake, river, or underground aquifer, so, too, is its ultimate destination (unless you live by the ocean, in which case it may be headed there). But, this isn't a direct loop if you live in the developed world. Unless you draw water directly from a well or spring, the freshwater you consume is treated before it comes to your home, and some of it is treated after it leaves your home. However, water run-off into storm drains usually isn't treated, so any drainage from your lawn, driveway, or street that is contaminated with oil, antifreeze, or trash passes untreated into the waterways—and back into the water supply.

The thought of drinking from a river without using a filter probably gives you pause, but the fact is there was a time in the not-so-distant past when drinking directly from streams, lakes, and rivers was not completely out of the question. These water sources can naturally contain microorganisms that cause human sickness, but before the days of intensive agriculture (think cow manure in the creek), this was not such a big threat. And today, we are concerned about numerous additional contaminants in the water, such as chemical-laced runoff from farm fields, industrial effluent from factories, and human sewage from towns and cities. Add to this all of the nonbiodegradable and toxic substances people rinse down their household drains (e.g., cleaning and grooming products), and you can see why the U.S. Environmental Protection Agency estimated back in 2000 that 40% of the rivers in the country were too polluted for fishing and swimming—let alone drinking (EPA, 2012b).

The fact is that water treatment doesn't actually remove all of these contaminants. Many of them make their way into natural bodies of water where they become hazards for wildlife first, and then for humans as the water cycles back through the system. Ironically, some water treatment processes themselves end up contributing to water pollution. For example,

> [The] unintended side effect of chlorinating water to meet federal drinking water regulations creates a family of chemicals known as trihalomethanes. . . .
>
> Scientists suspect that trihalomethanes in drinking water may cause thousands of cases of bladder cancer every year. These chemicals have also been linked to colon and rectal cancer, birth defects, low birth weight and miscarriage.
>
> (Sharp & Pestano, 2013)

And don't kid yourself by thinking well water or spring water are safe alternatives; if you are drinking from a well or spring, whatever has managed to seep into the underground aquifer will end up in your glass—or your bottled water. The imminent water shortages are not just due to disruptions of the hydrologic system through overextraction from aquifers, but also abuse of the water we have.

Food

Here is one more water statistic for you: Worldwide, 70% of freshwater consumption is for crops and livestock (United Nations, 2014), and that number reaches 80% in the United States (U.S. Department of Agriculture, 2013). Industrial agricultural practices, such as those which dominate American farming, are water-intensive, energy-intensive, and dependent on fossil fuels, not just to power farming equipment, but for manufacturing petrochemical pesticides and fertilizers. When it developed in the first half of the twentieth century, "industrial agriculture was hailed as a technological triumph that would enable a skyrocketing world population to feed itself"; but now, "a growing chorus of agricultural experts—including farmers as well as scientists and policy makers" recognizes that industrial agriculture as currently practiced won't work long-term (Union of Concerned Scientists, 2012). Growing single crops, such as corn or soybeans, in the same fields year after year taps the soil of the specific nutrients the plants need, thus requiring the addition of chemical fertilizers. Chemical pesticides are also necessary because **monocultures** are less pest resistant than biologically diverse plantings. There are natural ways to fertilize soil and deter pests, but the massive scale of industrial farming makes these methods impractical.

Monoculture plantings not only deplete the soil of nutrients, they also make fields susceptible to erosion by water and wind. This means that chemical fertilizers and pesticides don't just affect the fields to which they are applied, they affect plants, animals, and water in the region as they run off, or blow off, the farm fields. With the runoff also goes the **topsoil**, the top few inches of soil, which is the home of the organic matter necessary for growing food. Once topsoil is lost, it can only be replenished by the breakdown of more organic matter. When fields are never left fallow (unplanted, with decaying remnant crops and weeds), and organic fertilizer, such as manure or compost, is never applied, there is no topsoil restoration. Topsoil loss is a serious problem today. Half of the topsoil in the world has been lost in the

Source: Reprinted with permission of Kirk Anderson.

last century and a half (World Wildlife Fund, 2014). Some experts say that at current rates of loss without replacement, the world has about 60 years of topsoil left (World Economic Forum, 2012).

Perhaps you are thinking, "Okay, I get the point that I could probably reduce my consumption of electricity, transportation fuel, and nonessential water, but I gotta eat!" True, we have to eat to live, but we make choices about *what* we eat, *where* we get it, and how much of it we *waste*.

Do you know where your food comes from? How often do you eat food that is out of season or does not grow in your region? One study estimated that, on average, food in the United States travels more than 1,500 miles from its source to our plates (Pirog & Benjamin, 2003). Perhaps you have heard the increasingly popular catchphrase "Eat Local." In truth, distance transported from farm to table is only one factor in a food's environmental impact, and it is a relatively small contributor compared to the production phase (Weber & Matthews, 2008).

Although industrial agriculture in general has negative environmental impacts, such as those described above, industrial production of *meat* is the most detrimental. In industrial meat production, animals are prepared for slaughter by confinement for several weeks in large **concentrated animal feeding operations** (CAFOs). Some CAFOs are outdoor "feedlots" devoid of vegetation, and some are windowless buildings. Because of dense crowding, accumulation of wastes, and unnatural diets, diseases and other health problems flourish; so, farmers must use chemicals and antibiotics to keep (most of) the animals alive. Routine use of antibiotics ultimately produces strains of antibiotic-resistant bacteria, as these are the bacteria that survive and reproduce unchecked by the medications. Animals in CAFOs are fed a grain-based diet (mostly corn) to fatten them up, even though grain is not the natural food of cows, pigs, or poultry. In fact, about 80% of all corn grown in the United States is consumed by livestock, poultry, and farmed fish (National Corn Growers Association, 2013). This means that meat production relies indirectly, but heavily, on the use of chemical pesticides and fertilizers, and contributes indirectly, but significantly, to the loss of topsoil.

Livestock farming produces more greenhouse gases than transportation. For some perspective, consider that researchers have estimated the amount of CO_2 emitted in the industrial production of just one kilogram (2.2 pounds) of beef to be the equivalent of the amount emitted by an average European car driven 155 miles (Bittman, 2008). Half of the methane and two-thirds of the nitrous oxide that humans release into the atmosphere can be attributed to crop and livestock production (Food and Agriculture Organization of the United Nations, 2013). These greenhouse gases are, respectively, 25 and 300 times more potent than CO_2 (Worldwatch Institute, 2011). At the same time, the planet's ability to absorb CO_2 emissions is being reduced by deforestation for the purpose of ranching. The tropical rainforests, which play a major role in global **carbon sequestration**, the removal and storage of atmospheric CO_2, are rapidly disappearing. Ranchers in South America routinely clear areas of rainforest to create cheap pastureland, which is only usable for a limited time, thus requiring continued clearing. Eighty percent of the rainforest in the Brazilian Amazon has been cleared to support cattle ranching (Rainforest Alliance, 2012).

Worldwide meat consumption has tripled since the 1960s, increasing by 20% in just the first decade of this century, and this is largely due to the spread of industrial meat production methods across the globe (Worldwatch Institute, 2011). Meat production and consumption is expected to continue to rise by at least 30% by midcentury, with the most dramatic increases happening in the world's two most highly populated countries, China and India (Deutsche Welle, 2014).

Industrial food production includes more than the agricultural phase. Seventy percent of the average American diet consists of **processed food**, manufactured products that you could not make with the same ingredients in your home kitchen (Warner, 2013). Processed food has undergone energy-intensive (and sometimes polluting) steps that involve the addition of chemicals to enhance flavor, preserve texture, and increase shelf-life. At least 5,000 known food additives are used in the United States, and this figure is generally considered an underestimate because it is left up to the food industry itself to tell the Food and Drug Administration about new ingredients; in other words, there is no mandated testing of chemical food additives in the United States and the only regulation is "self-regulation" (Warner, 2013). Food additives are in nearly everything available in a typical grocery store and in most restaurants, even many items that do not look "processed," such as the whole chicken breast on your sandwich (Warner, 2013). Health implications of processed foods are discussed in Chapter 9.

Earlier in this chapter "consumption" was divided into actual use and waste. When it comes to food, this distinction is particularly poignant: About 39% of the edible food supply in the United States is wasted (Stokstad, 2009). The amount of food U.S. citizens waste *each day* could essentially fill the 90,000 seat Rose Bowl stadium (Bloom, 2010). This amounts to 36 million tons per year, nearly all of which ends up in landfills (U.S. Environmental Protection Agency, 2013b). Worldwide, that number is about 1.3 *billion* tons, which is about one-third of the edible supply. Waste varies across food categories and income levels; for example, in high income regions of the world, about 67% of *meat* is wasted (FAO, 2013). Food waste is obviously a concern in terms of human welfare and economics, but it is also a significant problem from an environmental perspective. In the first study to examine food waste in terms of its ecological footprint, the Food and Agriculture Organization of the United Nations stated that, with regard to worldwide CO_2 emissions, "food wastage ranks as the third top emitter after the USA and China" (FAO, 2013).

Food waste makes up nearly 15% of total municipal solid waste in the United States (EPA, 2014d). Add to this the containers and wrappers in which food (especially processed food) is packaged and served, most of which are not reusable or recyclable. Containers and packaging constitute another 30% of the municipal solid waste in the United States. Of course, not all of this packaging is for food items; a good portion of the packaging waste is from material goods, the topic to be addressed next.

Material Goods

The United States produces more solid waste than any other country, about 30% of the global total; this amounts to more than 4.5 pounds of garbage per person per day (Rogers, 2005). The amount of trash Americans generated in 2009 could circle the earth 24 times (Keep America Beautiful, 2013). Perhaps you are skeptical that you personally generate nearly five pounds of solid waste per day. Keep in mind that your daily garbage does not just include the things you physically throw out; it includes the waste created in the manufacturing, packaging, and distribution of products you consume, before those products ever get into your hands. It takes as many as 70 cans of garbage to make the stuff that we throw out in one garbage can (www.storyofstuff.org).

Where does waste go after we throw it away? Well, there really is no "away," even though people act as if there is. Instead, most solid waste goes to **landfills** or **incinerators**, both of which pose ecological threats. Not only do landfills claim acreage, they release emissions that cause air pollution and leach toxicants that pollute groundwater. According to the Environmental Protection Agency (2013c), landfills are the third-largest source of human-caused methane emissions in the

United States.[6] When precipitation percolates through the contents of a landfill, it creates **leachate**, a liquid mixture that carries contaminants to local bodies of water. Today, the negative effects of leachate are being reduced somewhat by the use of liners or sites that are geologically impermeable, but these methods have not eliminated the problem. When waste is incinerated, chemical materials combine and produce new chemicals that are released into the air. These chemical compounds are known to pose a variety of human health risks including reproductive and developmental abnormalities, immune system damage, and cancer. More specific information on how toxicants affect human development and functioning is presented in Chapter 9.

Earlier in this chapter, you learned about the hazardous waste created in the production of nuclear power, but there are many other substances and materials used every day that also qualify as hazardous, and should never make their way into landfills or incinerators, or be poured down the drain, dumped on the ground, or washed down storm sewers (Environmental Protection Agency, 2014b). These include paints, household cleaners, batteries, lawn chemicals, used motor oil, antifreeze, and compact fluorescent light bulbs (because they contain mercury). How often do you think these **household hazardous wastes** are improperly discarded? And what would represent "proper" disposal of them long-term?

Not all solid waste ends up contained in landfills or incinerators; some ends up as litter in the environment. Have you ever encountered litter in surprising places? We authors have seen plastic bags clinging to the cacti in the Sonoran desert, plastic bottles scattered over the slopes of Greek islands, and unidentifiable bits of plastic among the leaf litter in forested wilderness. Indeed, plastic is a particularly pervasive and pernicious form of solid waste (Royte, 2006). Evidence for this can be seen in the oceans where floating plastic waste is accumulating by getting caught in natural currents, creating large moving gyres of **nonbiodegradable** material (Goldstein, Titmus, & Ford, 2013). Over time, plastic does degrade into smaller and smaller particles; however, even at the molecular level, nonbiodegradable polymers never completely reintegrate into nature and thus are considered contaminants. It remains to be seen what the exact consequences of these moving "garbage patches" are, but there is evidence that aquatic animals are ingesting the smallest particles and birds mistake the colorful plastic for food, filling their bellies with bottle caps (NOAA, 2010; see Figure 1.6). In addition to the plastic debris found in the oceans, "microbeads" of plastic from personal care products such as facial scrubs are also accumulating at high concentrations in freshwater ecosystems such as the Great Lakes (Eriksen, et al., 2013).

Of course, some plastic and other materials are being recycled. However, not everything is currently recyclable, and much waste that could be recycled is not being diverted from landfills, incinerators, or the litter stream. For example, in 2009, Americans threw out nearly nine million tons of glass, enough to fill a line of tractor trailers stretching from New York to Los Angeles and back; and, in 2010, Americans threw away enough paper to cover 26,700 football fields three feet deep (Keep America Beautiful, 2013).

FIGURE 1.6 *Birds among plastic waste washed up on the beach.*
With permission from Shutterstock

Although many components of "**e-waste**" (discarded electronics) can be recycled, most are not. Of the more than two million tons of electronic gadgets discarded by U.S. consumers in 2009, only 25% were collected for recycling, including 38% of computers, 17% of televisions, and only 8% of mobile devices such as cell phones (EPA, 2012c). Recycling lessens the environmental impact from toxic metals being buried and burned, and reduces the demand for energy to mine and manufacture virgin material; yet, even e-waste that is reclaimed poses problems. Much of it is exported from industrialized countries to be recycled elsewhere in the world by impoverished workers in unregulated conditions, in spite of the fact that this practice is prohibited by United Nations conventions. For example, in Guiyu, China, where a massive amount of the world's e-waste ends up, crude methods of recycling contaminate groundwater and soil with mercury, lead, and other hazardous substances, and pollute the air with toxic fumes (Watson, 2013; see Figure 1.7).

Of course, the problem of solid waste starts with consumers. Material consumption has increased dramatically in the past century, and this is not just due to population growth. For example, from 1910 to 2010 in the United States, the use of raw materials (other than fossil fuels and food) rose 2.8 times more than the population increased (Center for Sustainable Systems, 2013). Although family size has decreased in the United States, the average size of single-family homes being built today is more than double what it was in the 1950s, and of course, these larger homes are not sitting empty. They are filled with stuff, more than twice the amount per person than in 1950 (Taylor & Tilford, 2000).

FIGURE 1.7 *Child sits among e-waste in Guiyu, China.*
Courtesy of Greenpeace/Natalie Behring.

The dramatic increase in material consumption in the industrialized world has coincided with the technological and social developments that you will read about in Chapter 2. Changes in lifestyle have been responses to, and have driven, changes in the types and amount of goods available. For example, the advent of industrial manufacturing in the nineteenth century introduced mass-produced consumer goods, allowing those who aspired to climb the socioeconomic ladder to practice **conspicuous consumption**, the display of power and prestige through the accumulation of unnecessary luxury items (Veblen, 1899). With the introduction of plastic products in the middle decades of the twentieth century came the concept that material possessions could be disposable or **single-use**. In an effort to recharge the U.S. economy after World War II, consumer goods engineers became enamored with the concept of **planned obsolescence**, intentionally designing products to have a limited lifespan so that consumers will have to intermittently replace them (Packard, 1960); this is still standard practice today. And, of course, just as common today is **perceived obsolescence**, consumers feeling that a perfectly functional item needs replacing simply because it is dated or out-of-fashion. Do you ever discard useable things just because a different design, color, or style has come out?

Many material possessions that are common today in industrialized cultures *did not exist* 100, 50, 25, or *even 5* years ago. How many do you own? Try mentally walking through your home, your school, your car, your workplace. How many of your possessions are *not* reusable, recyclable, or biodegradable? How many of them do you plan to keep for the rest of your life? How many will you pass on to future generations? How many do you consider essential? One way to come up with an answer to this last question is to take a global perspective. There is a wide disparity between the amount of material goods consumed in industrialized countries and developing countries. Yet, as developing countries aspire to the industrialized life-style, material consumption and associated resource use and waste production will continue to increase.

CONCLUSION

So, how ya doin'? Ready to go hide under a rock yet? This litany of what's wrong in the world is not meant to scare you; rather, it is our hope as the authors of this text that it will inspire you. Change can happen, and some recent data suggest people in the United States are moving in the right direction. For example, earlier, you read that the United States uses about 20% of the world's supply of oil; just a few years ago that number was closer to 25%. That's an improvement. Still, other parts of the world (e.g., China) are increasing their fossil fuel consumption in the process of becoming more industrialized, so the fundamental global problem of fossil fuel consumption is not abating. And fossil fuel use is a prime example of a behavior that is simply not **sustainable**.

What does that mean, "sustainable"? Certainly, you have heard this term; it's pretty popular these days. You may also have heard people argue that it is diffi-cult, perhaps even impossible, to define. Nonsense! The meaning is actually quite straightforward: A sustainable behavior is one that *can be sustained*; that is, main-tained or continued at a certain rate or level. Clearly, many of our current practices, particularly in the industrialized world, do not fit this definition: reliance on finite energy sources, agriculture that depletes topsoil and water, and creation of waste that cannot be reabsorbed into natural cycles come to mind. And, we must not overlook the impact our actions are having on other species; extinction is obviously an unsustainable pattern.

How, then, do humans move in a direction that will ultimately lead to a sus-tainable balance? As Einstein's well-known dictum puts it, problems can't be solved from the same mindset that created them. It is this mindset that is addressed in the next chapter. In order to know where to go from here, it is valuable to understand how we got here in the first place.

NOTES

1. Our thanks to Deborah Winter for her permission to retain some phrasing from the first edition of this book, *Ecological Psychology: Healing the Split between Planet and Self* (1996).

2. In general, the use of the plural pronouns "our" and "we" in this text refer collectively to human beings, especially those of us living in the Western industrialized world.
3. Special thanks to John Du Nann Winter for this teaching example.
4. In the United States, the Recommended Daily Allowance guidelines still use a 2,000 calorie standard, but in 2010, U.S. residents consumed an average of 2534 calories per day. And this is less than the U.S. food supply *produces* per person, which was 3900 calories per day in 2006 (Center for Sustainable Systems, 2013). Forty percent of the edible food produced in the United States is wasted (Gunders, 2012).
5. There are four countries where the average ecological footprint for individual people is higher than for individuals in the U.S.: Qatar, Kuwait, United Arab Emirates, and Denmark; but the combined population of these four countries amounts to only 20 million people, about 6% of the size of the U.S. population.
6. There is an effort in some places to capture methane release from landfills for energy; as of July 2013, 621 such projects existed in the United States. See more at http://www.epa.gov/lmop/basic-info/#a02.

CHAPTER 2

How Did We Get Here?
From Western Thought to "Wise Use"

After reading Chapter 1, you might be thinking, "Wow, we've gotten ourselves into quite a mess," and you may be wondering, "How did this happen?" There is no simple answer, but part of the explanation involves inherent traits, some of which we share with other species and others that are distinctly human. A characteristic shared with other species is an evolved predisposition to maximize our survival. Genetic inheritance is the product of evolutionary processes that have selected for the most **adaptive** features: i.e., those that enhanced survival and reproduction, enabling the transmission of those features to offspring. All animals, therefore, possess the motivation and ability to procure food, fend off predators, and conquer competitors. Among the qualities that are uniquely human is the capacity to dramatically modify our environment. Our drive to survive combined with our ingenuity has allowed us to develop technologies to harness water, land, plants, animals, and energy sources for our benefit, wipe out diseases, increase our lifespans, reduce infant mortality, build climate-controlled and illuminated habitats for ourselves, efficiently travel vast distances, and expand our territory into inhospitable regions. The crux of the matter is that, *in recent history, our technological innovations*

outpaced our understanding of their ecological impact. In fact, it has only been within the last few decades that humans have come to realize that what appeared to be wonderfully adaptive improvements to human quality of life may ultimately prove so maladaptive as to be our species' downfall.

An essential part of understanding how we got here is getting a grasp on the timeline. One goal of this chapter, therefore, is to raise your awareness about how relatively recently the problems described in the previous chapter emerged. Environmental damage has occurred throughout human history; humans have caused geographically isolated species extinctions and resource declines all over the world for centuries (Diamond, 2005; Redman, 1999). But, the point here is that humans are rapidly approaching planetary limits in carrying capacity because industrial development has involved unprecedented, large-scale exploitation of nature and disruption of ecological systems. In the past couple of centuries, humans have devised technologies that have made it possible for the average person in the industrialized world to live at a level of material comfort that the richest person a few centuries ago could hardly imagine. Unfortunately, these very same technologies have a destructive potential far deeper and broader than any that preceded them. Some scholars even propose the designation of a new geologic era, the "Anthropocene," to signify the scale at which human activity is altering the planet (Zalasiewicz, et al, 2008).

The technological developments of recent centuries have been inspired and supported by a way of thinking, or **worldview**, that some consider endemic to Western civilization (e.g., Pirages & Ehrlich, 1974). Psychologists study behavior at the individual level, but an analysis at the individual level is not adequate to completely explain, or remedy, our predicament. It is crucial to consider cultural "programming." Who people are, what they believe, and how they behave are greatly influenced by the social context in which they develop. No matter how much we attempt to alter individual-level behaviors, things in general will not turn in a different direction unless the cultural *mindset* changes.

© MARK ANDERSON, WWW.ANDERTOONS.COM

ANDERSON

"What's that boy?! A paradigm shift?!"

Everyone develops a worldview. Worldviews are rarely questioned by those who hold them; in fact, it is difficult for us to even perceive our own worldviews because we experience these implicit sets of assumptions and beliefs as givens, as *just the way things are*. If you have traveled or spent time with people from a different culture, you may have been surprised by the fact that not everyone shares the same values, priorities, goals, or ideas about how the world works. It can be eye-opening to contrast the ideas to which you've become accustomed with a different perspective, especially when the new perspective leads you to question notions that you have always taken for granted.

As an example, consider your ideas about what constitutes *progress*. Recall the questions asked at the beginning of Chapter 1: Do your visions for your own future include the acquisition of more and better possessions? Do you expect to generate a higher income than you do now? Do you assume that your wealth will continue to increase (even beyond inflation) over the course of your career? Do you anticipate having a standard of living that exceeds previous generations? If your answer to these questions is "yes," then you, like most of us living in the Western industrialized world, have internalized the notion that equates progress with *growth*. In fact, this idea is so ingrained in us that it can be difficult to imagine how one could proceed into the future without growth as the goal.

Equating progress and growth is one of the ways of thinking that comprises a worldview identified by Pirages and Ehrlich (1974) as the **Dominant Social Paradigm (DSP)** in industrialized culture. Although any culture may have a "dominant social paradigm," in this chapter, the term is used to refer specifically to the way of thinking that has spurred industrial development in the Western world. Within the DSP, economic growth is always good (and always possible), human beings should use natural resources however we can for our benefit, individuals have the right to develop land for the purpose of accumulating personal profit, and science and technology will solve any problems that may arise as a result. This set of axioms is labeled "dominant" because it has been pervasive among people in Western societies for the past few centuries, and continues to be endorsed to varying degrees by many people today.

Contrast the DSP with the notions that there are limits to growth, humans are not exempt from natural constraints, nature has intrinsic value beyond what it provides directly to humans, and potentially catastrophic human-caused environmental changes are looming. This second set of ideas represents a more recent way of thinking informed by environmental science, which has been able to pinpoint the errors in our industrial ways. The problem is that scholarly awareness of ecological realities has not yet significantly impacted the fundamental assumptions that drive and guide decisions and behaviors in industrialized culture. This chapter is therefore intended to give you a sense of the juncture at which we find ourselves. It begins with an overview of the intellectual history that shaped the DSP, followed by a more in-depth look at how the DSP influenced development, and inspired resistance, in the twentieth-century United States.

THE NATURE OF WESTERN THOUGHT

The Dominant Social Paradigm is a product of a centuries-long trail of Western intellectual and cultural history shaped by the Greek philosophers, the Judeo-Christian tradition, the Enlightenment thinkers, the Scientific Revolution, European colonialism, and the Industrial Revolution. (Whew!) But, in terms of overall human history, it is a *new* perspective. According to the fossil record, anatomically modern humans (Homo sapiens), with our enlarged frontal lobes, first appeared about 200,000 years ago; the DSP has been around for only about 250 years.

To get a sense of how recently this worldview emerged, imagine the 200,000 years of human history reduced to a year. Modern humans came into existence on January 1. Nearly the whole year, January through mid-December, was spent in small bands of nomadic hunting–gathering tribes; the innovation of agriculture and the subsequent establishment of the first settled societies in the Near East did not begin until December 12. The Greeks, to whom Western civilization owes much of its heritage, did not begin their celebrated Classical period until 10 a.m. on December 27. The Scientific Revolution, responsible for a massive change in the way Europeans viewed nature, happened after 2 a.m. on December 31, at the same time European civilization was spreading to the rest of the world through colonialization. And the Industrial Revolution of the late eighteenth and early nineteenth centuries did not begin until about 3:30 that afternoon. So, as you can see, even the most ancient roots of the predominant Western worldview occurred very recently in human history.

The following coverage of the historical roots of the DSP will be limited to four of its fundamental assumptions:

1. Humans are separate from, and superior to, nature

2. which can and should be controlled

3. by individuals exercising their right to seek maximum personal economic gain

4. which can be expected to grow progressively over time.

But, before proceeding, please keep in mind three caveats:

First, critically examining suppositions that have functioned seemingly successfully for generations can feel threatening. Worldviews influence not only our beliefs, motivations, goals, decisions, and actions, but also our values and definition of self. To challenge the DSP is to strike at the heart of many people's cultural, political, and religious identities, and cherished lifestyles. The DSP is an enticing worldview that offers a sense of freedom and opportunity unknown in most parts of the world, and that promises a life of unparalleled, and unlimited, material abundance. And, indeed, this worldview wouldn't have become dominant if it hadn't proved useful in contributing to improved quality of life for many people, at least for a time.

Second, this historical account highlights some scholars whose work inspired major shifts in Western thought that ultimately influenced the DSP; but, this is not an indictment of these intellectuals or their contributions. For example, you will read about Sir Francis Bacon, the father of the scientific method. As psychology researchers, we authors are deeply indebted to Bacon and dedicated to the scientific method. However, this does not mean that we are unable to recognize that along with scientific thinking came a distinctly different manner of relating to the natural world. Pointing to some of our species' greatest thinkers is not an attempt to blame them for current conditions, for they lived in very different times and places. Instead, the following sections tell a story about how ideas have emerged and embedded themselves in the way that many people, many generations later, in completely different circumstances, continue to think.

Third, this history is, by necessity, brief and selective. All history is told with some purpose, which sculpts what part of the story is articulated, and what other parts are omitted.

Humans Are Separate from Nature

In preindustrial societies, people live in small groups, deriving a sustained subsistence from the land, either through hunting and foraging, or from hand- or animal-based agriculture (Sabloff, 2001). Shelters and tools are constructed of natural materials. Basic routines of sleeping, eating, working, and playing are in sync with natural patterns of sunlight, weather, and changing seasons. What we in the industrialized world call "nature," these cultures call "home."

Daily, immersive interaction with nonhuman nature is often accompanied by a belief that it is not only alive, but also imbued with the same qualities as human beings (Manes, 1996). Other species are perceived as neighbors and family. In most indigenous cultures anthropologists study, **animism** and **anthropomorphism** inspire rituals, art, and daily practices that honor the spirit of nonhuman nature. In the words of Anishinaabe writer, activist, and former U.S. Vice-Presidential candidate, Winona LaDuke,

> Native American teachings describe the relations all around—animals, fish, trees, and rocks—as our brothers, sisters, uncles, and grandpas. Our relations to each other, our prayers whispered across generations to our relatives, are what bind our cultures together. . . . These relations are honored in ceremony, song, story, and life that keep relations close—to buffalo, sturgeon, salmon, turtles, bears, wolves and panthers. These are our older relatives—the ones who came before and taught us how to live.
>
> (LaDuke, 1999, p. 2)

In aboriginal societies, because of a belief that the natural world is alive and imbued with a spiritual essence, people live in kinship with it.

Similarly, early Western history shows allegiance to a vitalistic conceptualization of the natural world. Plato in his *Timeas* proposed that the world possessed a soul, which he called "anima mundi." Likewise, the Stoics of third-century B.C. Athens and A.D. first-century Rome saw the world as an intelligent organism. Modern life in the industrialized Western world involves a strikingly different orientation, one in which many people feel separate from, superior to, and entitled to use nonhuman nature. How did this shift in perspective develop?

Plato's student Aristotle proposed that all beings are arranged in a single continuum, a "scala naturae," based on their amount of soul, which determines how close they are to God, who sits at the top. And who is right up there next to God? Man. In Aristotle's version of the cosmos, God reigns over men, who rule over women, children, animals, plants, and inorganic matter, in that order. This ordering makes humans more important than animals, men more important than women, mammals more important than insects, plants more important than dirt, and organic matter more important than inorganic matter. The **anthropocentrism** (human-centeredness) evident in this hierarchical ordering foreshadows the thinking of Enlightenment Era scholars whose ideas furthered the perceived separation between humans and nonhuman nature.

Among these scholars, none was more influential than French philosopher René Descartes (1596–1650). Perhaps you have learned about Descartes's "mind-body dualism." In Descartes's view, everything in the universe, once created by God, operated strictly according to fixed laws—everything, that is, except the human mind, which was imbued with soul and was, therefore, distinctly different from

the rest of nature. By the middle of the seventeenth century, Descartes's ideas were widely discussed and increasingly substantiated. For example, Sir Isaac Newton (1642–1727) validated Descartes's view of a **mechanical universe** by providing mathematically verifiable predictions about the movement of stars and objects.

This transformation from perceiving the natural world as a dynamic, alive, souled entity to viewing it as an orderly, mechanical, and clockwork machine has been called "the death of nature" by historian Carolyn Merchant:

> In 1500 the parts of the cosmos were bound together as a living organism; by 1700 the dominant metaphor had become a machine. . . . Mechanisms rendered nature effectively dead, inert, and manipulable from without.
>
> (Merchant, 1980, pp. 214–215).

In other words, the view of nature as inanimate served as the foundation for the DSP belief that humans are separate from nature. And, importantly, this view of nature also led to widespread acceptance of the idea that nature can feasibly—and ethically—be manipulated and controlled.

Nature Can and Should Be Controlled

English philosopher Sir Francis Bacon (1561–1626) argued persuasively that philosophy had been unproductive because it was based on speculation, rather than on fact. He asserted that humans must bring the world to light through *scientific observation*. As the founder of empiricism, Bacon recognized that scientific rigor sometimes requires the exercise of control over one's subject matter. For example, we cannot conduct experiments if extraneous variables are allowed to interfere with the specific variables under study. So, it is no surprise that when it came to the scientific study of nature (which Bacon saw as female), he suggested that scholars should observe,

> not only nature free and at large (when she is left to her own course and does her work her own way),—such as that of the heavenly bodies, meteors, earth and sea, minerals, plants, animals,—but much more of nature *under constraint* and vexed; that is to say, when by art and the hand of man she is forced out of her natural state and *squeezed and moulded* . . . seeing that the nature of things betrays itself more readily under the vexations of art than in its natural freedom.
>
> (italics added; Bacon, 1620/1955, p. 447)

Bacon was absolutely correct that natural science benefits from the constraint of nature; however, with the emergence of natural science came increased confidence about humans' ability to control nature more generally beyond the exploratory scientific domain. And this was followed by the idea that if we can control nature, indeed we *should* do so for our own benefit.

Control over nature is perhaps most evident in the possession and cultivation of land for agriculture. John Locke (1632–1704), a British philosopher and political theorist, stressed the importance of land ownership as a foundation for democracy, and believed that the right to vote should be granted only to men who owned and cultivated land. Locke's emphasis on land use as a prerequisite for democratic participation helped promote the notion that unused land is wasted land. Consider his analysis of Native American land use:

> it is labor indeed that puts the difference of value on everything. . . . There cannot be a clearer demonstration of anything than several nations of the Americans are of this, who are rich in land and poor in all the comforts of life; whom nature, having furnished as liberally as any other people with the materials of plenty—i.e. a fruitful soil, apt to produce in abundance what might serve for food, raiment, and delight; yet, for want of improving it by labor, have not one hundredth part of the conveniences we enjoy, and a king of a large and fruitful territory there feeds, lodges, and is clad worse than a day laborer in England.
>
> (1690, in MacPherson, 1962, pp. 196–197, or Locke, 1690/1939, p. 419)

By the eighteenth century, this attitude toward nature had become part of the prevailing Western worldview. Of course, from an eighteenth-century perspective, utilizing nature and developing it for human benefit made perfect sense; at that time, no one could have predicted that this attitude, combined with tremendous advances in technology and dramatic population growth, would ultimately lead to impending ecological collapse.

Individuals Have a Right to Maximum Economic Gain

Philosopher Thomas Hobbes (1588–1679) proposed that there is no "good behavior" or "bad behavior"; instead, there are various ways that people behave because they are locked in a continual state of competition with each other for resources and for power. Hobbes saw nature as chaotic and dangerous, and said humans must fight for their own survival against nature and against each other.

> The life of man [is] solitary, poor, nasty brutish, and short . . . the condition of man, . . . is a condition of war of everyone against everyone; in which case everyone is governed by his own reason.
>
> (Hobbes 1651/1962, pp. 100, 103)

Hobbes believed that at the core of human nature is competitive self-interest; therefore, it is not natural for humans to join communally or behave cooperatively for the greater good. Indeed, Hobbes did not believe such a thing as the "greater good" existed, given that different individuals want and seek different things.

At first blush, this idea may resonate with you because it sounds like an early version of "survival of the fittest," but as sociobiologists tell us, natural selection does not occur only at the level of the individual; cooperation and altruism (particularly among kin) are innate behaviors that serve the purpose of group selection in humans and other species (Wilson & Wilson, 2007). Biologically, we are predisposed to act in the interest of the group, rather than ourselves, under certain conditions.

In any case, Hobbes's radical individualism was toned down a bit by several thinkers who followed. For example, Adam Smith, the Scottish economist–philosopher (1723–1790) argued that the government should leave individuals alone to amass their material wealth, not because self-interest is the natural state of things as Hobbes claimed, but because what is good for the individual is eventually good for the community. English–American political activist Thomas Paine (1737–1809), persuasively argued for the American revolution based on his belief that governments should not interfere with individuals' natural rights, and indeed, the U.S. Declaration of Independence later made the same point: "All men are endowed by their Creator with certain unalienable Rights that among these are Life, Liberty, and the pursuit of Happiness." Ultimately, primacy of the individual became a hallmark of the new democratic government. No longer did people have essential moral obligations to their clan or community. Instead, individuals were responsible for their own wellbeing. Lives were to be lived as individuals, competitive and separate, pursuing personal material wealth through the rights of individual freedom and noninterference from the state.

The discussion of how materialistic individualism became a fundamental part of the dominant Western worldview would be incomplete without mentioning the contribution of the Protestant reformers who settled the United States. Calvinists believed work was a divine calling and wealth was an indication of God's approval for hard labor; however, these people also believed that wealth should not be used for luxury or self-indulgence. The only appropriate thing to do with profit was to reinvest it. In this way, Calvinism ultimately fostered industrial development, and thus helped capitalism to flourish.

Progress Equals Growth

In the early days of the United States, the country stretched out into what must have seemed infinite wilderness, and the lure of the West soon attracted settlers and industrialists with visions of personal fortune. Although westward expansion was debated, many saw it as inevitable and desirable, at least those who believed in "manifest destiny."

> [The American claim] is by the right of our manifest destiny to overspread and to possess the whole of the continent which Providence has given us for the development of the great experiment of liberty and federative self-government intrusted to us. It is a right such as that of the tree to the space of air and Earth suitable for the full expansion of its principle and destiny of growth.
>
> (O'Sullivan, 1845)

The phrase "manifest destiny," coined by the *New York Morning News* editor, whose words appear above, succinctly captured a widespread sense of entitlement to land ownership and entrepreneurial conversion of that land. In the expanding United States, the idea of manifest destiny affirmed the link between progress and growth; the way forward for the nation was to occupy more land and use more resources. Stop for a moment and consider how this perspective was tied to time and place. Land and resources truly seemed unlimited in the mid-nineteenth-century United States. In this context, equating progress with growth made sense. But progress can't equal growth when there is no room to grow and when resources are tapped to their limits.

Yet today, when we are reaching (or even exceeding) those limits, as described in Chapter 1, those of us living in the industrialized West continue to expect that over time, growth will occur. Our economic models are based on this assumption. Most people in Western cultures perceive their lives as a linear sequence of events marked by change and development; to return to a previous state is to go backward, to regress, to fail. We continually strive for improvement and enhancement in our lifestyles (larger incomes, more possessions). The problem is that thinking of progress in terms of perpetual growth and change devalues sustainability as stagnation.

Divergent Voices

Just because the DSP is pervasive does not mean that everyone who lives in Western industrialized culture subscribes to it. Even in the nineteenth century, there were some individuals who expressed a contrary point of view. The minority perspective was voiced most eloquently by transcendental writers such as Ralph Waldo Emerson (1803–1882) and Henry David Thoreau (1817–1862). American transcendentalists rejected materialistic values and the utilitarian view of nature. Their writings were characterized by spiritual descriptions of nature. They praised wilderness and simple living, and they respected the interconnectedness of all life.

Consider the following quotes from Thoreau's writings in the 1850s, including his now classic book, *Walden; or, Life in the Woods*: "Nature is full of genius, full of the divinity; so that not a snowflake escapes its fashioning hand" (Thoreau, 1856/2009, p. 356), "Shall I not have intelligence with the earth? Am I not partly leaves and vegetable mould myself?" (Thoreau, 1854/1995a, p. 90), "Most of the luxuries, and many of the so-called comforts of life, are not only not indispensable, but positive hindrances to the elevation of mankind" (Thoreau, 1854/1995b, p. 8), and "In Wildness is the preservation of the World" (Thoreau, 1862/2015, n.p.). It is in the works of the transcendentalists that we can see the first glimmers of the type

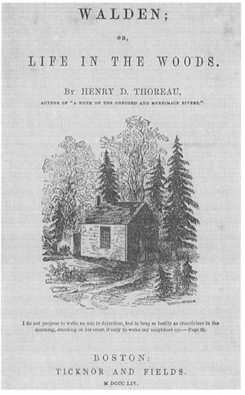

of ecological thinking that would challenge the DSP and inspire the environmental movement in the United States.

ENVIRONMENTALISM IN THE UNITED STATES

During the latter part of the nineteenth century, the use of natural resources, guided by the assumptions described in the preceding sections, ratcheted up to an unprecedented level in the United States. The increased demand was partly due to the spread of settlers into territory previously occupied only by Native Americans, who, as noted above, left the land basically undeveloped. The mid-century's California Gold Rush drew people to the west coast, and quickly thereafter, the middle of the country began filling in. Population also played a role in the last decades of the nineteenth century via a surge of immigration, which not only increased the number of people in the nation (especially in its rapidly growing cities), but increased the availability of cheap labor. Logging, mining, manufacturing, and large-scale agriculture thrived, aided by the advent and growth of the railroads, as well as changes in laws that made the formation and merger of corporations easier.

Preservation and Conservation of Wilderness

Although the natural bounty of the American wilderness may have seemed vast and abundant at the start of westward expansion, by the turn of the twentieth century, it was becoming clear to some that development was progressing at an alarming pace, seemingly without restraint. One such individual was the naturalist John Muir (1838–1914), who had emigrated from Scotland as a child, grew up in Wisconsin, and ultimately found his calling in the Yosemite Valley of California. Muir was a unique fellow who once described himself in a letter to a friend as a "self-styled poetico-tramp-geologist-bot. and ornith-natural, etc.!!!" (McFarlane, 2004). He is well-known for having founded the Sierra Club, which originally began as a hiking group, but morphed into an organization dedicated to environmental protection, and remains one of the most influential of such organizations today.

Muir is an important figure in the history of U.S. environmentalism, not just because he founded the Sierra Club, but because he represented a minority perspective focused on the **preservation** of untouched wilderness. Influenced by the transcendental works of Emerson, Muir valued pristine wilderness as a place of spiritual refuge, restoration, exploration, and recreation for people. His preservationist attitude contrasted with the more prevalent view that nature's value lay in the utilization of its resources; however, even among those who viewed nature as a supply of resources, there were some who cautioned that use should be monitored and managed in the interest of **conservation**.

The conservationist approach was advocated by President Theodore Roosevelt (1858–1919), who created the U.S. Forest Service, established 51 Federal Bird

Reservations, four National Game Preserves, 150 National Forests, and five National Parks, effectively putting about 230,000,000 acres of public land under government control (National Park Service, 2013). Muir had great influence on Roosevelt. For example, after spending three days camping with Muir in the Yosemite wilderness in 1903, Roosevelt left persuaded to make Yosemite Valley and the Mariposa Grove of giant sequoias part of Yosemite National Park (see Figure 2.1). Still, it was ultimately the conservationist perspective that prevailed legislatively after a contentious battle over whether to dam and flood the picturesque Hetch Hetchy valley in Yosemite National Park so as to provide water to the burgeoning city of San Francisco.

The fight over Hetch Hetchy lasted for 12 years, spanning Roosevelt's two terms in office and the term of his successor, William Howard Taft. During this time, it was

FIGURE 2.1 *President Theodore Roosevelt and John Muir in Yosemite, 1903.*
As founder of the Sierra Club, Muir was an ardent advocate for the preservation of untouched wilderness.
With permission from Yosemite Research Library (National Park Service)

FIGURE 2.2   *President Theodore Roosevelt and Gifford Pinchot, 1907.*
As first Chief Forester of the United States, Pinchot argued for managed conservation of natural resources.
Courtesy of North Carolina Theatre Conference

Gifford Pinchot (1865–1946), first Chief Forester of the United States (appointed by Roosevelt), who served as the leader of the conservationist cause (see Figure 2.2). Pinchot had been trained in scientific forestry in Europe and believed ardently in the managed use of natural resources. In his 1910 book, *The Fight for Conservation*, Pinchot argued that natural resources were there to be used, but used wisely, in a manner that minimized waste, would not exhaust the supply, and would benefit many rather than just profit a few. Ultimately, in 1913, Congress passed the Raker Act, and President Woodrow Wilson signed it into law, thus allowing San Francisco to build a dam and reservoir that flooded Hetch Hetchy valley.

Muir's relentless campaign to protect Hetch Hetchy is regarded by some as the first example of grassroots lobbying on behalf of the environment (Sierra Club, 2008), but this is debatable. More accurately, it is the most *visible* example of citizens banding together in an effort to influence environmentally relevant legislation in the early twentieth century—visible because of how environmental histories identify it as the pivotal event in the contest between conservationists and preservationists.

In truth, there were other instances of citizen resistance to environmental degradation happening at the same time.

For example, you may have heard of the National Audubon Society, an organization dedicated to wildlife protection and ecosystem restoration. Named for ornithologist John James Audubon (1785–1851), the society started in Massachusetts as an offshoot of one of the numerous ladies' clubs that were common among nineteenth-century middle- and upper-class women with philanthropic impulses and activist leanings. At the time, women could not vote or hold public office, so their influence tactics were necessarily more indirect than those of male peers like John Muir, who had the ear of the President. They worked through their own social networks, through persuasive appeals to husbands, sons, and fathers, and through writing and public speaking (when they were afforded the opportunity). The first Audubon group was "established in 1896 by Founding Mothers Harriet Lawrence Hemenway and Minna B. Hall, who persuaded ladies of fashion to forgo the cruelly harvested plumage that adorned their hats" (MassAudubon, 2014; see Figure 2.3). Other state-level Audubon groups soon sprung up and, in 1901, they banded together to persuade President Roosevelt to establish the first National Wildlife Refuge in the United States, Pelican Island, in 1903 (National Audubon Society, 2014).

Grassroots campaigns led by women's clubs were also responsible for numerous accomplishments in forestry protection, including the creation of the first congressionally mandated national forest, in Minnesota in 1902 (Merchant, 1995). In his autobiography *Breaking New Ground*, Gifford Pinchot wrote of the significance of the Chippewa National Forest legislation:

> Its passage was a victory for public opinion and the public interest against local opposition and the fierce hostility of the lumbering interests, who had long ruled the roost in northern Minnesota. Without the farsighted and patriotic support of the Minnesota Federation of Women's Clubs, it would have been impossible. With it, *the Bureau of Foresty emerged from the period of mere advice to actual control, for the first time, of timber cutting on public land.*
>
> (Pinchot, 1947/1998, p. 205, italics added)

Although both the conservationists and preservationists questioned unbridled and greedy development, the conservationist perspective was less of a challenge to the Dominant Social Paradigm. Advocates like Pinchot and Roosevelt did not question development in general and, consistent with the prevailing Western worldview, they believed that science and technology were the keys to successfully pursuing it.

The World Wars and Modern Living

While the preservationists and conservations duked it out over wilderness areas, an analogous debate was happening in the country's rapidly growing cities. Instead of loss of wilderness and wildlife, the first **urban environmentalists** were

JANUARY/FEBRUARY 1996

SANCTUARY

THE JOURNAL OF THE MASSACHUSETTS AUDUBON SOCIETY

THE MOTHERS OF CONSERVATION

1896 • CENTENNIAL ISSUE • 1996

FIGURE 2.3 *Centennial issue of* The Journal of the Massachusetts Audubon Society *(1996).*
The National Audubon Society originated with grassroots efforts by women to restrict the slaughter of birds and sale of their feathers to the millinery industry.
Courtesy of Massachusetts Audubon Society

concerned about the emerging problems of water contamination, air pollution, lack of adequate sanitation, solid and hazardous waste disposal, and occupational dangers faced by factory laborers who worked with toxic substances. Like the preservationists, the **reformers** fundamentally questioned the industrial enterprise. Like the conservationists, the **professionals** trusted that science and engineering could cope with any threats caused by industrial development (Gottlieb, 2005).

FIGURE 2.4 *Ellen Swallow Richards.*
In the 1880s, Richards conducted a comprehensive survey of public water supplies for the Massachusetts State Board of Health, which led to the establishment of the first water-quality standards in the United States. She is credited with introducing the term "ecology" into the English language.

One professional whose work focused on sanitary chemistry was Ellen Swallow Richards (1842–1911), the first woman to attend the Massachusetts Institute of Technology. While in school, she earned a reputation as the nation's leading water scientist, and her work led to the creation of the first water quality standards in the United States (Vassar Historian, 2005; Murphy, 2014; see Figure 2.4). Richards is an important figure in this environmental history because she is the person credited with introducing the word **ecology** into the English language (Clarke, 1973). Richards adopted the term and used it in her own writing after reading the work of German zoologist Ernst Haeckel, who coined the label (*oecologie*) in the mid-nineteenth century to describe the study of the relationship between an organism and its environment (Egerton, 2013). Richards was concerned with the human-environment relationship, and so her version of ecology bears more similarity to today's discipline of *human ecology* than to ecology in general. Still, Swallow's use of the term played a part in introducing a systemic perspective in environmental science.

In the decades leading up to World War I, industrial development was going full tilt and had resulted in a number of significant innovations (e.g., the automobile, the first fully synthetic plastic, the first commercial airline). Many aspects of modern living that we now take for granted—and now realize have negative ecological consequences—were just beginning to be developed and/or were being made more accessible and affordable to the general population (e.g., indoor plumbing, electricity, central heating, refrigeration). It is difficult to overstate how radically people's daily habits were changing in the first decade-and-a-half of the twentieth century.

The years of World War I interrupted the industrialized world's trajectory and shook people up in a way that we can only imagine today. The end of the war ushered in the "Roaring 20s," a decade that witnessed the demise of many nineteenth-century social roles and customs and the birth of the modern era. For example, for the first time, the majority of people in the United States lived in cities, young people went out on unchaperoned dates (in cars!), skirts rose to mid-calf and some women even donned trousers, jazz music inspired the first dirty dancing, department store chains were launched, commercial radio broadcasts brought advertisements to the masses, and widespread consumerism flourished as never before. In the midst of this heady decade, environmental concerns were not on most people's minds.

And then, in 1929, the stock market crashed. The roar of the 1920s was silenced by the Great Depression. The depression was an unparalleled global economic downturn. Most dramatically affected were areas dependent on heavy industries such as mining, logging, manufacturing, construction, and farming. Production and consumerism plummeted. After rising to a high of 25%, unemployment hovered above 15% for the duration of the 1930s (Temin, 2000). Conditions were made even worse by the Dust Bowl, a several-year period of extreme drought and erosion of topsoil across the great plains of the Midwest, exacerbated by industrial agricultural practices that had not included safeguards to conserve water or guard against wind.

Just as the exhilarating prosperity of the 1920s distracted attention away from environmental concerns, so too did the harrowing struggles of the 1930s. But, while the general public was desperately focused on putting food on the table, some conservation-minded individuals found the impoverished conditions motivating. In the words of science writer Randolph Fillmore,

In the 1930s the American experience with the Dust Bowl of the Midwest showed that the destruction of the soil by reckless farming could permanently affect the land. By this time, the harmful effects of industrialization and urbanization also became apparent. . . . [T]here was growing tension between those who wanted unrestrained exploitation of nature and those seeking to conserve it.

(Fillmore, 2009, p. 5)

Indeed, even President Franklin Delano Roosevelt, who was inaugurated in the midst of the Depression in 1933, was prompted to create what some consider "the single greatest conservation program in America," the Civilian Conservation Corps (CCC Legacy, 2013).

Roosevelt, who was Teddy Roosevelt's fifth cousin, created the CCC as part of his New Deal programs to generate economic recovery. Roosevelt saw an opportunity to help unemployed men while also benefitting the nation's parks and forests. By 1935, there were over half a million men employed by the CCC, living in 2,650 camps across the country (CCC Legacy, 2013; see Figure 2.5). Their varied jobs included, but were not limited to, building fire towers and roads, improving streams, protecting natural habitat, and planting trees (the total number of which is estimated to be about three billion). If you have visited any national parks (or even some state and regional parks), you have likely seen the characteristic stone and log buildings and picnic shelters that were constructed by the CCC.

While Roosevelt's CCC crews were industriously occupied grooming parks and national forests, a small grassroots group was banding together for the purpose of protecting the nation's roadless and primitive areas. The instigator was a man named Bob Marshall and the organization he formed in 1935 was the Wilderness Society, which is still going strong today.

No one could accuse Marshall of being uneducated on his subject. He had a Masters Degree in forestry from Harvard and a Ph.D. in plant physiology from Johns Hopkins University. He had first presented his ideas about wilderness

FIGURE 2.5 *Civilian Conservation Crew member planting a tree, ca. 1933–1942.*
Courtesy of Docsteach.org

protection in a 1930 article in *Scientific Monthly,* in which he defined wilderness as an area that,

> possesses no possibility of conveyance by any mechanical means and is sufficiently spacious that a person in crossing it must have the experience of sleeping out. The dominant attributes of such an area are: first, that it requires any one who exists in it to depend exclusively on his own effort for survival; and second, that it preserves as nearly as possible the primitive environment. This means that all roads, power transportation and settlements are barred.
>
> (Marshall, 1930, p. 141)

In the article, Marshall pointed out that when Lewis and Clark conducted their famous 1804–1805 expedition, it was possible to cross two-thirds of the continent without encountering "any culture more advanced than that of the Middle Stone Age" (Marshall, 1930, p. 142). He further lamented that Lewis and Clark's expedition,

> inaugurated a century of constantly accelerating emigration such as the world had never known. Throughout this frenzied period the only serious thought ever devoted to the wilderness was how it might be demolished. To the pioneers pushing westward it was an enemy of diabolical cruelty and danger, standing as the great obstacle to industry and development.
>
> (p. 142)

The Wilderness Society was instrumental in the passage of the 1964 Wilderness Act, the legislation that established the National Wilderness Preservation System. At the time he wrote his article in 1930, Marshall warned that "today there remain less than twenty wilderness areas of a million acres" (p. 142), none of which was federally protected. Thanks to his legacy, at the time of this writing, the total amount of federally designated and protected wilderness in the United States is nearly 110 million acres, which includes the over-one-million-acre Bob Marshall Wilderness Area in Montana (University of Montana Wilderness.net, 2014).

Of course Marshall did not found the Wilderness Society all by himself. One of the cofounders, who figures prominently among challengers to the Dominant Social Paradigm, was forester and philosopher Aldo Leopold (1887–1948). Leopold is perhaps best known for his book of essays *A Sand County Almanac.* It is in this volume that Leopold introduced the idea of a "land ethic" to guide how humans relate to the land and its other inhabitants. As you have learned, within the DSP, land is perceived in terms of its economic and utilitarian value, and therefore, it is considered ethical to use, alter, and control it. Inspired by an ecological mindset, Leopold proposed a different perspective in the foreword to his book,

> We abuse land because we regard it as a commodity belonging to us. When we see land as a community to which we belong, we may begin to use it with love and

respect. There is no other way for land to survive the impact of mechanized man. . . . That land is a community is the basic concept of ecology, but that land is to be loved and respected is an extension of ethics.

(Leopold, 1949, viii–ix)

With his focus on ecology and his premise that unspoiled nature is valuable for its own sake (not just as a retreat for humans, as Muir proposed), Leopold was truly ahead of his time. Although his book is now considered an environmental classic, it was not an immediate best-seller. Consider that when it was published, the American public was once again in the throes of postwar recovery.

Just as the country was getting back on its feet after the Great Depression, the Japanese bombed Pearl Harbor in 1941 and the United States entered World War II. Besides the fact that environmental protection was once more sent to the bottom of the priority list, the war proved ecologically detrimental because it yielded advances in technology and manufacturing that again dramatically affected Americans' lifestyles and daily habits.

Shortly before the war, big changes were already brewing. For example, the first home televisions were manufactured in the late 1930s and the Federal

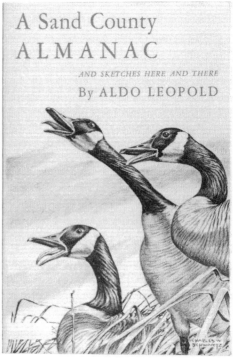

Communications Commission (FCC) legalized commercial TV advertising in 1941. But it wasn't until the 1950s that televisions became mainstream. According to the FCC,

> Between 1945 and 1948 the number of commercial (as opposed to experimental) television stations grew from 9 to 48 and the number of cities having commercial service went from 8 to 23. And, sales of television sets increased 500%. By 1960 there were 440 commercial VHF stations, 75 UHF stations, and 85% of U.S. households had a television set.
>
> (FCC, 2005)

Take a moment to contemplate how television changed people's lives. Consider how it affected their values, their material desires, their sense of identity, and their connection to popular culture. Think about how it influenced their leisure time pursuits, level of physical activity, and the amount of time they spent outside. (And then reflect on how many other forms of "screen time" exist today.) Ecologically speaking, the innovation of television was no-less-than momentous, and not in a good way.

Another ecologically significant change in postwar America was the birth of car culture. Automobiles had been introduced to the general public at the dawn of the century, when Henry Ford manufactured his Model T with the hope of putting a car in every garage, but by the late 1940s about 40% of Americans did not yet own one (Chase, 2010). In the postwar years, however, the car became an essential part of the "American Dream." Now the pursuit of happiness was guided by the iconic image of a nuclear family (headed by a returning vet and his homemaker wife) living in a single-family home, located in a newly created suburban development, which required a car to get to work, school, church, shopping, and so on. As transportation by car became more common, car-dependent businesses were created, including the drive-in restaurant, drive-in movie theater, and suburban shopping mall (surrounded by a vast parking lot). And commuting by car was made increasingly accessible by the creation of the National Interstate Highway System in 1956 (Chase, 2010). At the outset of the 1950s, there were 25 million registered automobiles in the United States; by 1958, that number had risen above 67 million (Reid, 2004). According to the 2012 U.S. Census Statistical Abstract, that number is now hovering around 250 million.

After the war, the nation's factories shifted from producing wartime necessities (e.g., bombers) to consumer products. These included myriad household appliances that are now standard (e.g., dishwashers and clothes dryers). In fact, the manufacturing and marketing of household appliances was an integral part of a bigger effort to promote the role of homemaker to women who had worked in factories and offices or served in the military during the war; the vets needed jobs upon their return, so women needed to go back home. During the war, the government had produced propaganda urging women to work for the war effort (and plant

Victory Gardens, and salvage and recycle); after the war, mass media was deluged with images of deliriously happy homemakers surrounded by modern labor-saving devices and products.

Among the products marketed to housewives were the first mass-produced plastic household goods, including dishes, countertops, floor tiles, and countless other gadgets and knick-knacks. These represented the introduction of the chemical products polyvinyl chloride, polystyrene, and polyethylene into American homes. Chemical household cleaners, fertilizers, and pesticides also made their debut. Along with the latter two came the promotion of a new ideal for the domestic lawn, a contrived landscape that we now consider normal: a lush green expanse of a single species of grass, most likely not native to the region and chemically dependent for its survival. One pesticide that proved particularly significant in provoking challenges to the Dominant Social Paradigm was *dichlorodiphenyltrichloroethane*, otherwise known as DDT (see Figure 2.6).

FIGURE 2.6 *Man spraying DDT on sheep to kill ticks, ca. 1948.*

Silent Spring and the Green Decade

Seventy years after Ellen Swallow Richards introduced the term "ecology" to the English language, two-and-a-half decades after English botanist Sir Arthur Tansley coined the term "ecosystem," and nearly a decade and a half after Aldo Leopold made his impassioned plea for a land ethic, the ecological awareness of the U.S. public was finally raised by the publication in 1962 of a ground-breaking book called *Silent Spring*. The author was scientist and nature writer Rachel Carson, who wrote it while battling the cancer that ultimately claimed her life in 1964. Carson's intent was to raise public concern about the negative health effects to humans, birds, and other species when chemical pesticides, which were routinely sprayed in neighborhoods, parks, and schoolyards, accumulate in the soil and in living tissues. And raise concern she did. Although the chemical industry questioned Carson's credibility as a scientist and attacked her with gendered accusations that she was "hysterically overemphatic," had an unseemly "mystical attachment to the balance of nature," and was a lesbian (perhaps you can figure out the relevance of that one), the book received enormous attention from other scientists, policy makers, politicians, and the public (Gottlieb, 2005). Carson's work was instrumental in bringing about a ban on DDT in 1972.

FIGURE 2.7 *Rachel Carson.*
Carson's groundbreaking 1962 book, *Silent Spring*, raised the public's awareness of the health hazards of commonly used pesticides.

Environmental historians consider *Silent Spring* the catalyst that sparked the second wave of U.S. environmentalism, the first wave having been the efforts of Muir, Pinchot, and allies. No doubt the book's success was facilitated by the endorsement of President John F. Kennedy, who referenced "Miss Carson's book" in a 1962 press conference, and ordered the President's Science Advisory Committee to investigate the issues the book raised. Another factor that enhanced the impact of *Silent Spring* was timing. The sixties were a period of cultural upheaval. Mainstream society was being challenged on many fronts. The Civil Rights Movement was at its apex, Vietnam War protests were widespread, and the Women's Liberation and U.S. American Indian Movements were launched. Hippies were following the advice of LSD-advocate Timothy Leary to "turn on, tune in, and drop out." The cultural conformity that characterized the 1950s was giving way to an antiauthoritarian attitude and a sense of empowerment to act on it.

In the midst of this decade, the U.S. government Bureau of Reclamation, which manages water in the Western states, made plans to build dams on the Colorado River that would have flooded sections of the Grand Canyon National Monument. These plans were withdrawn in response to a huge public outcry spurred by the innovative tactics of intrepid environmental crusader David Brower (see Figure 2.8). A former mountaineer, Brower had served as the first executive director of the Sierra Club since 1952. He had fought prior dam battles, with some success and some failure. This time around, he masterminded a novel approach to garnering grassroots support. In June of 1966, he took out full-page ads in the *New York Times* and the

FIGURE 2.8 *David Brower.*
In the 1960s, as Executive Director of the Sierra Club, Brower led a successful campaign to prevent dam building and flooding in the Grand Canyon.
With permission from Getty Images

Washington Post saying: "Now Only You Can Save the Grand Canyon from Being Flooded . . . for Profit" and "Should We Also Flood the Sistine Chapel So Tourists Can Get Nearer the Ceiling?" The public response was tremendous. Membership applications flooded into the Sierra Club and letters of protest bombarded the office of then-Secretary of the Interior, Stewart Udall. In the end, Congress passed legislation prohibiting dams anywhere in the Grand Canyon. Brower went on to found the Friends of the Earth and the League of Conservation Voters, both in 1969, followed by the Earth Island Institute in the 1980s.

In the late sixties, young adults were particularly eager to embrace counter-cultural ideas and were energized to confront the status quo. Wisconsin senator Gaylord Nelson devised a way to harness this youthful enthusiasm for the benefit of the environment. He sponsored a nationwide teach-in about environmental issues in 1970: the first Earth Day (see Figure 2.9). As it turned out, the participants in Earth Day included not just students (from thousands of colleges, universities, secondary, and elementary schools across the country), but also adult citizens who demonstrated in their communities, protesting the deteriorating environmental conditions and lack of regulation of industrial pollution (EPA, 2013a). The total number of participants is generally estimated to have been about 20 million Americans, some of whom vividly expressed their outrage by marching down Fifth Avenue in New York wearing gas masks, pounding cars with sledgehammers, and pouring sewage on corporate carpets (Hayes, 1990). Never before had there been

FIGURE 2.9 *Denis Hayes.*
Wisconsin Senator Gaylord Nelson recruited the Harvard University student to coordinate the first Earth Day, 1970.
With permission from Press Association Images

such a large-scale expression of proenvironmental sentiment. And the impact was significant.

Environmental historians refer to the 1970s as the "Green Decade," because it was shortly after Earth Day that some major policy changes took place, including (but not limited to) formation of the Environmental Protection Agency, and passage of the Clean Air, Clean Water, and Endangered Species Acts. One of the animals that appeared on the first Endangered Species list was the bald eagle, the U.S. national bird. These events all occurred during the Nixon Administration. It is unclear whether Nixon was personally a supporter of environmental protection, but he was at least politically savvy enough to recognize that the public was demanding change. Although many people consider the environmental record of Nixon's successor Gerald Ford rather lackluster, he was the president who signed the 1976 Toxic Substances Control Act, which initiated regulation on polychlorinated biphenyl (PCB) products.

Ford was the only president to have served as a National Park Ranger and during his term, he added 18 areas to the National Park System (National Park Service, 2006).

Ford's successor, Jimmy Carter, had a strong proenvironmental record.[1] Throughout his presidency (1977–1981), Carter proposed and lobbied for numerous environmental protections. Among other accomplishments, his administration oversaw the formation of the Department of Energy, as well as passage of the Soil and Water Conservation Act, the Surface Mining Control and Reclamation Act, the Superfund Act (which gave the EPA the charge of cleaning up toxic sites), and the Alaska National Interest Lands Conservation Act, which set aside an unprecedented 104 million acres. And, in 1979, to set an example for his constituents, Carter had a 32-panel solar water heating system installed on the White House. At the dedication he said,

> In the year 2000 this solar water heater behind me, which is being dedicated today, will still be here supplying cheap, efficient energy . . . A generation from now, this solar heater can either be a curiosity, a museum piece, an example of a road not taken or it can be just a small part of one of the greatest and most exciting adventures ever undertaken by the American people.
>
> (Biello, 2010)

As it turns out, it was a road not taken.

Perhaps it surprises you that President Carter was promoting solar panels more than 35 years ago; did you know solar panels existed way back then? Would it further surprise you to learn that the first wind farm was built in 1980? Carter's plan was to have 20% of the nation's energy generated by renewable sources by the year 2000, a somewhat more ambitious target than the 3.5% actually achieved (Roberts, 2011). So what happened?

As you've seen, the 1960s and 1970s seemed to represent a turning point, a surge in ecological awareness and concern. But, for the majority of the American public, the underlying (dominant social) paradigm had not shifted. The worldview assumptions discussed earlier were essentially unchallenged. For example, a lot of people recognized that pollution was a problem, but instead of quelling the practices that created waste, the solution was to "give a hoot, and don't pollute," (i.e., throw your trash in the landfill, not on the road). There were, however, exceptions to this, in the form of David-and-Goliath-type confrontations where relatively powerless citizens challenged the practices of powerful industries. For example, housewife Lois Gibbs became a household name across the country for her tireless work documenting unusually high rates of serious physical ailments (e.g., miscarriages and birth defects) among her neighbors in the Love Canal development of Niagara, New York; the suburban development had been constructed on a site that a chemical company had previously used as a toxic dumping ground (Breton, 1998; see Figure 2.10). After years of persistent effort by the homeowners, led by Gibbs, Love Canal was eventually deemed a **Superfund Site**, a region designated to receive federal clean-up funds.

FIGURE 2.10 *Love Canal protester, ca. 1978.*
Under the leadership of Lois Gibbs, residents of the Love Canal housing development in Niagara, New York succeeded in getting their neighborhood declared a Superfund Site. The development had been built on a chemical company's former dumpsite.

Although the green decade had witnessed numerous major developments in environmental regulations and policy, a fissure was developing along political party lines. Teddy Roosevelt was a Republican. Nixon and Ford were Republicans who worked with a Democratic congress. Jimmy Carter's successor, Ronald Reagan, was a Republican, too, but his attitude toward conservation and environmental regulation was decidedly negative, a position that has influenced the party to this day.

Professional Environmentalism, Direct Action, and Wise Use

Reagan campaigned on the promise of restoring America to the economic boom era of the 1950s and claimed environmental regulations were hurting the economy. Days before the 1980 election, during his final debate with Carter, Reagan asked Americans:

"Are you better off than you were four years ago? Is it easier for you to go and buy things in the stores than it was four years ago?" Reagan's campaign platform rested on the assumptions of the DSP as we have described them, and a majority of U.S. voters found this appealing; he beat Carter by a landslide. Once elected, Reagan cut research and development budgets for renewable energy, eliminated tax breaks for wind and solar, issued more leases for oil, gas, and coal development on national lands than any other president before or since, and removed the solar panels from the White House.

When Reagan was elected, environmental advocates were mobilized like never before. Most notably, large organizations including the Sierra Club, Wilderness Society, and the National Audubon Society recognized that they were going to have to up the ante to play against Reagan. Environmentalism was becoming a formal profession and the organizations matured into multimillion-dollar non-profits, complete with offices in Washington, D.C., boards of directors, and staffs of lobbyists, analysts, attorneys, and marketers (Gottlieb, 2005). At the same time, five self-proclaimed "rednecks," including Dave Foreman, a disillusioned employee of the Wilderness Society, founded the radical EarthFirst! movement with the rallying cry: "No Compromise in Defense of Mother Earth!" The movement's **nonviolent direct action** consisted of more extreme tactics than any that had previously been employed in the name of environmental protection; these included civil disobedience (e.g., forming human blockades on logging roads to prevent trucks from passing) and "monkeywrenching" (e.g., damaging logging equipment so as to render it inoperable) (Zakin, 1993).

In response to the explosive growth of the environmental movement in the early 1980s, antienvironmental constituencies also became more organized and came together to promote what they called the **Wise Use Agenda**. The 25 goals of this agenda, which were officially published in 1988 by the Center for the Defense of Free Enterprise (www.cdfe.org), included:

- Immediate development of the petroleum resources of the Arctic National Wildlife Refuge in Alaska.

- Logging three million acres of the Tongass National Forest in Alaska.

- Exempting from the Endangered Species Act "Non-adaptive species such as the California Condor and endemic (locally based species) lacking the biological vigor to expand in range . . ."

- The right of prodevelopment groups "to sue on behalf of industries threatened or harmed by environmentalists."

- Opening seventy million acres of federal wilderness to commercial development and motorized recreational use.

- Opening all public lands "including wilderness and national parks" to mining and energy development.

(Helvarg, 2004, pp. 45–46)

Ron Arnold, the man who gave the Wise Use movement its name, borrowed the term from the work of Gifford Pinchot. In 1991, Arnold told *Outside* magazine that he chose the name because it was short and "marvelously ambiguous," not because he shared Pinchot's belief in conservation (Krakauer, 1991).

In a 1996 essay called "Overcoming Ideology," Arnold wrote about the Wise Use Movement:

> As the environmental debate developed during the late 1980s, the "dominant Western worldview" gained an organized constituency and advocacy leadership: the wise use movement . . . [which] centered around a hodgepodge of property rights groups, anti-regulation legal foundations, trade groups of large industries, motorized recreation vehicle clubs, federal land users, farmers, ranchers, fishermen, trappers, small forest holders, mineral prospectors and others who live and work in the middle landscape . . . It came as a shock to environmentalists. The "competing paradigm" [i.e., environmentalism] unhappily found itself confronted with a competing paradigm [i.e., Wise Use]. . . . [which] advocated unlimited economic growth, technological progress and a market economy.

(pp. 18–19)

Besides the ramping up of environmentalism in the 1980s, the other motivating factor that compelled the Wise Use movement was fear inspired by the election of Republican George H. W. Bush in 1989, who campaigned on the promise that he would be "the environmental president." It was a hollow promise.

Partisan Policies and a Persistent Paradigm

It wasn't until Democrat Bill Clinton and his unambiguously proenvironmental Vice President Al Gore entered the White House in 1993 that the federal policy pendulum started swinging back toward where it had hung in the 1970s. The Clinton-Gore administration strengthened the Clean Drinking Water Act, tightened regulations on air pollution, accelerated the clean-up of Superfund Sites, raised environmental reporting standards for chemical manufacturers, and preserved millions of acres as national parks, monuments, and designated wilderness.

Then, in 2000, Republican George W. Bush was elected. His administration proceeded to weaken the Clean Water, Clean Air, and Endangered Species Acts; reduce funding for toxic waste clean-up programs; decrease the enforcement authority of the Environmental Protection Agency; lower toxics reporting standards for industry and mercury reduction requirements for power plants; approve mountain-top removal for coal mining; open millions of acres of wilderness to mining, oil and gas drilling, and logging; and, perhaps most significantly, announce that the United States would not sign the United Nations Kyoto protocol for limiting carbon emissions from industrialized nations because it did not put limits on developing nations, such as China. The stated concern was that U.S. jobs might be outsourced (Goldenberg, 2009).

Then, in 2008, Democrat Barack Obama was elected. Under Obama's leadership, the EPA raised emissions standards for cars and light trucks, and proposed the first standards for heavy trucks; introduced mercury emissions controls for coal- and oil-fired power plants; cut allowed soot emissions by 20%; developed a National Ocean Policy; and proposed the first greenhouse gas emissions standards for power plants, and the first regulations for storing and disposing of coal ash waste. In his second inaugural address, Obama vowed: "We will respond to the threat of climate change, knowing that failure to do so would betray our children and future generations."

Are you feeling like you're on a teeter-totter? As you can see, what commenced with the election of Reagan was a partisan polarization in the realm of environmental regulation, a polarization that was predicted by perceptive social scientists nearly a decade earlier (Dunlap & Gale, 1974) and has increased in the last dozen years (McCright & Dunlap, 2011). Importantly, however, even the ostensibly more environmentally friendly Democratic Party in the United States does not include in its platform any serious challenges to the Dominant Social Paradigm. For both parties, the primary agenda item has been, as famously stated by Bill Clinton's campaign strategist, "the economy, stupid"—an economy based on the capitalist ideals of personal profits and perpetual growth.

CONCLUSION

It is likely that real challenges to the Dominant Social Paradigm won't happen at the level of government. Counterculture radicals are rarely elected to public office. Consequently, if there is to be a shift in our cultural worldview, it will have to start with *individuals thinking differently*, individuals like Henry David Thoreau, John Muir, Bob Marshall, Aldo Leopold, Rachel Carson, David Brower, Gaylord Nelson, Denis Hayes, Lois Gibbs, Dave Foreman . . . and you.

By now you have a grasp on what *unsustainable* living entails. The next chapter provides a template for an alternative, ecologically harmonious modern worldview. It describes some basic principles of natural systems that can guide sustainable behavior, and offers examples of these principles being applied successfully to human systems.

NOTE

1. Having completed some graduate work in nuclear physics, Carter was interested in the option of nuclear energy to reduce dependence on oil and coal. His only real concern with the technology seemed to be the threat that plutonium byproducts would be used to create nuclear weapons, not that the waste would be an ecological hazard.

CHAPTER 3

Where Do We Go from Here?
Developing an Ecological Worldview

> Look deep into nature, and then you will understand everything better.
>
> Albert Einstein (2005, p. 60)

By the time you entered college, you had already formed the worldview that drives and directs your behavior. If you are from a Western industrialized culture, perhaps you recognize that your beliefs include some or all of the assumptions discussed in the previous chapter. That is to be expected because, after all, these assumptions are part of the *dominant* paradigm.

The assumptions that make up our worldviews are so deeply embedded and generally feel so *true* that they can be very difficult to see, and even harder to question. For example, the first edition of this book was written by psychologist Deborah Winter. While she was working on it, she had a friend over to dinner who asked her what she was writing. Deborah described some of the ideas addressed in the previous chapter, emphasizing in particular how many people in Western industrialized cultures see themselves as superior to the rest of nature, which is subject to their control. As Deborah was describing this idea, a carpenter ant crawled across the counter and she mindlessly smashed it. "Is that a problem?" her friend asked. "Yes," Deborah answered, still thinking about the view of nature, "because it allows people to unconsciously manipulate or harm anything in the name of human convenience." "No" her friend said, "I mean is *that* a problem?" pointing to the smashed ant. "Yes, those damn things are destroying our house." "But isn't that *the* problem?" persevered the friend. Deborah looked up in puzzlement and then realized what she had done. Even though she had spent all day writing about people's assumption that we have the right to control nature, she hadn't realized she was acting on that very assumption when she automatically smashed the ant. The assumption was so deeply ingrained that even as she consciously critiqued it, she still unconsciously accepted it.

The majority of our daily actions happen without much thought. This means they are being driven by automatic habits. It is a good thing that we can behave automatically, because if we had to think carefully about our every decision we'd be stuck in analysis paralysis! So, we end up relying heavily on mental shortcuts that allow us to act quickly and decisively and move on (more detail to come in Chapter 6). These mental shortcuts are based on things we have learned from our culture about how the world works and what's most important. For example, most of us ask ourselves "Is it worth the money?" when considering a purchase. We have learned through advertisements ("Lowest prices in town!"), store titles ("Pick 'n' Save"; "Best Buy"), and hearing others talk ("Was it on sale?") that economics trump other considerations when shopping.

Unfortunately, as you read in the previous chapter, the assumptions of the Dominant Social Paradigm (DSP) in Western industrialized culture are causing ecological demise. If we are going to survive as a species on this planet, we cannot continue to rely on habits that stem from these pervasive but ecologically incompatible ideas. In order to align our future choices with resource realities, we need to develop a different worldview; we need to think more like ecologists!

Here is something to ponder. Before humans developed the technologies that have proven detrimental to ecosystems, it wasn't necessary to understand ecology, or think like an ecologist, in order to behave in an ecologically harmonious way; there simply was no alternative to living within the bounds of natural systems.

Behaviors were ecologically compatible by default. Even very recently, technological innovations have made some behaviors more ecologically detrimental than they previously were. For instance, before the creation of nonbiodegradable materials such as plastic (which, as you learned in the last chapter, was only a few generations ago), throwing things away did not have the same consequences. Consider a common practice in India that occurs when people eat during their daily train commutes. Historically and today, after finishing their meals, riders simply toss their food wrappings out the train windows (Bhave, 2014). When food was wrapped in paper, one would encounter discarded bits of this and that along the tracks, but the trash consisted of materials that *eventually* broke down and were reabsorbed. In contrast, consider the consequences now that paper wrappers have been replaced by plastic containers; imagine the tons of inert materials now lining the tracks, blowing hither and yon to perch on tree limbs and settle into waterways. The behavior isn't so different, but the consequences are, because people now produce things that do not follow nature's rules.

ECOLOGICAL PRINCIPLES

Because it is not the standard way to do things today, *behaving in an ecologically compatible way requires conscious intention* (at least until those behaviors become automatic habits). Given this, how can we learn to think in ways that will lead to sustainable behaviors? The best teacher is nature itself. "After 3.8 billion years of research and development, failures are fossils and what surrounds us is the secret to survival. . . . The more our world looks and functions like the natural world, the more likely we are to endure" (Benyus, 2010, p. 34). In other words, the structure and function of natural systems provide models for ecologically harmonious human behavior and criteria against which we can evaluate the sustainability of our choices.

When ecologists study earth systems, they generally do not judge organism-environment interactions as being good or bad, they just are what they are. However, many of the ways humans currently interact with the natural environment are undeniably problematic. To move in a sustainable direction, we need to better align our thinking and behaviors with the ways natural systems function. Therefore, the following five fundamental principles, grounded in ecology, will give you a taste of an *ecologically informed worldview*. Depending on your upbringing and experiences, some of these ideas may be familiar to you, others not so much. In any case, you will be able to see how they contrast with the DSP assumptions described in Chapter 2. These tenets are drawn from a wide range of classic writings and contemporary research, and represent just a sampling of the full array of concepts that might comprise an ecological worldview. The chapter concludes with some big ideas and real examples to give you an idea of what our world can look like when we aspire to ecological harmony.

All Life Is Interdependent

We all belong to the system of Planet Earth. And in any working system, all parts are interdependent. Over a century ago in *My First Summer in the Sierra*, naturalist and preservationist John Muir wrote: "When we try to pick out anything by itself, we find it hitched to everything else in the Universe." (1911/1988, p. 110). In other words, while it is not always clear to us, *everything* we are, do, and have depends on a resource or function provided by natural systems and their nonhuman inhabitants.

Ultimately a complex system survives to the degree that it serves its component parts—ALL of its component parts. Thus, we ignore the state of ecosystems and our role in their healthy functioning at our own peril. Take the case of the honeybee. In recent years, honeybee colonies have been dying off at unprecedented rates. In

fact there are half as many colonies now as there were in the 1940s. It is thought that this phenomenon, called Colony Collapse Disorder, is due to a mix of new pathogens and parasites, overcrowded hive conditions, contact with incapacitating chemical insecticides, and habitat stressors such as lack of plant diversity, and reduced access to clean water (USDA, 2013). Human behavior has a hand in all of these causes.

It may be hard to care much about the plight of honeybees when there are bills to pay and exams to study for. However, if you like your morning coffee, your almond trail mix, and your crisp veggie salads, listen up: in the United States, about one out of every three bites of food we eat depends on honeybees for pollination (USDA, 2013). While there are other native pollinators, most commercial fruit, vegetable, and nut crops are highly dependent on the honeybee. Furthermore, many of the other native pollinators, including bumblebees and monarch butterflies, are also on the decline due to habitat loss, climate change, insecticides, and the interactions between these stressors (Potts, et al., 2010).

Some human communities, most notably in Southwest China, have already inadvertently wiped out their pollinator species. Orchard workers must now hand-pollinate every blossom on their apple and pear trees using paintbrushes (Partap & Ya, 2012). Needless to say, there are not enough humans on the planet to replace the work of pollinators. Besides, the financial cost of this kind of human labor would render much of our diet unaffordable. To continue to think of ourselves as separate is to ignore the stark reality of our fundamental interdependence with all other parts of nature, and in doing so, threaten our very survival.

Small Actions Can Cause Big Consequences

Ecosystem interdependence means we can't just alter one thing in isolation. Even one seemingly minor change can ripple through a system, sometimes slowly and sometimes more quickly, sometimes dramatically and sometimes subtly, and often in ways that couldn't be foreseen. While science constantly advances our understanding of complex ecosystems, this knowledge is far from complete. As environmental scientist and systems thinker Donella Meadows (2008, p. 86) put it, "our models fall short of representing the world fully. That is why we make mistakes and why we are regularly surprised."

For instance, although humans have harnessed coal for heat since the Stone Age, the widespread use of it to generate electricity dates back only to the late 1880s. Unlike our Victorian era predecessors, however, we now know that burning coal releases particulate waste containing heavy metals such as mercury into the atmosphere. In fact, "coal-burning power plants are the largest human-caused source of mercury emissions to the air in the United States, accounting for over 50 percent of all domestic human-caused mercury emissions" (EPA, 2014c). Minute mercury particles are distributed via rain into rivers, lakes, and streams where **biomagnification** occurs: Plankton digest particles, small fish eat plankton, big fish eat small

fish—and we love to eat the big fish like tuna and halibut, which now have elevated levels of mercury in their delicious fatty tissue (Lavoie, Jardine, Chumchal, Kidd, & Campbell, 2013). As you will read in Chapter 9, mercury is a significant health hazard for humans. By burning coal for energy, we have inadvertently poisoned a food source that otherwise supports healthy brain development and functioning.

Although you may not know exactly how your daily decisions affect your ecosystem, you can be sure that they do. Being unintentionally (or willfully) ignorant of the causal relationships between our actions and other parts of the ecosystem can lead us to behaviors that seem preferable in the moment, but ultimately have significant, negative consequences. Rather than working from the overconfident assumption that we can fully control nature, we would do better to try and understand the system in which we're embedded, examine carefully the extended links likely to be affected by our behaviors, and then act with humility and caution, knowing things may not turn out as we planned.

Life Systems Are Circular

Not only are we living in a complex system, we are living in a *closed* system which currently contains all of the natural materials it ever will. Even though resources are finite, natural systems work because they *cycle* in ways that "create conditions conducive to life" (Benyus, 2010). Animals eat other animals and plants, producing waste that builds soil, which provides nutrients to plants, which are then eaten, and

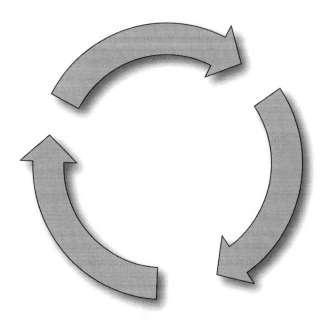

so on. Water evaporates, travels in the form of clouds, precipitates as rain, fog, and snow, migrates back into bodies of water, aquifers, and glaciers, is consumed and employed by all living species somewhere along the way, only to eventually evaporate again, over and over. In natural cycles such as these, resources are continually replenished, there is really no such thing as "waste," and the process can continue indefinitely without compromising the system.

Some human systems are also circular in nature. Until the mid-twentieth century, most family farms were self-sustaining, using practices focused on building soil by cycling a variety of nutrients back into the system. With circular practices, farmers can continue to cultivate the same fields year after year. In contrast, industrial systems tend to function *linearly*: Nutrients are extracted from the soil repeatedly until it can no longer provide any more, at which point synthetic chemicals must be added or the now exhausted field is abandoned in favor of one that still has some life left. When we think in terms of extractive lines, rather than restorative circles, we consume materials faster than they can regenerate and remove materials without taking the necessary steps to replenish them.

Not only are many of our systems linear, having a starting point (extraction) and an ending point (waste), our knowledge of them is often truncated. As consumers, we are unaware of steps that came before we obtained the resource and oblivious to steps that occur after we are finished with it. Food serves as a good example here, too. Picture yourself buying groceries: You are probably thinking about how appetizing your choices are, not about the soil and water that created them or, for that matter, the farmers and farm workers who grew and harvested them. So, for you, the grocery store is the beginning of the linear food system. After dinner, you scrape your left-overs into the trash. And, at this point, the system ends for you. This is a common U.S. perspective: We buy food, we eat, and we dispose of the leftovers. But, where does that food go after we scrape it into the trash? Do the nutrients return to the soil to help create more food (as nature would have it)? Not if they end up sitting with other trash in a landfill, they don't. This means that by scraping food into the trash, we have removed precious organic matter from the system. How long can we keep removing nutrients without replenishing them? How long can this linear practice endure before the system is so depleted that we can't grow enough food?

To be sustainable, our food production must be circular and, by modifying some key behaviors, we *can* reintegrate into the natural system. However, some human-made systems are simply ecologically incompatible (and therefore unsustainable) because they are inherently linear rather than circular. These are systems which rely on nonrenewable inputs and/or produce wastes that cannot be used by the system to produce new materials. Can you think of examples?

There Are Limits to Growth

In 1971, the year after the first Earth Day, beloved children's author Theodore Geisel (aka Dr. Seuss) published his book *The Lorax*. Perhaps you read this book as a child. Read it as an adult and you will be confronted with a not-so-subtle message about the consequences of exploiting limited natural resources in pursuit of personal gain. The story is told retrospectively by the Once-ler, a former entrepreneur who developed a method for knitting the tufts of Truffula trees into multipurpose "Thneeds." He reflects,

Then, Oh baby, Oh how my business did grow

Now chopping one tree at a time was too slow

So I quickly invented my Super-Axe-Hacker

which whacked off four Truffula trees at one smacker...

I meant no harm I most truly did not

but I had to grow bigger so bigger I got. . .

I went right on biggering selling more Thneeds

And I biggered my money which everyone needs

The Once-ler continues with a description of what happened when he was challenged by the angry Lorax, who scolded him for his production of "smogulous smoke" and "Glupitty-Glup" which was destroying the habitat of the Swomee-swans and Humming-Fish,

Well I have my rights, sir

And I'm telling you

I intend to go on doing just what I do

And for your information you Lorax, I'm figuring

on biggering and biggering and biggering and biggering. . .

And at that very moment we heard a loud whack

From outside in the fields came a sickening smack

of an axe on a tree

Then we heard the tree fall

The very last Truffula tree of them all

One belief that gave rise to the Industrial Revolution (and apparently inspired the Once-ler's zealous logging) was the assumption that natural resources are inexhaustible. This idea seemed accurate from the seventeenth to twentieth centuries, but is clearly out of date in the twenty-first century. Even industrial leaders are reaching this conclusion. For instance, back in 1997, Robert Shapiro, the CEO of Monsanto (the multinational, agro-chemical company, widely known as a pioneer in biotechnology such as genetically modified organisms), stated, "What we thought was boundless has limits, and we are beginning to hit them" (as quoted by McDonough & Braungart, 1998, p. 82). One example of these limits is the worldwide collapse of wild tuna and salmon populations. Over 50% of our planet's fisheries have already been fully exploited and another 30% are endangered. And, the World Wildlife Fund estimates that the global fishing fleet is two to three times larger than what we need (WWF, 2014). The consequences of overuse are reversible, but not without penalty. "The land recovers, but at some reduced level of complexity, and with a reduced carrying capacity for people, plants, and animals" (Leopold, 1949, p. 257).

Many of the problems we now face stem from the fact that finite resources must be *shared*. Acting in self-interest by taking more than one's fair share hurts the common good (consider the Once-ler's use of Truffula Trees and creation of "smogulous smoke" and "Glupitty-Glup"). If each individual uses the resource for maximum personal benefit, everyone ultimately suffers as the resource is overused and degraded (more on dilemmas like this in Chapter 5). Take a moment to brainstorm why a community would overfish to the point of losing a fishery altogether. What do individuals have to gain? Who bears the burden when the fishery is depleted? What impulses do we have that make us behave even more aggressively when resources become scarce? In the meantime, consider that while there are ways to use our shared natural resources without risking their long-term viability, the Dominant Social Paradigm emphasizes the right of the individual to pursue personal happiness, and promotes material acquisition as the way to achieve it. Although pursuing personal happiness is not inherently at odds with sustainability, it *is* unsustainable to continue to equate happiness, progress, and success with unrestrained resource use.

Diversity Equals Resilience

What happens when one part of a system fails? Hopefully there's a back-up, a Plan B, a safety net. When there is potential for multiple system members to do the same job, one functioning member can compensate for another malfunctioning member and a larger system fail is averted. Organisms with redundant functions create ecosystem flexibility and resilience (Wohl, Arora, & Gladstone, 2004). In contrast, systems of reduced complexity become vulnerable and the damage from a single component failure is more likely to ripple through the interdependent network. Consequently, another key ecological principle is that system integrity depends on *diversity*.

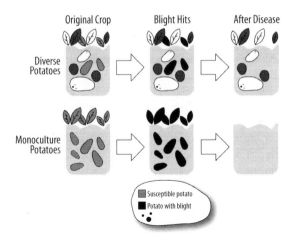

Original Crop   Blight Hits   After Disease

Diverse Potatoes

Monoculture Potatoes

■ Susceptible potato
■ Potato with blight

FIGURE 3.1 *Diversity of plantings creates crop resilience.*
As illustrated by the example of the potato blight, when disease strikes a monoculture, the whole crop is wiped out. Diverse plantings protect the crop because some species may be resistant to specific threats.

Our monoculture-reliant food system is an example of a simplified system, and its lack of diversity puts our food security at risk (Tuomisto, Hodge, Riordan, & Macdonald, 2012). Monocultures have reduced resistance to stressors and are at greater risk for total failure. When there is little variety in what is grown, an entire farm (which is a system) will fail if its single most important plant fails to grow. This was vividly demonstrated by the devastating Irish potato famine of the 1840s (see Figure 3.1). More recent monoculture failures have included North American corn crop devastation due to fungus and significant loss of Russian wheat due to extremely cold temperatures.

If choosing vegetables based solely on price, we will most likely purchase from a few, often distant, enormous monoculture farms rather than from small, local producers. Price pressure puts local farmers at a disadvantage, forcing some of them to go out of business. Each time a system loses a member, resilience is diminished. With fewer farmers, if there is a crop failure, the consequences have a much farther reach. At the time of this writing, it is predicted that, due to extreme drought in California's Central Valley, about one-quarter of its crops will be lost. Much of the United States now depends on California for a significant amount of its produce. This simplification of the food system might affect you directly: You may notice reduced availability of common fruits and vegetables, as well as higher prices, due to drought or flooding in distant places.

There are additional problems with putting all of our proverbial eggs in one basket. For instance, when contamination occurs in large monocultures, it affects more people. In 2006, an e-coli outbreak originating from one field in California affected people in 26 states. In contrast, people who grew their own spinach, or bought directly from farmers elsewhere in the country, could eat with confidence throughout the crisis.

In sum, complexity is important for overall system functioning. Complexity of the ecosystems that support our lives is reduced by overuse of resources, over-reliance on a few system members to the exclusion of others, and outright eradication of species. Because of human encroachment on, and despoilment of, natural habitats, one third of the species in the United States are classified as vulnerable or endangered (Wilson, 2006). For most of geologic history, species extinction occurred at the same rate as speciation, the emergence of new species; the extinction rate is now estimated to be 100 times that of speciation, with the rate accelerating as people harvest more resources from the few remaining biodiverse areas of the planet (e.g., the Carribean & Madagascar; Myers, Mittermeier, Mittermeier, daFonseca & Kent, 2000). Biologist E. O. Wilson (1992) has argued that loss of biodiversity is the most harmful aspect of environmental degradation, because it creates a disastrous feedback loop. As species are lost, their ecosystems become simplified and are more likely to fail, resulting in more species loss and extinction.

> The variety of organisms . . . once lost, cannot be regained. If diversity is sustained in wild ecosystems, the biosphere can be recovered and used by future generations to any degree desired and with benefits literally beyond measure. To the extent it is diminished, humanity will be poorer for all generations to come.
>
> (Wilson, 1992, p. 35)

In other words, species extinction reduces an ecosystem's ability to withstand natural disasters and further human-instigated damage. Combine this with the principle of interdependence and it is clear that when we cause the loss of other species, we put ourselves at risk too.

The five natural systems principles described above provide the seeds for an ecological worldview. With these in mind, you are better equipped to identify optimal solutions to ecological problems. To further help you, here's a bonus principle:

Upstream Solutions Are Better than Downstream Solutions

> The residents who live in the village Downstream had built their community beside a river. Many years ago, they began to notice that growing numbers of drowning people were caught in the river's swift current. So they went to work inventing ever more elaborate technologies to save them. Talk to the people of Downstream today, and they'll speak with great pride about the hospital by the edge of the water, the fleet of rescue boats ready for service, or the large number of dedicated lifeguards ready to risk their lives to save victims from the raging waters. So preoccupied were these heroic villagers with rescue and treatment that they never thought to look upstream to find out why people were falling in the river in the first place.
>
> (Baxter, Boisvert, Lindberg, & Mackrae, 2009, p. 4)

SWIMS UPSTREAM

Because downstream is too mainstream

An ounce of prevention is worth a pound of cure (Franklin, 1735; as found in Keil, 2011). A stitch in time saves nine (Fuller, 1732; website). An apple a day. . .well, you probably get the point. Tackling the *root cause* of a problem is ultimately more efficient and effective than trying to relieve its symptoms. In other words, upstream changes make more of a difference than downstream changes.

Consider food waste. Composting uneaten food scraps so that they decompose and return to soil completes the natural cycle, but these days, most uneaten food ends up in landfills instead. Rather than trying to deal with where the waste ends up, upstream solutions eliminate the problem of leftover food in the first place. For example, we can make an effort to better estimate how much food is likely to be eaten so as to not prepare too much, or we can intentionally plan to take leftovers for lunch the next day. If we have too much food, we can invite friends over to share in the bounty. Here is another example: If your community has curbside recycling, you may have noticed how people will put out paper bags, neatly arranged, and filled with other paper bags. Clearly folks accumulate more paper bags than they need. Is it good that the paper bags are being recycled instead of being thrown away? Of course. But recycling is a downstream solution. It is after the fact, and it requires additional resources to transfer, process, and remake the material into new products. A more efficient solution would be to prevent the accumulation of paper

bags in the first place by taking a canvas bag to the store, or by not using a bag at all for just a few items. This example illustrates *the priority of reduce and reuse over recycle*. Clearly, we don't want people to stop recycling, but recycling is one of our least efficient options.

Here's a detailed example that is literally an upstream/downstream issue. Perhaps you have heard that every summer, a zone appears in the Gulf of Mexico where aquatic creatures cannot survive because the area is devoid of oxygen. This dead zone is not small; in 2013, it grew to be the size of Connecticut, nearly 6,000 square miles (Main, 2014). The cause of this lifeless region lies upstream. The Gulf of Mexico is where the mighty Mississippi River reaches the ocean. By the time it gets there, it is laced with excessive amounts of nutrients coming from fertilizers that have washed into the water from lawns, gardens, and farmland all along the river's enormous watershed (which includes tributaries from 31 states and two Canadian provinces sandwiched between the Rocky and Appalachian mountains). While "nutrients" might sound like a good thing, they actually create large algea blooms, and the trouble begins when the algae die. Their decomposition consumes oxygen from the water, lowering the concentration below the level necessary to support most animal life; this is called **aquatic hypoxia**.

There isn't a good way to address hypoxia downstream. To prevent these annual die-offs, and the negative impact they are having on fisheries in the Gulf, people who live in the states and provinces that drain to the Mississippi will need to reduce or halt their use of the problematic fertilizers. This does not mean foregoing beautiful landscaping or healthy farms, but it may mean planting species that are more biologically suitable to the climate, and transitioning away from monoculture farms that require heavy doses of supplemental nutrients.

Upstream solutions are sometimes a tough sell. The most upstream approach, prevention of problems before they have even manifested, can seem costly in terms of lost opportunity. For example, it can feel like a waste to not cut down those trees that are just standing there, or to not mine the plentiful, thus cheap, coal. Once problems are already present, and it is too late to prevent them entirely, upstream interventions require retrofitting, reworking, rethinking, and sometimes going back to square one; in other words, they are expensive, time-consuming, and effortful. And, when we alter our modus operandi, we may be left with infrastructure that is now useless. Think of the hospital built by the people of Downstream. If they fix the problem upstream so that people stop falling into the river, what will they do with the expensive building? What will happen to the employees who have developed specialized skills? Similarly, if we switch from fossil fuels to nonpolluting wind and solar energy sources, we will have vacant coal plants and unemployed oil rig workers on our hands. There will always be economic pressure to maintain the status quo rather than pursue upstream solutions.

Thinking ecologically is only the first of many steps needed to actually move in a sustainable direction. Knowledge must be translated into ideas, ideas into plans, and plans into action. But what kinds of action? What changes should people be striving for? What will it really mean to live sustainably? The next part of the chapter offers a sampling of inspirational ideas and exemplary innovations to help you begin to envision a sustainable future. Perhaps they will also serve as the seeds for your own decisions, designs, and dreams.

LIVING THE DREAM OF A SUSTAINABLE WORLD

We share just one planet, yet most of us live as if there is some other source for meeting our needs and fulfilling our desires. The following sections revisit two components of our ecological footprint described in Chapter 1: food and material goods. Looking through the lens of an ecological worldview, you'll see that sustainability will require radical changes to our own lifestyles and to the larger systems in which we live. In other words, it will not be enough to individually behave less consumptively and wastefully. We need to collectively reenvision the big picture. It is a grand opportunity for creativity.

Food

We know a lot about what we eat: We count calories, look at nutritional labels, and we have favorite brands. What we don't tend to think about is where our food comes from and how it is grown. Understanding the impact of what we eat is a very basic but essential part of the journey toward sustainability. Each meal is an opportunity to have a say about our future.

Interconnection. There are many methods, old and new, for creating sustainable food systems. One common strategy for farming and gardening is to leverage interconnectedness through **companion planting** techniques, basically intermingling plants with complementary qualities. A very simple ancestral example is the Three Sisters: corn, beans, and squash (Lewandowski, 1987). Corn stalks provide a natural pole for beans to climb, the bean vines in turn make the corn stalk more durable. Beans also fix nitrogen in the soil to enhance the fertility for future crops (Postma & Lynch, 2012). Squash leaves shade the stalk roots from sun, reduce moisture evaporation, and deter pests with their prickly spines. The three plants are also nutritiously complementary, providing carbohydrates, protein, and vitamins.

More recently, ecologists Bill Mollison and David Holmgren introduced **permaculture**, a method of designing spaces that focuses on the *relationships* between people, animals, plants, activities, *and* buildings (Holmgren, 2002). The goal is to create permanently productive, multifunctional, self-sustaining, low-maintenance spaces. Permaculture principles, such as observing patterns and using system feedback, are applied uniquely in every situation. A permaculture space generally includes a mix of perennial fruit trees, berry bushes and herbs, bees, small animals like chickens and fish, heirloom vegetables and legumes, and compost and rain capture systems, which together provide a synergistic network of functionality.

The reality for many of us is that we don't grow our own food. However, knowing about the interdependencies within food systems, such as the relationship between our food and pollination, can inspire a great variety of behaviors that support bees, birds, bats, and butterflies. Planting a window box with a variety of pesticide-free, native flowering plants, supporting neighbors who apply for apiary permits, and buying honey from urban beekeepers are among the many ways people have begun to engage with sustainable food production (PBS, 2014).

Circular Systems. Recall the CAFOs described in Chapter 1, and consider a circular alternative. Joel Salatin, author of *Everything I Want to Do is Illegal*, is a farmer in the Shenandoah Valley of Virginia (see Figure 3.2). Despite raising cows, pigs, chickens, turkeys and rabbits, he says he doesn't farm animals; he considers himself a *grass* farmer. But as the name of his farm, Polyface, suggests, it is not your average turf. It is a mix of 20 perennial varieties that utilize soil and sun to provide a diverse "salad bar" for his cows. Salatin relies on the natural cycles of plants and animals to run his farm: Cows graze a field and then move on to the next one; three days after the cows have cleared out, Salatin brings in a mobile chicken coop carrying his feathered "sanitation crew" to pick away at the larvae that develop in the cow manure. As the chickens digest the high-protein grubs, they disperse nitrogen throughout the field by excreting as they move freely about the field. Additional soil is built by microorganisms and soil biota digesting the roots of the cow-eaten grass. After six weeks of sun and rain, the grass has matured again, the pasture is ready for the cows and the cycle continues. In two generations, Salatin has transformed a farm once nutrient depleted by monoculture farming into one of the most fertile and fecund farms in the eastern United States (he produces over 100,000 pounds of meat each year on just 100 acres of land; Pollan, 2006).

FIGURE 3.2 *Joel Salatin, owner and steward of Polyface Farm.*
Courtesy of Polyface Farms

We can all contribute to closing the loop in our food systems. Salatin relies heavily on manure and other sources of compost for the soil nutrients needed to grow food, but nonfarmers can compost too. Prior to modern sanitation practices, which mix organic waste with all types of nonorganic waste, food scraps were fed to animals or composted by default. Some cities are revisiting this paradigm and providing curbside pickup of compostable material alongside trash and recycling. Even apartment dwellers without such services can create a circular system with under-cabinet **vermiculture**, using a tub of worms to transform food waste into soil. The soil can then be used to grow houseplants and pots of herbs and tomatoes in balcony gardens (Janagan, Sathish, & Vijayakumar, 2003).

Where else do food nutrients go besides the trash? Into our bodies, of course, by virtue of the food we eat. In a closed cycle, we would excrete whatever our bodies don't use in a way that feeds the system (see Figure 3.3). However, this is not the way we've set things up. Clearly, the introduction of the flush toilet was a critical step in reducing pandemic spread of disease in crowded cities. But why poop in perfectly drinkable water, especially when we then have to pay to separate it again using our elaborate sewage treatment systems? Drinking water is a precious and increasingly scarce resource worldwide, and human waste is chock full of nitrogen and other soil-building nutrients. Conventional sanitation makes them both unusable. There are lots of good alternative ideas floating around, ranging from composting toilets to buses that run on methane harvested from citywide poop "digesters" (March, Gual, & Orozco, 2004). The point is that thinking cyclically about food can have powerful results.

FIGURE 3.3 *The human nutrient cycle, intact and broken.*
Composted human waste is used to fertilize food-producing gardens in permaculture systems (from Jenkins, 1999, p. 29).

Limits to Growth. We've previously noted how various *wild* foods like salmon are a limited resource that is overtaxed by current fishing practices. But what do *domesticated* crops have to do with limited resources? Growing global demand for commodities such as meat, palm oil, coffee, and tea have contributed to extensive deforestation (in this case, rainforest in places like Brazil and Madagascar rather than the imaginary Truffula Trees). This means we can make a difference by examining our dietary impact and changing our relationship with food commodities, especially meat. The variables that shape the environmental impact of meat are numerous and complex, as are the potential responses. Some people decide to use meat as a side dish or only in a few meals each week, while others eat only the meat they hunt, catch, or raise. Some decide to eat only meat from carefully chosen sources (e.g., farms like Polyface). Others choose to forego meat altogether. While there are many reasons why millions of Americans choose **vegetarianism** (ranging from religious, to health, to ethical, to environmental) and many forms it can take (pescetarian, vegan, macrobiotic, raw), a plant-based diet generally uses substantially less land, water, and fossil fuels (Pimentel & Pimentel, 2003).

However, just replacing beef with tofu will not solve the problem because soybean plantations also contribute significantly to deforestation (Barona, Ramankutty, Hyman, & Coomes, 2010). Palm oil is quickly surpassing soybeans in popularity and impact on rainforests. In fact, over one third of the world's vegetable oil is now produced from oil palm (Fitzherbert, et al., 2008). This commodity is appealing because it is a far more efficient plant than soybeans. Nonetheless, the increased global demand for processed foods made with these oils, such as cereal, crackers, and cookies, continues to cause deforestation. Opting for rainforest-certified products is something consumers can do to mitigate their personal contributions to deforestation. Individuals can also use their voices to change the way food companies source their oils. Lucia von Reusner did just this. As a shareholder advocate for Green Century Capital Management, von Reusner phoned into a Kellogg earnings conference call and asked the CEO: "As a company publicly committed to sustainability, how will Kellogg ensure that the Kellogg brand is not associated with illegal deforestation?" Since that conference call, Kellogg has set up new supply chain criteria, only purchasing palm oil from companies that demonstrate sustainable growing practices. There has even been a ripple effect, inspiring policy and process changes among some of the most notoriously unsustainable palm oil suppliers (Aubrey, 2014).

Upstream Solutions. You've already read about the relative merits of composting and eliminating food waste earlier in the chapter. We can make many dietary choices even further upstream that reduce the demand for resources such as supplemental water and nonrenewable energy. For one thing, there is food all around us, but most of us haven't been taught to recognize it. Even in cities there are greens, vegetables, fruits, beans, berries, seeds, nuts, grains, and sources for sweeteners, flours, and oils growing wild in parks, vacant lots, and residents' backyards (Brown & Brown, 2013; Thayer, 2006). **Foraging** for wild food may sound like something relegated to wilderness camping and TV shows like *Survivorman*; however, it is growing in popularity and foraged wild foods are making their way into grocery stores, gourmet restaurants, and people's pantries. In addition, **urban food forestry** initiatives in dozens of U.S. cities have demonstrated the enormous potential in planting, mapping, and harvesting from city trees and shrubs (Clark & Nicholas, 2013). This requires transforming our mindset from one that assumes urban green spaces are merely for decoration or recreation to one that recognizes they can also provide goods (e.g., food, wood) to city dwellers (McLain, Poe, Hurley, Lecompte-Mastenbrook, & Emery, 2012).

Most of the food you consume is probably not wild; but within the realm of domestic agriculture, there is significant opportunity to reduce our impact. It may satisfy our senses to eat fresh strawberries whenever we want, however, eating them where they cannot grow easily or when snow is falling outside one's window means that they have been on a long, resource-intensive journey that creates a lot of CO_2. One method for mitigating this kind of impact is to buy **carbon offsets**, which are investments, typically in renewable energy and energy efficiency programs, intended to compensate for greenhouse gas emissions associated with one's lifestyle, including transporting food long distances (Peck, 2008). However, this is a down-stream solution that doesn't really address the core of the problem. Upstream alternatives are to eat **seasonal meals** (such as those that highlight apples throughout the fall, citrus and root vegetables during winter, asparagus in spring, and tomatoes in summer) and to be a **locavore**, eating what is regionally grown. Both of these strategies generally mean less intensive water and petroleum use for storage and transportation (Brodt, Kramer, Kendall, & Feenstra, 2013). **Slow Food** refers to an emerging global trend of mindfully growing, preparing, and sharing nutritious, sustainably produced, seasonal, local cuisine, an approach that benefits both planetary and human health (more on this in Chapter 9).[1]

Another potentially effective tool for reducing the energy used to transport food from farm to plate is to purchase locally grown food directly from farmers. Neighborhood **farmers' markets** are popping up across the nation; the U.S. Department of Agriculture (2014) estimates there are over 8,000 currently operating in the United States. Farmers' markets are one method for connecting citizens with regional producers. Another mechanism is **Community Supported Agriculture** (CSA): Consumers buy "shares" prior to the growing season and in turn receive regular deliveries of produce, meat, and dairy directly from a nearby farm (King, 2008). Even those of us who do not cook can lobby those who do to buy from local farmers. Indeed, many school cafeterias are finding it cheaper, healthier, and more sustainable to do so.

Diversity Is Resilience. Currently, just a handful of multinational companies conduct most of the food trade worldwide. They grow, buy, ship, and sell food, and increasingly own patents on the genetically modified seeds that produce it. What do you think this means for the resilience of our food systems? Local food is important, not only in terms of its role in reducing environmental impacts; it also enhances the local farming community and its economy by directly supporting the people who grow the food. But what if a neighborhood does not have a farmers' market, or access to any fresh food at all?

Many American cities have minimal green space and unreliable access to fresh, whole foods. In fact, these neighborhoods are referred to as **food deserts** and are a result of people's reliance on a few corporate farms and chain grocery stores. When a corporate decision maker decides a chain grocery store is not profitable, the store is closed down, leaving only fast food and convenience stores in its wake (Smith, Miles-Richardson, LeConte, & Archie-Booker, 2013). Many individuals are working to create solutions for the problem of food deserts. Perhaps the most well-known is Will Allen, whose nonprofit organization, Growing Power, employs standard business practices to market and sell fresh produce in the city, and uses the proceeds to train urban adults and children how to grow their own food (see Figure 3.4).[2] He has perfected the art and science of **aquaponics**, using a combination of fish and gravity to create fertile vegetable gardens in small urban areas (Graber & Junge, 2009). Allen has instigated and inspired a variety of models for urban agriculture including

FIGURE 3.4 *Will Allen, founder of Growing Power.*

planting over asphalt lots, growing vertically to save space, and repurposing abandoned buildings for year-round growing in cold climates.

Sometimes change emerges from an individual act of civil disobedience. Ron
Finley was tired of driving 40 minutes to buy fruits and vegetables, so he planted
a garden on the boulevard patch in front of his South Los Angeles home (see
Figure 3.5). He was cited and fined by the city, which demanded he pull up the
garden. He literally stood his ground, obtaining hundreds of signatures in support
of his edible boulevard garden. He has since started a movement of volunteers
who **guerilla garden**, planting food in the unlikeliest places in Los Angeles.
His motivation is aesthetic (he's an artist), practical (his mantra is "growing your
own food is like printing your own money"), and hopeful, with vision to make
acts of urban self-sufficiency culturally acceptable; he wants "gangsta" to mean
"gardener" (Hochman, 2013).

Even in wealthy communities where food security is not questioned, there are
barriers to engaging directly in food production. Many housing developments are
governed by homeowners' associations that require homeowners to sign "cove-
nants" restricting practices that are perceived to stray too far from manicured lawns.
This often means that no composting or food gardens are allowed. Look around
your own neighborhood. Where could food be grown? How could resilient food
systems be fostered?

FIGURE 3.5 *Ron Finley, urban garden activist.*
With permission from Ron Finley

Material Goods

We rely on the earth's resources for all of our tangible goods. However, as you read in Chapter 1, current extractive and manufacturing practices in the industrialized world surpass the earth's capacity to replenish the necessary raw materials, and they create waste with nowhere to go. Consumer habits exacerbate these problems. The following sections describe some alternative approaches to production and consumption that can significantly reduce, and in some cases even rectify, the negative ecological impact of our clothing, tools, shelters, and myriad other possessions.

Interconnection. Generating material goods affects everything essential to life: water, air, land. The business world is recognizing this fact, and is working with industrial ecologists to develop decision-making tools such as **Life Cycle Assessments** (LCA) that aim to minimize environmental impacts by better understanding the scope of possible effects (EPA, 2006). LCA follows products and processes from the extraction of raw earth materials to eventual disposal with the goal of minimizing impacts across *all* ecological systems. During the process of examining the real environmental costs of doing business, leaders generally find that using virgin materials and creating toxic waste are financial losers in the long run (Anderson, 2005; Hawken, 1993).

Circular Systems. Drawing too much raw material from our natural systems and creating waste in the process of using them are two problems that can be solved by developing circular systems. Some visionary leaders have engaged this concept in earnest, redesigning whole industries so that "humanity works with nature, where technical enterprises are continually reinvented as safe and ever renewing natural processes."[3] In this model, called **cradle to cradle**, "wastes" must become food, just as they do in natural cycles in which organic wastes serve as fertilizers for the next generation of growth. In other words, natural resources do not end up in sewers and landfills, but are used in new production cycles (Hawken, Lovins, & Lovins, 1999; see also Krupp & Horn, 2008; McDonough & Braungart, 2002).

To transform linear systems into circular ones, traditionally resource-intensive manufacturers, making everything from photocopy machines to carpeting to cars, have been switching from sales to service models. Renting products rather than selling them creates an incentive for making durable and reusable parts. When consumers are done using a product, instead of sending it to the landfill, it's returned to the manufacturer who has produced it for its recyclability (Anderson, 2005; Savitz & Weber, 2006). On an even larger scale, industrial communities based on sharing resources, maximizing efficiency, and utilizing each other's waste materials are spreading worldwide. These **eco-industrial parks** are modeled on the original site in Kalundborg, Denmark, where a fish farm, a power plant, a manufacturer of gypsum wallboard, a pharmaceutical company, a cement factory, local farmers, and an oil refinery each use by-products of another's production. The collaborators enjoy reduced costs associated with energy use, waste, and material acquisition (Tudor, Adam, & Bates, 2007).

Individuals can close linear systems by donating, consigning, sharing, and buying used material goods. It has long been acceptable to buy used autos and houses and rent vintage apartments. Can you think of other items, even some that are not as expensive and durable, that could be purchased secondhand? The next generation of circular systems supplants recycling, which can lead to lower quality material, with **upcycling**—using dated or tainted materials for new, useful, and higher-quality purposes (McDonough & Braungart, 2002). This is a significant step in the direction of a **regenerative design culture** in which the material flow from system members is cleaner, healthier, and of generally superior value than the material inputs (Cole, 2012; Tainter, 2012).

Limits to Growth. To many of us, the idea of using fewer resources evokes images of hardship and sacrifice. But instead of presupposing that a sustainable future will be Spartan and severe, consider the possibility that with increased human attention, *less can be more*. For instance, a large amount of our resource use is just plain sloppy. Consider the current state of the automobile: 80% of the energy from the gasoline it burns is used to run and cool the engine and deal with its exhausts. Only 5% of the remaining 20% is used to move the person; the rest is used to move the car. Thus, only 1% of the total energy in gasoline goes to moving the driver; the rest is wasted (Hawken, et al., 1999). The mantra of less is more can inspire enhanced efficiency without necessarily losing convenience or style.

Use it up, wear it out, make it do, or do without was a popular saying during World War II when money was tight and every available scrap of metal and cloth was channeled toward the war effort. Today, **voluntary simplicity** is a social movement spreading through Europe and North America in which people choose to downscale their material possessions in order to live "consciously, deliberately, while not being distracted by consumer culture . . . taking charge of a life that is too busy, too stressed, and too fragmented, [while] consciously tasting life in its

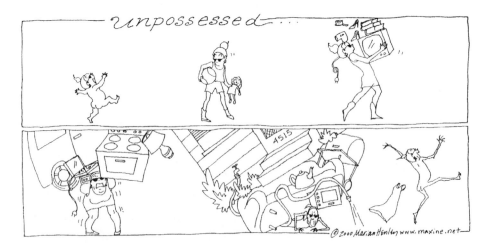

unadorned richness" (Elgin, 2000). By taking the time to intentionally recognize their present abundance and sufficiency, people learn to appreciate and experience more from less.

People choose simplicity for a variety of reasons ranging from religious beliefs to achieving more balance between work and other aspects of their lives. Reduce and reuse, share and repair, repurpose and recycle; you can find modern movements for all of these methods of cutting our use of resources. Sharing has shown particularly enthusiastic growth in recent years. The Internet has revolutionized carpooling. Bike sharing and garden sharing programs are popping up all over the United States. Car sharing programs can be found in most large, North American cities and on many college campuses, claiming over a million members. In the United States, each shared car averages about 70 users (Shaheen & Cohen, 2013). Car sharing enables people to shed their own personal cars and rely more heavily on alternative modes of transportation, while still having convenient access to an automobile. This reduces traffic, emissions, and resources needed for auto production.

Upstream Solutions. Much of the clothing manufactured in the industrialized world is **Fast Fashion**, mass produced imitations of recent designer creations. In an effort to keep up with the runway, retailers continually turnover their inventory and consumers buy new styles, even whole new wardrobes, every six months. Just when we've bought the flared jeans, skinny jeans come into vogue. This uses significant energy and resources and leads to massive amounts of waste, as the cheap retail prices allow merchandise to seem disposable. U.S. residents discard about 14.3 million tons of textiles annually, only about 15% of which are recaptured for further use (EPA, 2014d). In contrast, **Slow Fashion** has emerged to counter the race to replace by shifting the goal from quantity and novelty to quality and timeless style. This eliminates overproduction and excessive waste before it's created. Perhaps you have examples of Slow Fashion in your own wardrobe: Do you have a favorite jacket, hat, or pair of shoes that you've had for many years, your go-to items for when you just want to be you? Slow Fashion is about finding one's own personal panache, rather than dressing according to the dictates of designers. It's about sticking with clothing that suits you, regardless of fleeting trends (Shor, 2002).

Diversity Is Resilience. Imagine a tailor in every neighborhood, someone who personally knows the residents and can design and create unique outfits for each individual, someone who can fix those outfits when they wear out or become damaged (Shor, 2002). It used to be commonplace for communities to have a tailor, a cobbler, a woodworker, and assorted other specialized craftspeople that make and repair material goods, but in recent decades, their numbers have been dwindling. Returning to this model, and expanding it to all kinds of products and services, will tap into the diverse skill sets present in every neighborhood, fostering entrepreneurship and creating new jobs (Dearie & Geduldig, 2013). Ultimately, this approach will produce a winning combination of sustainable results: reduced consumption and waste, and thriving, interdependent communities.

The Functions of Individual Behavior

Individual behavior, in many forms and many domains, matters for sustainability. As you have read, individuals can innovate, initiate, and influence changes that move society in a sustainable direction. Lifestyle choices are the most obvious place to begin. Our lifestyles include not only numerous, small, daily decisions, but also big-picture things like choosing where to live, what to study, and how to earn money. However, while private behaviors are necessary, they will not be sufficient for slowing or reversing the massive scale of our global impact. We also must behave in ways that lead to larger, systemic change.

For example, individual behavior can take the form of exercising economic clout. As seen with Lucia von Reusner, corporate shareholders have enormous power to effect change given that corporations must respond to their concerns. Consumers, too, can compel corporations to be more sustainable by practicing **conscious consumption**, which involves not only reducing how much one buys and consumes, but also choosing environmentally sound products. Voting with our dollars is one of the public's most powerful tools. At the collective level, customer **boycotts** have led to corporate change: McDonald's ceased using Styrofoam containers for their hamburgers, Burger King no longer imports rainforest beef, and Starbucks became a leader in promoting fair-trade coffee. A recent example of consumers exerting influence is the *Carrotmob* movement that uses social media to generate support for store ***buycotts***. In this model, small business owners (generally food vendors like grocery stores, coffee shops, and pizza parlors) commit a percentage of their sales to sustainability initiatives and, in return, customers show up in droves. Carrotmob campaigns have provided funding for projects related to energy efficiency, bicycle parking, solar panels, and green spaces.[4]

Individual willingness to financially support and adopt new regenerative technologies as they become available can also help drive sustainability. However, we will not be able to simply buy our way out of our unsustainable lifestyles. Political advocacy for the greater good is as crucial as conscious consumption. Individuals must participate in public decision making and speak out about legislation, because people can't easily make ecologically responsible choices when the larger systems work against them. Recall how Ron Finley was fined by the city of Los Angeles for his boulevard garden. Also in the state of California, despite a variety of environmentally progressive laws and policies, it is currently a violation of construction codes to build houses that are completely consistent with permaculture principles (Warren Brush, personal communication). Similar constraints on sustainable behavior can be found in the form of rules, regulations, and laws around the globe. You've probably heard the dictum "Think Globally, Act Locally." Grassroots efforts of individuals working for systemic change in their own communities are *essential* to achieve sustainability.

CONCLUSION

Individual-level behavioral change is where all social movements begin. We encourage you, therefore, to be mindful throughout your reading of this book about how individuals like yourself can effect change for sustainability. Think not only about personal lifestyle changes, but about exerting systemic influence within families and circles of friends, in the service of organizations (such as school, workplace, or church), within neighborhoods and local communities, and while engaging in regional, national, and even global, politics.

Try to notice when your thinking is influenced by the DSP, and take the opportunity to counteract those thoughts with what you now know about ecological principles. In the end, our long-term success will depend on viewing the natural world differently and acting in ways that are consistent with scientific information about how natural systems operate (Amel & Manning, 2012). If we are mindful about maintaining "conditions conducive to life," we have a good shot at achieving ecological, economic, and social wellbeing, the three dimensions of a society that is truly sustainable.

Now that you've read these first three chapters, you should have a good sense of what the problems are, how they developed, the forces currently acting against change, and some of the alterations in mindset and behavior needed to foster sustainability. In the next chapter, you will see how the discipline of psychology, the science of behavior, can play a crucial role in bringing about those changes.

NOTES

1. http://www.slowfood.com
2. http://www.growingpower.org
3. Quoted from the website, http://www.earthomeproductions.org/context.html; see also information about sustainable designs at http://www.mbdc.com/cradle-to-cradle/c2c-framework/
4. http://www.carrotmob.org

PART 2

Psychology for a Sustainable Future

The goal of this section of the book is to educate readers about theory and research in psychology relevant for understanding and changing unsustainable behavior. Chapter 4 describes psychology as a sustainability science, and reviews methods used by both basic and applied researchers in this field. Chapters 5, 6, and 7 represent psychological facets that together affect behavior: situational and social influences, thinking processes, and individual differences. Chapter 8 presents theories that explain how these facets combine to motivate behavior.

Psychology Can Help Save the Planet

Psychology is defined as the study of behavior and mental life. And although psychological theory and research has had myriad impacts on social issues in the United States, psychology's relevance to sustainability is not immediately obvious to most people (Koger & Scott, 2007). One reason for this, as described in Chapter 1, is a failure to recognize that "environmental" problems are actually human behavior problems. Another reason is that the discipline of psychology is commonly misunderstood. Our culture's predominant stereotype of psychology is that it's all about identifying people's neuroses and treating mental illness through therapeutic techniques such as dream analysis. Trade bookstores place psychology books alongside self-help books and books on the occult, making psychology look like a do-it-yourself method for managing a divorce or experiencing past lives. Contrast this with the psychology classroom and research laboratories, which is where you'll find about one-third of all professional psychologists. These academic psychologists conduct scientific studies to illuminate basic behavioral and mental processes like learning, perception, motivation, and thinking. The majority of psychologists (about 60%) apply this knowledge to solving real-world problems: Half are trained professionals who, indeed, provide therapy to help clients improve their mental health; the other half apply scientifically tested principles within domains such as business, education, the military, sports, and law. As you will see throughout the remainder of this book, this same body of psychological knowledge about human thinking, social interactions, and motivation to act can also be leveraged to solve the behavioral challenges associated with our current environmental crises (Schultz, 2014).

As described in Chapter 2, natural scientists have been aware of, and alarmed by, negative human impacts on ecosystems since the mid-twentieth century. And since the 1960s, they have made concerted efforts, along with like-minded politicians, professional environmentalists, environmental educators, grassroots activists, and everyday concerned citizens, to motivate individual and societal change. But very little has changed. It has only been within the past 15 years or so that environmental scientists and practitioners have come to realize that inspiring large-scale human behavior change will require input from the people who are experts on human behavior: psychologists.

Certainly, sociologists, geographers, political scientists, and other social scientists study people and environmentally relevant behavior, but psychologists study behavior at its roots, at the *individual* level. Every family, organization, institution, business, community, and political body is made up of individuals; and *every decision and action ultimately stems from an individual*, or from the interaction of individuals. Individual behaviors happen at home, at work, in neighborhoods, and in Congress. Thus, to achieve a more sustainable world, change is needed in both private individual behaviors, such as our personal transportation and shopping habits, and in communal actions such as voting for ecologically-minded political candidates and initiating greener practices in one's congregation, school, and workplace. It is at the level of the individual that all behavior starts, and so it is at the individual level that behavior must change.

Changing individual behavior makes an impact because we are a system of individuals, and when one member of a system changes, other parts of the system adjust! When one small change is made by many individuals, or one individual makes many small changes, it begins to add up. Consider how the cumulative effects of what we each do as individuals can be either devastating (each American's post-workout bottle of water adds up to two million bottles sold every five minutes) or healing (e.g., if every household were to plant native species in their gardens, we would likely see a reduction in erosion, fewer water shortages, and recovery in the honeybee population). And, the more that people see others doing things a certain way, the more they come to accept it as the normal way to be and live. In other words, individual behavior change is socially contagious.

As you will see in the chapters to follow, psychology offers an extensive array of theoretical insights, scientific data, and useful strategies for building a more sustainable world by way of individual behavior change. You will also see that a wide variety of psychological perspectives are relevant; note that the four of us writing this book come from the disparate subdisciplines of social psychology, physiological psychology, industrial-organizational psychology, and cognitive psychology. In order to prepare you for the content to come, the goal for this chapter is to provide a primer on some of the methods psychologists use to ask and answer questions in the context of sustainability. However, before reviewing these research methods, first consider some of the developments in the field that represent psychologists' increasing engagement in environmental issues.

GROWTH IN GREEN PSYCHOLOGY

The earliest psychological research specific to environmental issues dates back several decades and originated in **Environmental Psychology**, an interdisciplinary, scientific field that examines human interaction with the physical environment, broadly defined to include both natural and built settings and elements (Gifford, 2014; Giuliani & Scopelliti, 2009). Much of the work of environmental psychologists is not related to sustainability per se, but among the research topics that are, some are obviously relevant (e.g., people's preferences for particular wild landscapes), while others are more indirectly related (e.g., the impact of airport noise on cognition). You will read more about the work of environmental psychologists in Chapters 9 and 10.

The 1990s witnessed the emergence of **Ecopsychology**, a perspective that focuses more on the mental health aspects of the reciprocal relationship between human and nonhuman nature. Ecopsychologists theorize that humans are not just physically but also psychologically dependent on the rest of nature. A fundamental premise of this perspective is that living physically separated from the rest of the natural world, as those of us in modern societies do, is psychologically distressing to us. This disconnection from nature may explain elevated rates of common problems like depression, anxiety, substance use and addiction, and other compulsive behaviors (such as shopping and checking our text messages) in industrialized cultures. In addition, ecopsychologists propose that many people experience personal distress when they witness environmental degradation. An example would be the sense of loss you might feel when you return to a favorite wild area where you played as a child, only to find it turned into a subdivision.

Traditionally, ecopsychologists have tended to be clinicians and therapists rather than researchers. These practitioners expanded the practice of psychotherapy to include **ecotherapy**, moving their practice from offices and hospitals into gardens and wilderness settings, with the goal of helping people nurture a sense of connection to their natural surroundings (Buzzell & Chalquist, 2009). Ecotherapy also involves taking clients' anxiety about ecological problems seriously (White, 1998). For the most part, however, ecotherapeutic techniques, and the theoretical ideas upon which they are based, have not been subject to scientific scrutiny.

Importantly, this means ecopsychology is often ignored by mainstream psychologists. You will read more about ecopsychology in Chapter 10.

In recent years, there have been efforts to bridge the divide between ecopsychologists, some of whom are skeptical about the value of Western empiricism (Fisher, 2002), and some scientific psychologists who have serious reservations about the philosophical, spiritual, and new age underpinnings of ecopsychology (Reser, 2003). One example of this bridging occurred in 2009, when a clinical psychologist with a background in wilderness therapy founded *Ecopsychology*, the first peer-reviewed academic journal to provide a venue for scientific ecopsychological research. Developments like this are encouraging because ecopsychology speaks to deeply felt experiences people have in relationship to the natural world (Kahn & Hasbach, 2012), and intersects with current needs for adaptation to our new environmental circumstances (e.g., coping and decision-making responses to climate change; Reser & Swim, 2011). As trained experimental psychologists, the writers of this text see value in generating scientific data: It allows a critical examination of the claims of ecopsychologists, and thus puts validated ideas in more direct view of mainstream psychology and other sciences.

During the last decade, psychological research specifically focused on environmental sustainability has proliferated. Researchers from a variety of academic backgrounds have joined forces under the banner of **Conservation Psychology**, defined as the "scientific study of the reciprocal relationships between humans and the rest of nature, with a particular focus on how to encourage conservation of the natural world" (Saunders, 2003, p. 138). Conservation psychologists subscribe to the **scientist-practitioner model** in which rigorous research is accompanied by active application toward solving real problems. This is akin to how basic biology is put to work finding cures for disease and enhancing public health practices, how workplace psychologists create and test interventions to increase employee satisfaction and safety, and how conservation biologists work with an eye toward protecting species and their ecosystems. Conservation psychology overlaps to some degree with environmental psychology, but is distinct in its explicit mission to study human interaction with the *natural* environment for the purpose of advancing sustainability (Schultz & Kaiser, 2012). In addition to their work within psychology, you can also find conservation psychologists engaged in transdisciplinary fields, like **Human Ecology**, which combine expertise from many social science disciplines to address the interplay between humans and their varied environments.

The research described in the remaining chapters of this book will draw from these diverse theoretical perspectives to explain what we think of as the **psychology of sustainability**, a collection of psychological insights relevant to, and perhaps crucial for, creating a more sustainable world. The essential underlying feature of this upcoming content is adherence to the scientific method of inquiry. Let's now turn our attention toward what this means.

PSYCHOLOGY AS A SUSTAINABILITY SCIENCE

What we currently know about human behavior rests on the interplay of theory and empirical observation, the building blocks of psychological science. **Theories**, or explanations about how things work, can both represent what has already been discovered and provide ideas for new research. **Empiricism** refers to the discovery of knowledge via systematic data collection. In psychology, this includes direct observation of behavior, inferring mental processes using behavioral markers such as how quickly and accurately people respond, measurement of physiological functions like neural and hormonal activity, and asking people questions about intangibles like opinions and preferences.

Scientific knowledge is not based on a hunch. We all develop ideas, inspired by our own experiences, about what makes people tick and why we behave the way we do. However, science requires that even common sense notions be rigorously tested before they are accepted as probable. As it turns out, there are many examples of robust psychological research results that fly in the face of what most of us believe about ourselves. For instance, people generally don't believe they are likely to copy others' behavior, yet there exists a compelling body of evidence showing that what one sees others do is often the best predictor of one's own behavior (Nolan, Schultz, Cialdini, Goldstein & Griskevicius, 2008). In the pursuit of scientific knowledge, it is vital to maintain a skeptical mindset. This means playing the devil's advocate and attempting to debunk our own and others' assumptions. It also means being open to being wrong.

Because it is not possible to measure the whole of humanity, research psychologists sample subsets of people and use the principles of probability to statistically assess whether results are likely to apply more generally. "The truth" is never known for certain. Rather, conclusions are drawn regarding what is likely to be true. Thus, there is always the risk of making mistakes and mistakes do have consequences. A **Type I error** is to identify a result as real, or unlikely to be due to chance, when really there is no systematic pattern. Alternatively, to overlook a phenomenon when it actually exists is a **Type II error**. Take a moment to consider the relative drawbacks of each. Scientists traditionally adopt a conservative approach so as to avoid false positives (i.e., Type I errors), because making unwarranted claims is considered scientifically egregious, and because the error of a false positive can be costly. For example, we might spend time, money, and resources on implementing a behavioral change technique that does not actually work. However, failing to detect real effects can also be costly. For instance, researchers cannot confidently quantify the impact of chronic, low-level human exposures to many toxic chemicals, yet to continue to assume "innocent until proven guilty" could mean serious risks to human health. Under these circumstances, some decision makers, especially in Europe, utilize the **Precautionary Principle**, which encourages avoiding potential risks to human and overall ecological health, even if scientific proof of harm has not yet been fully established.

To minimize mistakes, psychologists try to build accurate tools, measure data from people with a lot of different characteristics, and eliminate as many alternative explanations for observed phenomena as possible. Despite limitations within any given research project, researchers become pretty confident over time if there are enough data from different labs using different participants, different measures, and different methods providing corroborating evidence. It is important to remember that research is a process, not an event. In other words, one project does not "prove" anything; instead, conclusions are based on a body of research results.

Psychologists follow a code of ethics when conducting research, which requires working responsibly and with integrity.[1] This means striving to be unbiased and dispassionate observers rather than trying to find particular research results. However, political, commercial, or economic forces can pressure researchers (in any field) to come to certain conclusions. This is why it is imperative to know who sponsors and funds research. Indeed, some research is underwritten by companies, organizations, and political parties, who then selectively report the results to further their cause. This kind of bias is to be avoided. Yet, as stated above, the goal of conservation psychology, or any applied science for that matter, is not value-neutral. Clearly, we know that a healthy environment is essential for healthy humans, and that this is worthy of our collective attention just as curing cancer and eliminating child abuse are. It might seem impossible for scientists who are trying to solve specific problems to remain neutral under the circumstances. However, engaging in research intended to solve a problem does not mean forgoing stringent empirical standards. In other words, although research may be conducted in support of an overarching goal, the study of any particular solution must be unbiased, must adhere to accepted scientific principles, and must be reproducible.

Throughout this book, you will encounter the standard language of psychological science, including vocabulary like "variables," "operational definitions," "hypotheses," and "participants." A variety of research methodologies are referenced, typically involving hypothesis testing with observations that are quantified and analyzed using statistics. Here are some of the basics that will support your understanding of the studies described in later chapters.

What Psychologists Study: Operational Definitions of Variables

The main ingredients of a scientific research project are **variables**, the factors under study. As the label implies, these factors *vary* across people and across situations. Consider the example of environmental attitudes: Not all people have the same attitudes, and an individual's attitudes may change depending on the context. Sometimes researchers measure multiple variables to establish whether they appear to be associated with each other, and sometimes one variable is intentionally changed to see how it will affect another. More will be said about these different approaches to studying variables in the section on research designs, but first, it is important to understand how researchers turn a variable that may be an abstract concept (such as "proenvironmental behavior") into something tangible that can be empirically studied.

In order to study something, it needs to be **operationalized**, or specifically defined in terms of how it is represented and measured. In psychology, many variables can be tricky to operationally define. To illustrate, just try asking 20 different people to define proenvironmental behavior. You are likely to obtain 20 loosely related but different definitions. This may be okay in everyday conversation, but science requires precision. So, can you think of some things that a researcher might use to systematically observe or measure proenvironmental behavior?

As you read about studies in the chapters to come, you will see examples where different researchers operationally define the same theoretical variable, or **construct**, in different ways (see Figure 4.1). In one study, the operational definition

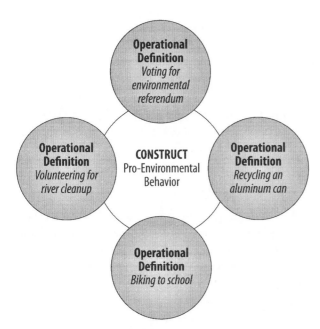

FIGURE 4.1 *Operational definitions.*
The same theoretical construct can be operationally defined in a variety of ways.

of proenvironmental behavior may be recycling a can when leaving the lab room; while in another, it might be circling "Yes" in response to a survey item asking whether the participant voted in favor of a particular environmental conservation referendum. Paradoxically, while very specific operational definitions enable the study of the variables of interest, each precise operational definition limits what can be learned about the conceptual variable. What might be missing when the operational definition of proenvironmental behavior is limited to recycling? This is another reason to rely on the outcomes of many studies to draw conclusions about behavior.

Whom Psychologists Study: Participants

All people make decisions that impact ecosystem health. Thus, studying any person is relevant and studying a variety of people is important. However, one of the primary challenges in psychological research is obtaining diverse samples of people that represent the general population. Workers, shoppers, and indigenous tribes are groups that researchers find hard to reach or entice into participating. In contrast, students taking psychology courses (particularly in the United States, Europe, and Australia) are convenient and form the basis for many research samples. Can you think of ways in which college students might differ from executives, families you find in the grocery store, or people from non-Western traditions? It's likely you can think of a dozen important differences right off the top of your head. Because different groups of people are likely to differ in their experience and circumstances, it is important to bear in mind that what we find out about one group cannot automatically be generalized to everyone else. If you've taken a Research Methods class, you might recognize this concept as **external validity**. Case in point: Scholars conduct a great number of conservation psychology studies in Europe. Do you think the results of these studies would also hold for U.S. residents? Maybe, maybe not. That is why it is so important to think critically about the samples used in studies, and to question whether a study's conclusions should be considered applicable to populations other than the one studied.

How Psychologists Study: Research Designs

Scientific methods are a means for testing **hypotheses**, which are predictions about the expected relationships among variables. For instance, you might propose that people will be more likely to enjoy their first ride on the city bus if they go with a bus-riding friend than if they go alone. You might hypothesize this because of the discomfort people experience when they do not feel competent or confident. In this case, a respected other can provide cues about where to stand, how to pay, where to get off, or how to read the bus map. Hypotheses can be inspired by theories, previous research, or real-world observations.

When little is known about a phenomenon, one of the most informative first steps is **naturalistic observation**. Observation of public behaviors as they naturally unfold helps researchers identify critical variables and suggest how they might be related. In one study involving zoo visitors, researchers observed how hundreds of small groups reacted to the animal exhibits (Clayton, Fraser, & Burgess, 2011). The most commonly observed verbal responses were calling another person's attention to and discussing the animals, while the most common nonverbal response was pointing to the animals. Since the human interactions frequently involved efforts to create or describe relationships with the animals, the investigators suggested that future research should examine the role of zoos in creating a sense of human–animal relationship and increasing support of habitat conservation efforts.

One main drawback of observation is that one can only see so much. Underlying psychological processes cannot be observed, only inferred. **Interviews** and **focus groups** are additional exploratory methods that allow researchers to move beyond inference to hear participants' feelings, beliefs, and attitudes about a topic. Like naturalistic observations, data collected through interviews or focus groups can provide a springboard for further research. Once a researcher has generated a specific hypothesis, the next step is determining how to test it.

Laboratory Experiments. The most highly controlled method of hypothesis testing is the laboratory experiment in which the goal is to isolate a cause and effect relationship. The researcher predicts that the **independent variables (IVs)** will impact the **dependent variables (DVs)**. In an experiment, a researcher **manipulates** the IVs and **measures** the DVs. For instance, suppose a researcher hypothesizes that the persuasiveness of an environmental message depends upon who the spokesperson is. The researcher could operationally define "environmental message" as a radio commercial about choosing dolphin-friendly fish, and "persuasiveness" as study participants' ratings on a 1 to 7 scale, where 1 = not at all persuasive and 7 = extremely persuasive. To test the hypothesis, the researcher could manipulate the source, meaning half of the study participants would hear a commercial featuring an expert from the Nature Conservancy, and the other half would hear the same message, but featuring a Hollywood celebrity. Everything else would be held **constant**, meaning that *nothing* else about the experience would vary between the two sets of participants. The only difference would be which spokesperson they heard. Why would this be important?

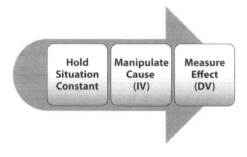

Another important feature of experiments is **random assignment** of participants to the different versions of the IV, called **conditions**. In random assignment, every participant has an equal chance of ending up in any of the conditions. Why should this matter? Imagine that a researcher wants to test the hypothesis that receiving detailed instructions will make people more willing to ride the bus for the first time. Now suppose that the researcher goes into a classroom to recruit participants. Right away, a dozen outgoing students eagerly volunteer to be part of the research study, but the researcher needs more than a dozen participants, so he or she waits until another dozen, somewhat more reserved, students also sign up. What would happen if the first dozen people were assigned to the condition receiving bus-riding instructions and the second dozen people ended up in the condition that didn't get instructions? If the instructions group was indeed more likely than the other group to jump on the bus, could the researcher be confident that this was just because they received instructions? What is an alternative explanation? In this example, personality is a **confounding variable**, another factor that varies between the conditions, and may actually be responsible for any observed bus-riding differences (DV) between the two groups (see Figure 4.2). While not perfect, randomly assigning participants to different conditions minimizes the risk of confounds like this one by increasing the likelihood that individual characteristics (personality, intelligence, expertise, and so on) will be evenly distributed among conditions; that is, if this researcher had randomly assigned participants, we would expect that the average level of extroversion of the two groups would be very similar at the outset.

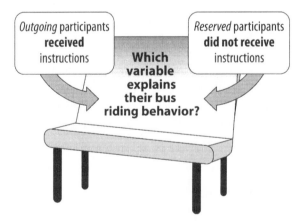

FIGURE 4.2 *Experimental confound.*
When participants are not randomly assigned to conditions, the independent variable manipulation may not be the only systematic difference between the groups. This makes it impossible for the researcher to be confident that observed differences in the DV (bus riding) are due solely to the IV (instructions versus no instructions).

Now you understand why experiments require random assignment. But what happens if *random assignment is impossible*? Can you think of any variables that might affect people's environmental behaviors that cannot be manipulated by researchers? One example might be hypothesizing that "political party affiliation" will affect participants' reactions to a news article about climate change. As researchers, we cannot manipulate political party selection. That is, we cannot randomly assign some participants to be Democrats and others to be Republicans—they already are what they are. The same thing is true of any demographic variable that might be of interest as a potential causal factor, such as gender, age, environmental attitudes, and so on. When it is not possible, either practically or ethically, to manipulate a causal variable, **quasi-independent variables** must be created by identifying participants based on preexisting characteristics. So, participants would be asked to identify their political party, and then their reactions to the news article would be compared based on their reported political affiliation. Now suppose the two groups did indeed respond differently. Could one conclude this was due to their different political affiliations? Nope. The difference could be due to something else that just happens to coincide with political party orientation. Can you think of any? When dealing with quasi-independent variables, confounds are always a distinct possibility. Researchers and readers alike must use caution when interpreting the results of such studies.

Sometimes researchers conduct experiments that include both quasi-IVs and IVs to which participants have been randomly assigned. This kind of design is good for testing how a particular IV manipulation might affect different kinds of participants. A clever example is a project that measured how people responded after receiving their ecological footprint (EF) results (Brook, 2011). The hypothesis was that EF feedback would affect people differently depending on how tightly linked their sense of self-worth was to environmentalism (ESW).

Brook (2011) asked a couple hundred students to take a computerized ecological footprint questionnaire. Then, regardless of their actual footprints, one random half of the participants received negative feedback: "IN COMPARISON to previous studies we have done with University of Michigan students, your footprint measures 140.23% of the footprint of an average University of Michigan student." The other half, assigned to the positive condition, received an identical message except for the footprint percentage, 55.23%. So, feedback condition was the true IV in this study.

The quasi-independent variable was level of ESW. Brook had pretested these participants and divided them into high-ESW and low-ESW groups based on their responses to an environmental self-worth scale. After they received their EF feedback, all participants were given an article to read about environmental degradation and solutions, so they would share the same minimum level of knowledge. Finally, participants were invited to write a letter about *a policy issue of their choice* and e-mail it to their governor. The dependent variable was whether or not they chose to write about an environmental policy issue.

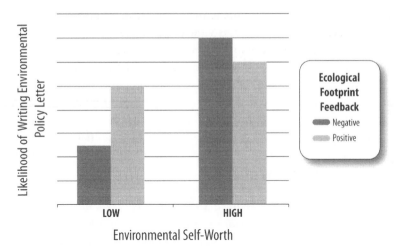

FIGURE 4.3 *Results from a study combining a true IV and a quasi-independent variable.*
Brook (2011) found that the effect of ecological footprint feedback on writing an environmental policy letter depended on the participants' environmental self-worth. Adapted from Brook (2011).

Brook's results indicated that for people with high ESW, receiving negative feedback increased the likelihood of sending an environmentally-focused letter, but for those with low ESW, the same negative feedback decreased their likelihood of sending an environmentally focused letter (see Figure 4.3). In this example, it isn't type of EF feedback alone that determines whether people are likely to communicate environmental concerns to politicians; there is an **interaction** between type of feedback and participants' preexisting ESW.

Testing for interactions is one way that researchers can try to capture some of the complexity of behaviors as they occur in the real world. Generally, however, even with more than one IV, what happens in a lab is a stripped down or contrived version of reality. This is the tradeoff for the high level of control. Also, we must assume that the laboratory heightens people's awareness of their own behavior in an unusual way that could change it. They may even try to guess the hypotheses so as to behave "appropriately" rather than naturally. Of course, researchers can make efforts to improve realism. In fact, **simulations** can be developed that provide realistic experiences in the laboratory. For instance, researchers recently created the Virtual Bus, a laboratory simulation that mimics everything from bus controls to road conditions. It is being used to design and test real time feedback, which can help drivers adjust their own behavior to improve fuel efficiency (Rolston, 2012). Simulations, while effective (and really cool), require extensive amounts of time and money to develop.

Sometimes the best way for researchers to add some realism to their laboratory experiments is to employ **deception**, which involves temporarily misleading participants to create more realistic situations. In the Brook study described above, the false feedback is an example of deception. Another example would be the use of a **confederate**, an actor hired by the experimenter to pose as another participant, a passerby, or some other pretend role. Can you think of ways confederates could be useful for studying people's environmental behaviors? Deception might be needed for an experiment to capture genuine responses, but it is always cause for concern within the research community. It doesn't feel good to be duped, so deceiving people increases the level of psychological risk to participants. Therefore, **debriefing**, revealing and explaining the purpose of the deception to the participant after the experiment is over, becomes critical.

Correlational Studies. A large amount of current research in conservation psychology takes the form of correlational studies in which no independent variable is manipulated and there is no random assignment. In fact, in these types of studies, any hypothesized causal variable is called a **predictor variable** rather than an IV (see Figure 4.4 for examples of predictor variables). Similarly, the term DV is reserved for experimental designs, so, in nonexperimental methods, it is replaced with **criterion variable**. Both predictors and criterion variables are measured *as is* by the researcher. In some correlational studies, the researcher does not even have a specific prediction about whether or how variables are causally related to each other. Instead, the researcher may be measuring multiple variables simply to explore whether they seem to be related to each other in any systematic way.

For example, a researcher might wonder whether the amount of time people spend in natural settings is somehow related to their environmental attitudes. The researcher could ask people how often they visit nature and assess their attitudes using a standardized set of questions. (Bonus point question: How might you *operationally define* a nature visit?) Then, the researcher could conduct a statistical

Personality	Age
Values	Gender
Thinking style	Occupation
Social identity	Socioeconomic status
Group affiliations	Geographical location
Political party	Upbringing
Religion	Life experiences
Interests	Lifestyle

FIGURE 4.4 *Examples of predictor variables in conservation psychology.*
Variables like these cannot be manipulated. Instead, researchers must measure them to see whether they correlate with environmental attitudes and behaviors.

analysis to determine whether there is a pattern present in the data. Suppose the researcher found a trend in which people with higher scores on spending time in nature also had higher scores on the measure of proenvironmental attitudes. Could the researcher conclude that spending time in nature *causes* people to develop proenvironmental attitudes? That might be what is going on, but can you think of an alternative interpretation? Maybe people with proenvironmental attitudes are more inclined to seek out experiences in nature, so the direction of cause and effect might be the other way around! Or, it may be that there is some other variable that explains *both*. For instance, maybe growing up in a family that took a lot of wilderness camping vacations influenced both a habit of seeking out nature and proenvironmental attitudes. In this case, spending time in nature and environmental attitudes are not actually influencing each other in either direction, but seem related because they are both systematically influenced by something else (how participants were raised; see Figure 4.5). We hope you now understand why researchers emphatically say **correlation does not equal causation**!

Every correlational finding represents a "chicken-and-egg" problem because the researcher measures all the variables at the same point in time. The researcher does not know which ones "came first," nor whether any variable is actually exerting an influence on any other. Still, correlational research is extremely valuable in cases where manipulating IVs is impossible. It is also valuable as a way to assess many variables for large numbers of people over great geophysical space, even spanning the globe (Mostafa, 2012).

For example, in an effort to understand when people will drive their cars rather than take alternative transportation, researchers used online surveys to study approximately 27,000 trips taken by 3,560 German university students (Klöckner & Friedrichsmeier, 2011). They measured many situational variables such as day of the week, time of day, availability of the different travel modes, purpose and length of trip, and the weather. They also measured many personal attributes of the participants including habits, attitudes, perceived control over their choices, knowledge of the consequences of individual car use, and intentions to use travel alternatives.

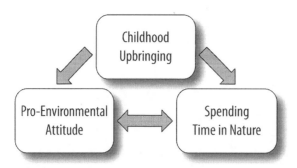

FIGURE 4.5 *Potential causal directions in correlational research.*

Analyzing the data, they found that personal and situational variables were *both* correlated with car use.

In this study, the researchers identified the situational and personal variables as predictors and car use as the criterion. However, it is still a correlational study and the researchers cannot definitively conclude that the predictors *cause* people to choose taking a car. Even so, knowing that both types of predictors were correlated with car use suggests, as the researchers conclude, that neither personal nor situational factors should be neglected as potential influences on car use. In other words, if the goal is to reduce people's tendency to choose cars over public transportation, then behavior change strategies should address both personal and situational variables.

Field Experiments. Field experiments occur in actual workplaces, schools, and public places with people going about their normal behavior. Here you have the best of both worlds with high control over experimental variables within a real-world context. These types of studies are possible when there is a captive audience such as in a meeting, classroom, or parking lot. (Really, you can learn a great deal about behavior in a parking lot!) A now-classic example is an experiment that examined how social influence affected littering behavior in a hospital parking ramp (Cialdini, Reno, & Kallgren, 1990). The researchers manipulated two independent variables: the descriptive social norm (that is, information about what most people would do in the situation; you'll read more about this concept in Chapter 5), and norm salience (how obvious the descriptive norm was to the participants).

Each unknowing participant returned to his or her vehicle to find a large flyer on it (and on all the other car windshields). The behavior in question was what the participant would do with the flyer. The descriptive norm pertained to what other people seemed to be doing with their flyers; it was operationally defined as the state of the parking garage floor (clean or littered with crumpled up flyers). Norm salience was operationally defined by whether or not a confederate littered in front of the participant. With two independent variables, this experiment included four conditions (see Figure 4.6). Each participant was randomly assigned to experience just one of these four conditions.

	Litters	Does Not Litter
Littered	Confederate litters in a littered parking garage	Garage is littered but confederate doesn't litter
Clean	Confederate litters in a clean parking garage	Garage is clean and confederate doesn't litter

FIGURE 4.6 *Research design using two independent variables.*
In this field experiment, norm salience was operationally defined as the confederate's behavior, and descriptive norm as the state of the parking garage floor (adapted from Cialdini, et al., 1990, p. 1017).

Before we tell you the results of this experiment, think about what you would predict. In which condition do you think participants were most likely to litter, and in which condition do you think they were least likely to litter? Have you made your predictions? Okay, now look at Figure 4.7 to see if you were correct.

In this experiment, there was a **main effect** caused by the descriptive norm, meaning that, overall, participants were more likely to throw their flyer to the floor in the littered setting than in the clean setting (as you probably guessed). But, there was also an interaction. Whether the confederate's behavior served as a model to be imitated depended upon whether the setting was already littered or clean. Specifically, participants who saw the confederate litter in an already cluttered garage were the most likely to follow suit and throw their flyers on the ground. However, participants who saw the confederate litter in a clean garage were the *least* likely to litter themselves. Are you surprised? This is an example of a **counterintuitive** finding, where the empirical evidence doesn't gibe with common sense. As noted earlier in the chapter, this is why it is necessary to study behavior scientifically.

The littering study is considered a true experiment, even though it took place outside a laboratory, because many contextual variables (time of day, presence of other people) were controlled, IVs were manipulated, and the IV manipulations were presented randomly to participants. Therefore the results have very few possible alternative explanations. The advantage over a contrived lab situation is that conducting the experiment in this field setting—a real parking garage using real people doing their real-life activities—makes the results both scientifically compelling and likely to represent real-life behavior.

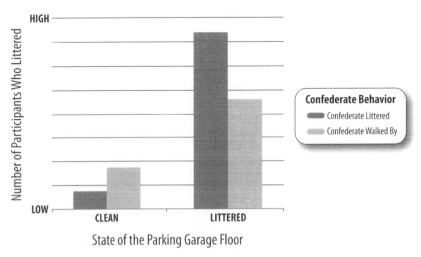

FIGURE 4.7 *Field experiment results.*
Participants were more likely to litter in an already littered environment, particularly when they saw a confederate do so. However, those seeing a confederate litter in a clean garage were the least likely to litter themselves. Graph adapted from Cialdini, R. B., Kallgren, C. A., & Reno, R. R. (1991).

Quasi-Experiments in the Field. Sometimes researchers choose to maximize realism by studying people in their real-world context, but are unable to conduct a true experiment (in which every possible variable in the situation is controlled and participants are randomly assigned to conditions). A recent example of this type of quasi-experimental design compared outcomes for 397 kids who attended camp located in either nature or urban environments (Collado, Staats, & Corraliza, 2013). Measurements were taken on the first and last day of camp to determine attitudinal changes over time. Kids attending nature camp demonstrated larger changes in proenvironmental feelings (e.g., finding solace in nature), beliefs (worldview), and willingness to behave proenvironmentally (e.g., conserve water and energy) than kids attending urban camp. Because families chose where to send their kids, the comparison groups were not randomly assigned, thus there could be some other systematic differences between the groups besides the camp locations. Any ideas? However, despite the design's lack of control, the results suggest that these different camp experiences *may* have had different impacts on the children, which deserves further research attention.

Some quasi-experiments are conducted for the specific purpose of trying out **behavioral interventions**, which are formal efforts to influence behavior (e.g., training a new skill) in real-world contexts. Due to logistical limitations in field settings, researchers may not have the luxury of making a comparison between distinct groups of people exposed to different conditions. When researchers can't randomly assign people to conditions (as in the littering study) and there are no comparable preexisting groups (as in the camp study), an alternative option is to try out a series of manipulations on the same group (or at least in the same setting) over time.

An exemplar of this kind of study involved a four-step intervention to encourage composting in a university cafeteria (Sussman, Greeno, Gifford, & Scannell, 2013). Researchers systematically varied the situation, first providing a compost bin next to the garbage. Then, after a baseline period, they added new features one by one starting with informational signs, then persuasive signs, and then confederates setting an example with their own behavior. Observers measured whether people composted all, some, or none of their waste as they exited the cafeteria. As predicted, people composted more after each type of sign was posted. Composting increased further when confederates modeled composting behavior. We can conclude that it is *possible* that signage and/or modeling improved composting behavior. However, because there was no comparison group (who contemporaneously didn't see signs or confederates), we cannot eliminate alternative explanations. For example, maybe there happened to be a high-profile media blitz touting the benefits of composting during that same period of time and that's what changed the behavior. Or maybe it was just natural change due to the passage of time. Further, because the confederates were added after the signs, it is not possible to know if the composting was ultimately being influenced just by a delayed effect of the signs or by the combination of the two factors. Signage and modeling are confounded, or inseparable in their effects. This kind of confounding is not unusual when quasi-experiments are carried out in a real situation and without a control condition. However, the advantage of maximizing realism by observing people in a real-world context often outweighs this risk of confounds and the inability to draw certain causal conclusions.

Running the Numbers: Statistical Analysis

Whether researchers are conducting laboratory experiments, correlational studies, or field experiments, they are collecting **quantitative** data. That is, they are employing operational definitions that convert human thoughts, feelings, and behaviors into numbers that can be statistically analyzed. Importantly, in psychology, each individual participant's responses are rarely considered separately. Instead, the data are aggregated from many individuals to determine what is typical, using measures of **central tendency** (mean, median, and mode). Analyses also determine "how typical is typical" by measuring **variability** (standard deviation), which indicates whether most people's scores hover around the average point or if they are widely spread out (see Figure 4.8).

No matter how spread out or clustered a group's scores are, there will always be a few people near the extreme ends of the continuum (unusually high or low). If they are really different from anyone else who was measured, they are called **outliers** and sometimes they are eliminated from further statistical analysis since keeping them in the data set can trick the statistics, masking a more general pattern. This does not mean, however, that outliers are ignored. Indeed, it is important to dig deep and try to understand why they are so different from everyone else.

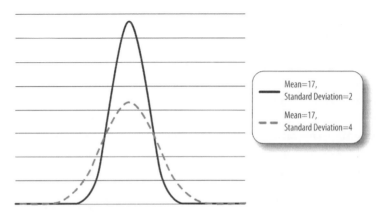

FIGURE 4.8 *Distributions of scores.*
Variability puts "what is typical" into context.

Inferential statistics are used for understanding differences between groups, and patterns among variables. They help estimate whether any patterns in the data are likely to be real rather than just random; that is, whether a finding is "statistically significant." Tests of **statistical significance** are essentially answering the question: "What are the odds that I would see this pattern in my sample of participants if no such pattern actually exists in the broader population of people (i.e., if the pattern were simply due to chance)?" It is customary to consider a pattern statistically significant if those odds are less than five in 100. More extensive and nuanced explanations of statistics are beyond the scope of this text, but suffice it to say that most published empirical research, and consequently this text, focuses on statistically significant results, meaning the data patterns are not likely to be a fluke.

Once researchers complete a large number of studies on a topic, a statistical method called a **meta-analysis** is often used to assess the big picture. In contrast to a traditional **literature review**, in which authors summarize and subjectively interpret large numbers of studies on the same topic, a meta-analysis involves *quantitative synthesis*. In other words, it is a technique that combines the numerical results of many studies into representative summary statistics called **effect sizes**. Effect sizes, labeled "small," "medium," or "large," indicate which IVs (or predictors) are most influential, and to what extent they reliably predict a DV (or criterion) (Cohen, 1992).

One recent meta-analysis examined studies about the relationship between materialism and environmental attitudes and behaviors (Hurst, Dittmar, Bond, & Kasser, 2013). The meta-analysts scanned relevant databases, reference and citation lists, and contacted prominent researchers in the field. They considered any study, published or unpublished, that included correlational data between materialism, defined as individuals' "long-term endorsement of values, goals, and associated beliefs that center on the importance of acquiring money and possessions that convey status," and environmental attitudes or behavior. There were 13 independent samples that were ultimately included in the analysis. The researchers concluded that there is a medium effect size for materialistic values as a predictor of both an attitude that the environment does not need protection and "environmentally-damaging" behavior.

Meta-analytic summaries are used to update theory and can serve as the foundation for subsequent research. For instance, now cited by more than 100 other published studies, Bamberg and Möser (2007) summarized research about the Theory of Planned Behavior based on results from 57 samples. They found that attitudes, a sense of control, and personal moral norms are the best predictors of behavioral intentions. They also conclude that intentions directly, but only partially, predict proenvironmental behavior (it's about 30% of the explanation, which means that a lot about the behavior is NOT explained by intentions). Their meta-analysis provides support for the Theory of Planned Behavior (described in Chapter 8), which continues to be a strong thread in conservation psychology research.

Measurement Tools for the Psychology of Sustainability

As described earlier in the chapter, one of the fundamental challenges of designing quantitative studies is operationally defining variables. Operational definitions of dependent variables in experiments, and all variables in correlational studies, require measurement tools. But rulers, balances, and graduated beakers aren't much use when one is trying to measure thoughts, feelings, and behaviors. In this section are some examples of the types of measurement tools most commonly employed when studying the psychology of sustainability.

Frankly, extensive work is required to develop strong measurement tools. The best measurement tools are **reliable** (yielding consistent results time after time), **valid** (measuring what they are supposed to be measuring), and **sensitive** to differences among people. Because research on the psychology of sustainability is burgeoning, with contributors from diverse backgrounds deriving ideas from a variety of literatures, there has been a proliferation of tools, sometimes without evidence for their **psychometric properties** (i.e., reliability, validity, and sensitivity). This "anarchy of measurement" (Stern, 1992, p. 279) makes it hard to build a systematic understanding of constructs and phenomena. There are signs of maturity, however, in recent, concerted efforts to develop psychometrically strong instruments (Milfont & Duckitt, 2010) and rigorous meta-analyses identifying best practices (Hawcroft & Milfont, 2010). Nonetheless, when you read research articles, your curiosity and skepticism should lead you to ask questions about the measurement. Have the authors provided strong evidence that they are measuring a variable accurately?

The most commonly used scale in conservation psychology is the New Ecological Paradigm (NEP revised: Dunlap, Van Liere, Mertig, & Jones, 2000). Cited by over 300 scientific papers, the NEP, which captures people's assumptions about the relationship between humans and the natural world, demonstrates acceptable reliability and predicts proenvironmental behavioral intentions (e.g., Cordano, Welcomer, & Scherer, 2003). Looking at Figure 4.9, can you tell which items represent the dominant social paradigm (DSP) worldview described in Chapter 2 of this book? Which ones might represent the NEP, an ecological worldview consistent with principles found in Chapter 3? Upon examining the measure, you might think there's a bit of truth in items from both perspectives. This is why responses are made on a five-point scale, rather than just asking "True or False." Specifically, when completing the NEP, people respond to each statement using a scale ranging from *strongly agree* to *unsure* to *strongly disagree*, allowing representation of a wide variety of beliefs. Some people lean one direction or the other, others see some value in both NEP and DSP ideas, and others yet land at the very extremes.

Many scholars criticize the use of this measure because of the many ways researchers adjust it, sometimes by changing the number of items, the wording of the statements, and the number of options on the response scale (Hawcroft & Milfont, 2010). These sorts of alterations can significantly affect a scale's psychometric

How would you respond to these statements using the following scale?
SA = strongly agree **MA** = mildly agree **U** = unsure **MD** = mildly disagree **SD** = strongly disagree

1. We are approaching the limit of the number of people the Earth can support.
2. Humans have the right to modify the natural environment to suit their needs.
3. When humans interfere with nature it often produces disastrous consequences.
4. Human ingenuity will insure that we do not make the Earth unlivable.
5. Humans are seriously abusing the environment.
6. The Earth has plenty of natural resources if we just learn how to develop them.
7. Plants and animals have as much right as humans to exist.
8. The balance of nature is strong enough to cope with the impacts of modern industrial nations.
9. Despite our special abilities, humans are still subject to the laws of nature.
10. The so-called "ecological crisis" facing humankind has been greatly exaggerated.
11. The Earth is like a spaceship with very limited room and resources.
12. Humans were meant to rule over the rest of nature.
13. The balance of nature is very delicate and easily upset.
14. Humans will eventually learn enough about how nature works to be able to control it.
15. If things continue on their present course, we will soon experience a major ecological catastrophe.

FIGURE 4.9 *Items from the New Ecological Paradigm (NEP) Scale.*
Adapted from Dunlap, R. E., Van Liere, K., Mertig, A., & Jones, R. E. (2000), p. 433.

properties and thus its comparability across studies. Researchers often have real reasons to modify scales such as time and space limitations. For instance, in a public
venue, one might only have a minute or two to hold people's attention so researchers may shorten a scale, maybe even simplify it to a single question, in an effort to
capture what they can. Though there may be good reasons for adjusting a tool like
the NEP, untested variations should be viewed with skepticism and results should be
corroborated with additional research using validated measures.

One example of a measure that was purposely designed to be brief—a single item, in fact—is Schultz's (2001) Inclusion of Nature in Self (INS; see Figure
4.10). The INS is a measure of the sense of connectedness people perceive between

Please circle the picture below which best describes your relationship with the natural environment.
How interconnected are you with nature?

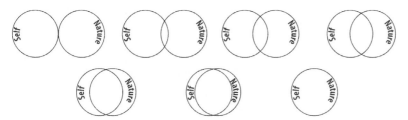

FIGURE 4.10 *Inclusion of Nature in Self (INS) scale.*
Adapted from Schultz, P. W. (2002b).

themselves and the natural world. Although the INS is conveniently short, it is predictive of ecological attitudes and behavior.

Growing quickly in popularity is the lengthy but comprehensive Environmental Attitudes Inventory (the EAI has 120 items; a shortened version, EAI-S, contains 72 items), designed to measure a diverse set of predictive constructs identified by previous research (including concepts originating in the NEP). The EAI reliably and validly measures 12 unique dimensions of environmental attitudes (Milfont & Duckitt, 2010; see Figure 4.11).

Though attitude measures like these are useful and informative, they are not necessarily correlated with actual behavior (Kollmuss & Agyeman, 2002). Behavior is the problem, and behavior will also determine future sustainability. Therefore, it is imperative to have good behavioral measurement.

Enjoyment of nature	Altering nature
Support for interventionist conservation policies	Personal conservation behavior
Environmental movement activism	Human dominance over nature
Conservation motivated by anthropocentric concern	Human utilization of nature
Confidence in science and technology	Ecocentric concern
Environmental frugality	Support for population growth policies

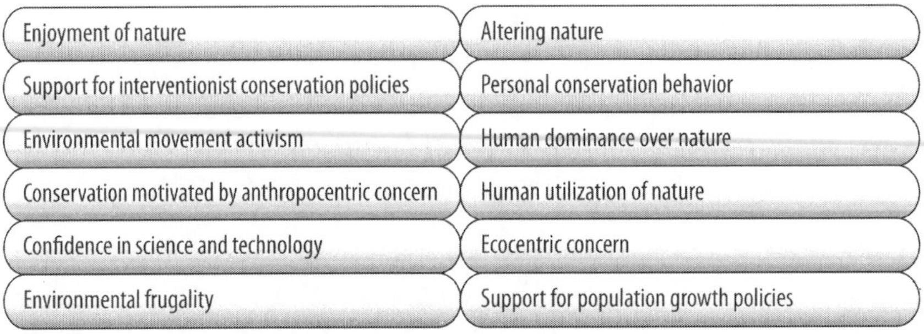

FIGURE 4.11 *Environmental Attitudes Inventory dimensions.*
Adapted from Milfont & Duckitt (2010), pp. 89–90.

It might seem that the best type of behavioral measurement would be **behavioral observation**, meaning actually watching and recording what people do. However, there are many drawbacks to this method. Often there is limited time or opportunity to observe the behavior of interest (we can't follow people around all day to see whether they turn lights off, recycle their cans, write a letter to their state representative, etc.). Even when there is an opportunity to observe, researchers can make errors in recording and interpretation. For example, suppose you saw someone throw a can into the trash bin instead of the recycling bin. Was the behavior on purpose or accidental? Was it typical for this person or unusual? Was it an example of "antienvironmental behavior" or was it just thoughtless habit? Because behavioral observation is so laborious and imprecise, the research literature is rife with studies measuring behavioral *intentions* like "how often do you think you will take the bus to work?" and proxies for behavior like "willingness to pay" (e.g., to subsidize public transportation). Although these types of outcomes are easier to assess than actual behavior, it is clear from research studies looking at how well (or poorly) intentions predict actual behaviors that the two are definitely not interchangeable.

Some researchers develop creative, albeit indirect, measures of proenvironmental behavior that don't require following people into their homes or workplaces. For instance, what would you do if a "student leader" approached you after a research study, asking you to suggest budget allocations to a variety of university programs, including one that was environmentally focused? In one such study, participants who had read about an oil spill from the perspective of a struggling animal were more likely than those taking a human view to allocate funding to the environmental program (Berenguer, 2007). Here's another idea. What do you think it means if some people spend more time than others looking at environmental pamphlets after a research project is over? Do you think it matters whether or not they take one away with them (Pahl & Bauer, 2013)? Given the popularity and ease of online surveys, why not track virtual behavior, such as whether or not participants click on a web link to seek information about an environmental issue or donate to an environmental organization?

Another way to get at actual behavior without the need for chronic surveillance of participants is to administer **self-report** measures, such as the Energy Efficient Behavior Scale (EEBS; Stragier, Hauttekeete, De Marez, & Brondeel, 2012). The EEBS is a 13-item scale that asks participants whether they perform any of a variety of energy-saving actions such as upgrading appliances and switching from machine drying to line drying clothes. A second example of a self-report behavioral measure is the 30-item General Ecological Behavior scale (GEB; Kaiser, 1998), which measures a more general constellation of environmentally related behavior ranging from consumer behavior to volunteering. This measure can be used to identify the relative difficulty of different behaviors (e.g., apparently talking to friends about environmental problems is generally harder than eliminating the use of fabric softener) and has been used to compare behavior across different cultural contexts (Kaiser & Biel, 2000). See Figure 4.12 for a variety of examples of how proenvironmental behavior can be operationally defined.

Self-report measures allow assessment of multiple behaviors over extended periods of time. However, because they are retrospective, participants' answers may be inaccurate due to memory foibles. Can you think of any other reasons self-reported behaviors might not be quite accurate? How eager do you think participants are to admit to behaviors that they think make them look bad, and how tempted might they be to overestimate behaviors that make them look good? Participants aren't always fully honest in self-reports because of the pressure they feel to give **socially desirable** responses. Another problem that can arise with self-report measures is that participants may feel frustrated trying to communicate their behavior quantitatively. It is not unusual for researchers to administer a scale to participants only to have them circle a number and then draw a line from that number to a paragraph, written in the margin of the survey, explaining their answer. In cases like this, the participants clearly want to tell the researcher more than can be communicated by a numerical quantity. This is where qualitative measurement serves a valuable function.

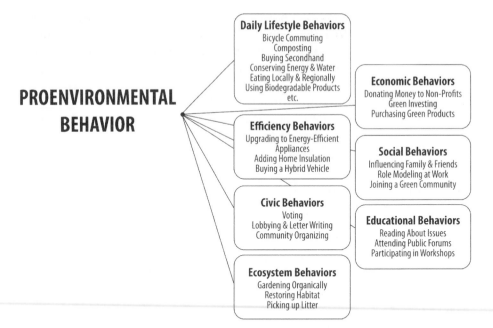

FIGURE 4.12 *Examples of environmentally relevant behavioral constructs with sample operational definitions.*
Researchers can assess people's intended or actual behaviors by way of observation or self-report.

Qualitative data are collected through surveys, interviews, and focus groups in which participants respond freely to open-ended questions. Earlier in the chapter, qualitative techniques like these were mentioned as a useful first step, when not much is known about a topic, and the researcher is trying to identify critical variables for future research. They can also provide the basis for building quantitative tools such as checklists or multi-item scales. And, they can complement quantitative measures by providing a means (other than scribbling in the margins) for participants to contextualize and elaborate on their quantitative responses.

In spite of the fact that qualitative techniques can elicit rich and complex responses from participants, many researchers avoid them. Two primary reasons are that they are extremely time consuming for both participants and researchers, and they require subjective interpretation that feels "unscientific." For qualitative measurement to be successful, participants need to be articulate about their thoughts and be given time and space for reflection. Researchers need to be well-trained listeners and transcribers. To analyze qualitative responses, researchers first scour the data to generate themes and categories. For instance, when analyzing interviews about what influenced people most to become involved in environmental work, researchers developed categories such as *role models*, *experience of natural areas*, and *formative books* (Chawla, 2006). The next step then is to code all of the individual responses into those categories (In what category would you place the influence of parents and friends? Where would you place remarks about fishing and hiking?). When coding qualitative data, results are only as reliable as the people categorizing observations; training is essential and agreement between coders, termed **interrater reliability**, should be high before the data analysis begins.

Limits to Empiricism

Contemporary psychology, conducted as a science, is reductionistic. It evolved through the modern Western worldview, which assumes that experience can be partitioned (reduced) into smaller elements and that we can understand people by studying individuals separated from each other, as well as from their communities, culture, and the rest of the natural world. This practice has its limitations (Kidner, 2001). Seeing the world as consisting of separate elements can lead to one-way causal thinking, in which specific causes are assumed to impact specific outcomes. This is certainly an oversimplification, because the chain of influence is likely to be longer and more complex than just one IV causing one DV in a unidirectional manner. In fact, a variable that we use as a DV in one study is often the very same variable we use as an IV in another. For instance, the physical environment of lighting on bike paths (IV) may influence perceptions of risk (DV). Perceptions of risk (IV) are likely to influence choosing to go by bike (DV). Choosing whether or not to go by bike (IV) might further influence perceptions of risk (DV).

Research on the psychology of sustainability will progress as theory becomes more integrated and measurement tools are developed for what was previously believed to be unmeasurable. On the other hand, there are limits to measurement, so it is important to remember that whenever something gets measured, something else gets missed. When something is quantified, we often assume it is understood, and potentially miss out on a deeper understanding because a concept has been reduced to a number. Moving forward, a combination of quantitative and qualitative approaches will likely be best for studying our interconnected existence within a complex system, and cause-effect relationships that work both ways.

CONCLUSION

As suggested in Chapter 1, labeling problems as "environmental" diverts our attention from their ultimate cause: human behavior. Calling them "environmental" problems also downplays the reciprocal relationship between the health of the ecosystem and our own wellbeing. Hopefully, you are now prepared to think about our ecologically related behaviors through the lens of a psychologist. You will learn more if you consider your own behavior, thoughts, and emotions as you read the rest of this book. Keep track of your own reactions, which constitute the raw data of psychology. These observations will enable you to judge the adequacy and relevance of the psychological research discussed in subsequent chapters. Psychological responses are what this book is about, so treat yours with respect. They will enhance your understanding of the material, your world, and yourself.

NOTE

1. http://www.apa.org/ethics/code/index.aspx

CHAPTER 5

The Power of the (Unsustainable) Situation

Most thoughtful people agree that the world is in serious trouble. . . . Traditional explanations of why we are doing so little are . . . that we lack responsibility for those who will follow us, that we do not have a clear perception of the problem, that we are not using our intelligence, that we are suffering from a failure of will, . . . Unfortunately, explanations of that sort simply replace one question with another. Why are we not more responsible or more intelligent? Why are we suffering from a failure of will? A better strategy is to look at our behavior and at the *environmental conditions* of which it is a function. There we shall find at least some of the reasons why we do as we do.

(Skinner, 1991, pp. 19–20; italics added)

While it may fly in the face of people's sense of free will, behavior is greatly influenced by forces outside of us. People have a tendency to assume the causes of behavior are internal things like attitudes, values, personality, and preferences (Ross, 1977), but in fact, how we behave is largely a function of whether our behavior is rewarded or punished, and whether it is consistent with social expectations and cultural conventions. The quote above comes from B. F. Skinner, one of the first psychologists to repeatedly relate the issues of resource depletion, pollution, and overpopulation to human behavior (Skinner, 1948; 1971; 1987). Skinner argued that to address the ecological crisis, it is necessary to understand how the environment shapes behavior. In this context, the word *environment* refers to the total physical, social, political, and economic **situation** in which a person behaves. As it turns out, situations are very powerful and often override internal influences on our behavior.

How often have you felt unable to make behavioral changes even when your attitudes, values, and preferences favor the new behavior? For example, maybe you would like to bike instead of drive because you know it is better for the environment (and better for your own health), but you still don't do it. What are the situational influences at work? Maybe biking takes extra time, planning, and preparation, or there isn't convenient bike parking where you are headed. Maybe no one you know rides a bike, or there is no shower at your destination, making a sweaty bike ride socially awkward. Maybe you tried it once and had a bad experience getting soaked by rain or harassed by honking automobiles, leading to the conclusion, "I won't be doing *that* again anytime soon!"

Psychologist Kurt Lewin famously stated that behavior is a function of the interaction between the person *and* the person's environment $B = f(P, E)$. This means that people are likely to behave in ways supported by the situation in which we find ourselves, and unlikely to behave in ways not supported by the situation. This chapter will draw on the subdisciplines of **behaviorism** and **social psychology**—the domains of Skinner and Lewin, respectively—to identify how and why the situation is so powerful.

BASICS OF BEHAVIOR MODIFICATION

Any situation includes multiple **contingencies**, external forces that affect the behavior of individuals in that particular situation. These contingencies include **antecedents**, cues that precede behavior, and **consequences**, outcomes that reward or punish. Antecedents can be anything from billboards and post-it notes to being buzzed by your smartphone or seeing someone else's actions. A student recently told one of us authors about an antecedent that guides her behavior: "The signs above the bathroom lights that say *Save a Light, Kill a Watt!!* . . . always remind me to turn off the lights when I leave." Additionally, every behavior has consequences. Another student recently started biking regularly as a mode of transportation due to a mixture of consequences: "I don't have to pay for gas money or borrow a friend's car" and "I get fresh air and exercise during travel time."

Skinner called rewarding consequences like these **reinforcers**, because they strengthen or increase the likelihood of behavior. As an example, several states have a "bottle bill" where the purchase price of products in glass bottles and aluminum cans includes a 5¢ deposit, refundable upon return of the bottles and cans. If a customer returns bottles and gets a refund, each nickel serves as a reinforcer. As a result, beverage container recycling is much higher than the national average in the ten states where this type of law has been enacted (Container Recycling Institute, 2013). In contrast, **punishment** decreases the probability of a particular behavior. If mom yells at you for throwing a returnable can in the garbage, that's punishment; presumably that behavior is unlikely to recur (at least in her presence).

Skinner elaborated by noting that both reinforcement and punishment can involve either *adding* or *removing* something to the situation. Both the five-cent refund and the scolding mentioned above are examples of adding something. If the added stimulus is pleasant, it will be a reinforcer; if it is aversive, it will be a punisher. Conversely, removing a pleasant stimulus is a punisher while removing an aversive stimulus is a reinforcer. In Skinner's terminology, an added stimulus is "positive" and a subtracted stimulus is "negative" (think mathematical addition and subtraction, rather than "good" and "bad"). For example, recycling is *negatively reinforced* if it terminates mom's scolding (see Figure 5.1). Both positive and negative reinforcement will potentially increase proenvironmental behaviors, like using bikes for transportation; however Figure 5.2 illustrates how punishers can simultaneously discourage people from making the leap into bicycle commuting.

Is a punisher always a punisher and a reinforcer always a reinforcer? The short answer is no. Consequences are subject to interpretation by individuals, and may be experienced differently depending on the situation. For example, using the bus might feel reinforcing in a large city where auto parking is expensive and difficult to find, but it might feel punishing in a rural location where the bus stop is a mile from

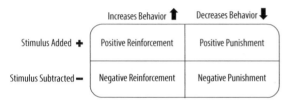

FIGURE 5.1 *Types of reinforcement and punishment.*

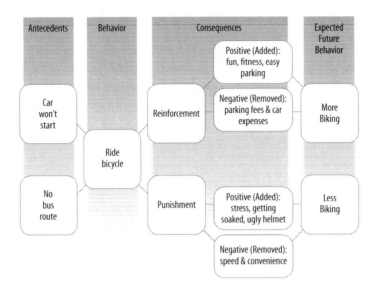

FIGURE 5.2 *Behavioral contingencies.*
Antecedents and consequences of bicycle commuting.

one's destination. Bus riding might be reinforcing to someone who feels stressed by driving, but punishing to someone who loves to get behind the wheel. Therefore, it is important to understand that whether a particular consequence is reinforcing or punishing is in the eye of the beholder.

You Catch More Flies with Honey

Although all four behavior-consequence relationships are relevant for thinking about behavioral change, Skinner determined that reinforcement is generally most effective. You may know from experience that punishment feels really unpleasant and often produces undesirable behavioral side effects such as aggression or avoidance of the person implementing the punishment. People like to perceive themselves as having free will, and generally they do *if* their behavior is controlled by reinforcement. Specifically, when reinforced, people usually perceive the freedom to choose their actions, while punishment creates a feeling of being constrained or even coerced and manipulated (e.g., DeYoung, 2000). Using the biking example, note that reinforcement increases the appeal of biking, yet allows one to take a car if needed. However, being punished for driving essentially reduces a person's perceived options. In that light, consider that "the principle modus operandi of [environmental] organizations is to frighten people rather than offer them a world to which they will turn because of the reinforcing consequences of doing so" (Skinner, 1991, p. 28). No wonder people reject their appeals for action. Instead, efforts to encourage environmentally responsible behavior should forgo punishing inappropriate behaviors or using scare tactics, and instead focus on reinforcing achievements.

Feelings are important for maintaining many behaviors, because emotions and thoughts associated with various contingencies also begin to operate as reinforcers or punishers that affect future behavior. For instance, when that student had a good experience bicycling to school, her behavior changed and she also experienced certain emotions such as joy. Feeling pleasure became associated with biking, and now when she considers the biking option, that same feeling recurs. Experiencing happy feelings continues to positively reinforce her conservation behaviors. The opposite is true, too, such that the shame and guilt associated with being punished further punish the individual.

Timing Is Everything

Rewards and punishers work best when they occur in close proximity with the behavior, so that their connection is clear. The pattern of consequences also makes a difference in the strength and durability of behavior change. As you might expect, people's actions tend to persist with continuous reinforcement. For instance, receiving money every time you return a beverage can is likely to get you to bring in your bottles regularly; however, if the refund is then discontinued, you will be likely to immediately stop bringing in bottles.

In contrast, **intermittent reinforcement** occurs when the pleasant consequence happens only *sometimes*. This type of reinforcement can take a while to have an impact but **extinction**, or the weakening of the behavior, will slow down too, meaning behaviors will continue for a while even after reinforcers are withdrawn. Intermittent reinforcement can be regular and predictable or irregular and unpredictable, and it can differ depending on whether consequences occur after some set amount of time, or after a certain number of behaviors (see Figure 5.3).

To understand how intermittent schedules of reinforcement work, consider another biking example: Imagine you are setting up a bike share program in your community. Once people sign up, you'd like to encourage them to use the bikes often and on a long-term basis. What kind of reinforcement should you use? Perhaps virtual coupons for the local coffee shop would be popular, but how should you distribute them? Giving out a coupon for *every* ride, continuous reinforcement, would work at first but not so much after the introductory offer expires. Using a

	Dependent on Time	Dependent on Behavior
Regular and Predictable	Fixed Interval	Fixed Ratio
Irregular and Unpredictable	Variable Interval	Variable Ratio

FIGURE 5.3 *Intermittent schedules of reinforcement.*

fixed ratio schedule, you could send out a coupon after every fifth ride. On a **fixed interval** schedule, participants would earn coupons once a week if they've used a bike, no matter how many times they actually rode in that week. On the other hand, using a **variable ratio** schedule, you would distribute a coupon after about five rides, sometimes fewer and sometimes more. And, giving a coupon every so often, no matter whether they've used a bike only once or many times, is a **variable interval** schedule.

Which schedule do you think would be the most effective at encouraging frequent and prolonged use of the bike share program? The answer is the *variable ratio* schedule, because it is behavior-dependent and unpredictable. (Slot machines employ this type of reinforcement schedule to keep people gambling.) In contrast, the fixed interval schedule is least effective, leading to procrastination until the deadline for earning the reward draws near.

The Short and Long of It

Sometimes the short- and long-term consequences of behavior differ. This is the case with many unsustainable behaviors, where the immediate rewards are not in line with distant punishers (Toates, 2009). For example, driving may ultimately be bad for people (air pollution and climate change), but the immediate reinforcement is so powerful (getting us where we want to go quickly and conveniently), we do it anyway. In other words, we are caught in a **contingency trap** (Baum, 1994). One way individuals can get themselves out of a trap like this is to deliberately change the short-term consequences to be more consistent with the long-term ones. For instance, people who want to decrease how often they drive to work could choose not to purchase a parking permit. With no permit, if they succumb to the temptation to drive (because it is immediately rewarding in terms of time and convenience), they will also likely experience immediate punishment (a parking ticket). By purposefully declining to buy a parking permit, they have specifically altered the contingencies so that the situation will support their new behavior.

Like the driving example, many environmentally relevant contingency traps pit self-interest against what is best for the greater good, in what are known as **social dilemmas**. A **commons dilemma** occurs when individuals are tempted by personal benefits to overuse a shared resource. Hardin's (1968) description of the *tragedy of the commons* is the classic example: If individual farmers who graze their cows on a limited piece of common land allow too many of their own animals to graze there, overgrazing will ruin the land for everyone. The immediate consequence benefits the individual, but the long-term consequence punishes the group. Similarly, individuals find themselves in a **public goods dilemma** when they must decide whether to contribute to a pooled resource, such as donating to a nonprofit organization that works to protect the local watershed.

In some social dilemmas, the consequences to the self and to others are separate. For example, if you are at a potluck and you take more than your fair share when you go through the line, it is the people behind you who suffer the consequences of your gluttony. However, in many environmental social dilemmas, the consequences to the group *ultimately affect the individual* who is a member of that group. In these situations, people behave selfishly to their own detriment. Consider climate change. Individual behaviors that contribute to the problem, such as driving, affect the whole planet, including all of us drivers. For many people, the fact that they are part of the greater whole that will suffer for their actions is not salient, so the immediate rewards easily outweigh the long-term punishers. However, when people are reminded of the personal relevance of their ecologically harmful actions, they are more willing to forgo immediate benefits and make contributions for the benefit of the group, because they recognize that acting for the common good *is* acting in self-interest (Milinski, Sommerfeld, Krambeck, Reed, & Marotzke, 2008; Ostrom, Burger, Field, Norgaard, & Policansky, 2007). Likewise, studies have shown that people will forgo immediate personal reinforcers for longer term group goals if they identify with the group and feel responsible toward it (Dawes, 1980; van Vugt, 2002).

Old Habits Die Hard

From a behavioral perspective, one of the biggest obstacles to proenvironmental behaviors may be that many environmentally destructive behaviors are strongly habitual. **Habits** are patterns of responding to particular stimuli, built through repeated association. If you have ever tried to modify your own behavior, you know that instant and permanent change just doesn't happen. (Think New Year's resolutions!) New behaviors are inconsistent and weak and are no match for long-reinforced habits; therefore, they require considerable attention and practice to become established (more on this topic in Chapter 6) (Neal, Wood, & Quinn, 2006).

Importantly, habits can cut both ways. In other words, weak, proenvironmental behaviors can eventually become consistent and durable if contingencies support them. For example, if you get involved in campus sustainability work, it can be embarrassing to be seen acting in unsustainable ways, such as throwing away compostables or driving for short distances. Your friends might harass or ridicule you for being hypocritical. Embarrassment and humiliation are punishing. Because you never know when someone might see you, you are likely to become very conscientious about what you do so as to avoid these punishers. The intermittent schedule keeps the rate of responding so high for such a long time that the behaviors eventually become habitual, and thus become very resistant to extinction.

As the example in the previous paragraph illustrates, contingencies sometimes take the form of social pressure. Things that other people do or say can serve as both antecedents and consequences. The next section describes in more detail some of the ways that interpersonal situational influences powerfully affect behavior.

BASICS OF SOCIAL INFLUENCE

Most people underestimate the extent to which their behavior, and the behavior of others, is subject to **social influence** (Cialdini, 2005). Especially in **individualistic cultures**, such as those in North America, Western Europe, and Australia, people feel autonomous. They think their behavior comes from within and fail to recognize how much it is shaped by the behavior, suggestions, requests, and reactions of others around them. This blind spot pertains to all behaviors, including environmentally relevant ones. For instance, one study showed that intentions to conserve energy correlated more highly with beliefs about neighbors' conservation practices than it did with the desire to save money, protect the environment, or benefit society in general (Nolan, Schultz, Cialdini, Goldstein, & Griskevicius, 2008). And yet, people in this study reported that their beliefs about neighbors' behavior had *less* impact than their desire to save money or the environment. Social influence is strong and ubiquitous, but when it comes to recognizing the causes of behavior, people "do not just fail to get [its] relative role right; they tend to get it precisely wrong (Cialdini, 2005, p. 159).

We Do as Others Do

Whether you are aware of it or not, you are constantly "reading" social settings to determine appropriate language, manner, gestures, and other behaviors. We learn a lot about how to behave by watching and imitating what others do. In fact, **modeling** others is a primary influence on behavior (Bandura, 1977), especially when situations are unfamiliar or ambiguous. Perhaps you remember feeling unsure about how to behave when you first came to college. As a first-year student, you likely looked to more experienced peers for examples. You noticed which behaviors seemed to work best for other students, and which ones to avoid. As you have read, consequences shape people's behavior, but individuals do not have to experience consequences first-hand to be influenced by them. When we observe others' behavior, we also sometimes *witness* punishing or reinforcing consequences of that behavior. Just seeing the consequences happen to someone else can be enough to push us toward, or away from, a behavior (Bandura, 1977).

Some behaviors that people imitate are witnessed from a distance. For example, many teenagers emulate the fashions worn by popular celebrities whom they do not personally know. People are most likely, however, to copy the behaviors displayed by members of more relevant and immediate **reference groups**, such as neighbors, friends, and family members. One classic study demonstrated that the best predictor of whether or not people purchased solar equipment was the number of acquaintances they had who owned solar equipment (Leonard-Barton, 1981). Similarly, a strong predictor of recycling is having friends and neighbors who recycle (Oskamp et al., 1991).

As reference groups change across different situations, so do people's behaviors. For example, a nonvegetarian might be less likely to order meat when dining out with vegetarian friends than when dining with fellow meat eaters. Someone who typically ignores litter on the sidewalk may become an avid trash collector when walking with friends who stop every few steps to pick something up. As in these examples, the actions of other people often exert a more powerful influence over behavior than personal preferences, attitudes, values, or habits. Of course, we do not imitate other people's behaviors all of the time. We do have minds of our own. Still, the pressure to **conform** to the group, to behave as others behave, can feel very strong. Perhaps you can recall a time when you found yourself going along with what others were doing, even though you weren't sure it was the right thing to do; maybe you thought it was silly, or even disagreed with it in principle. People feel compelled to do as others do because they don't want to stand out (in a bad way).

When other people around us serve as models, their behaviors sometimes communicate **social norms**, general guidelines about what sort of conduct is typical, expected, or correct in that situation. Norms are more subtle than rules or laws, though they can feel just as binding. You may not be formally penalized or arrested for violating a social norm, but you risk feeling embarrassed, ashamed, or ostracized. Theoretically, humans are evolutionarily predisposed to create and enforce norms as a means to maintain order within groups (e.g., Chudek & Henrich, 2011). When someone violates a norm, it may indicate that the person is an outsider, and, therefore, a potential threat. This is why norm violations are usually perceived negatively and why it feels so intimidating to perform them.

When people go along with the crowd, they are following the **descriptive norm**, a guideline about what most people typically do in a particular situation.

Descriptive norms can vary across similar situations. For example, in some of your classes, students probably sit and quietly take notes, while in others, the norm is to actively participate in discussion; however, in neither setting is it considered acceptable for a student to suddenly stand up and launch into an impromptu lecture. The prohibition against students spontaneously assuming the role of instructor is an example of an **injunctive norm**, which refers to what the culture as a whole *approves of* as the appropriate way to behave. As you might assume, these two types of norms often correspond to each other: In a given situation, most people are likely to be doing behaviors consistent with what the culture condones. But, sometimes there is a discrepancy: Although everyone knows the "right thing" to do, most people aren't doing it (Reno, Cialdini, & Kallgren, 1993). Such is the case for many environmentally relevant behaviors.

For example, littering behavior is generally frowned upon in the United States due to widespread anti-littering campaigns in the 1970s. Nevertheless, if you walk along any U.S. riverbank frequented by people, you are likely to see smashed cans, broken bottles, food wrappers, fishing line, and cigarette butts. In this example, the injunctive norm discourages littering, but the descriptive norm supports it. When injunctive and descriptive norms are in conflict, the injunctive norm may be less vivid in people's minds than the more immediate descriptive norm. For instance, in Chapter 4, you read about a field experiment in a parking garage where researchers varied whether the setting was clean or littered with flyers; this manipulation communicated a descriptive norm (Cialdini, Reno, & Kallgren, 1990). In keeping with the descriptive norm, people were more likely to litter in an already littered environment, even though they presumably were aware of the injunctive norm against littering. Recall, however, that researchers also manipulated the salience of the injunctive norm by having a confederate model relevant behavior; specifically, when the confederate threw his flyer onto the clean garage floor, his startling action reminded participants that littering is a no-no. This was the condition in which participants were least likely to litter, because the injunctive norm was on their minds.

As you can see, the communication of norms is often implicit rather than overt. Similarly, when others model behaviors, they often do so unintentionally. Sometimes, however, social influence is intentional. Friends give their buddies tips about how to fit in before introducing them to a new social group. Parents try to set a good example for their children to follow. In one study, researchers deliberately created a new norm among students to turn out the lights when leaving a classroom (Werner, Cook, Colby, & Lim, 2012). Just posting signs reminding students to turn off the lights was not effective. Creation of the norm required social interaction in the form of presentations (that included the presenter modeling how to turn off the lights) and group discussions (in which the conversation established the idea that it was acceptable to turn off the lights). More direct forms of social influence like this are addressed next.

We Do as Others Push Us to Do

One way to think about different forms of social influence is where they fall on a continuum from indirect and unintentional to direct and deliberate. On one end of the continuum are subtle social norms and inadvertent modeling; on the other end are directions, commands, and orders. Between the two extremes are suggestions, requests, and persuasive appeals (see Figure 5.4). Behavior change in response to the high pressure tactics on the far right of the spectrum is not necessarily voluntary; people feel obligated or forced to respond. Behavior change in response to the social antecedents in the middle is less compulsory. Whether people respond depends on how the pressure is exerted, and by whom, as described in the following discussion of persuasion and compliance techniques.

When someone tries to influence what you believe, that person is exerting pressure in the form of **persuasion**. Often, the ultimate goal of persuasion is behavior change. As you will read in upcoming chapters, people's behavior does not always follow from what they are thinking, but changing a person's beliefs and attitudes can be a step toward changing that person's actions.

Although formal persuasion is deliberate on the part of the communicator, people are sometimes persuaded inadvertently. For instance, imagine you are at a party listening to a debate between friends. No one is trying to convince you of anything, but as you listen to the dispute, you find yourself swayed by one side of the argument. In this situation, you are being vicariously persuaded. Because the opinions of reference groups matter to us, we are even more likely to find ourselves feeling persuaded when most or all of our acquaintances agree about something. For example, participants who heard their friends and colleagues favorably discussing the use of nontoxic household products were more interested in pursuing that behavior themselves, because the overheard conversation made them feel like their friends would approve (Werner, 2003; Werner & Stanley, 2011).

Deliberate attempts at persuasion take many forms, including advertisements, sales pitches, political campaigns, lectures, sermons, books, essays, and more. Among younger people, social networking is a primary format. For example, researchers are beginning to document teens' use of platforms such as Facebook and Twitter to persuade their peers to engage in proenvironmental political, social, and personal action (Allen, Wicks, & Schulte, 2013).

FIGURE 5.4 *Continuum of social influence.*

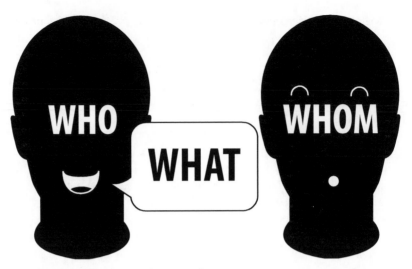

FIGURE 5.5 *Who says what to whom.*
Whether a communication is persuasive depends upon characteristics of the source, the message, and the audience.

Not all communications intended to be persuasive are equally effective. Obviously, the form and content of the message matter; some arguments are more convincing than others. In addition, all messages come from a *source* and are directed at an *audience*; characteristics of both affect how the message is received (Hovland, Janis, & Kelley, 1953; see Figure 5.5).

One of the primary factors that affect persuasion is perceived **credibility** of the source. In general, experts are seen as more credible than nonexperts. This is why people are influenced by sources such as the Union of Concerned Scientists; even the name sounds credible. But, sometimes expertise can be a bit alienating. For example, researchers have found that people may be more receptive to messages about environmental issues, such as climate change, when they come from familiar local figures rather than nationally recognized scientists (Cole, 2007; Leiserowitz, 2007). Even when local sources lack expertise, they can seem credible because they are "people like us" (Chess & Johnson, 2007).

Besides expertise, another thing that influences perceived credibility is trustworthiness (Wiener & Mowen, 1986). When audiences are suspicious of a communicator's motives, they may pay less heed to the message. In one classic study, New York City residents cut their electricity use significantly more when asked in a letter with New York State Public Service Commission letterhead than when the same letter was sent on stationary from Con Edison, the local utility provider. Apparently, people were wary of a utility company encouraging conservation, given its financial interest (Craig & McCann, 1978). More recently, researchers tested how people's responses to persuasive messages about climate change were affected by their perceptions of scientists' motives in delivering the messages. People trusted scientists more, and were more persuaded, when they believed that the scientists' intent was merely to inform, rather than persuade (Rabinovich, Morton, & Birney, 2012). In other words, when participants believed the scientists were *trying* to persuade them, they were less persuaded! Perhaps this resonates with you. We tend to feel threatened and defensive when we think that others are intending to influence us. In fact, their attempts to manipulate our behavior can lead to **psychological reactance**, a motivation to resist or rebel (Brehm & Brehm, 1981).

Some sources seem to be trustworthy experts because their arguments are dense with details. But more is not always better when it comes to persuasive communications. In a series of experiments, researchers presented persuasive messages about carbon dioxide capture and storage (CCS) technology as a method to help mitigate climate change. Participants were randomly assigned to read one of three messages: The first contained only highly relevant information (*By implementing CCS, approximately 90 percent of the CO2 emissions released by the burning of fossil fuels can be captured*); the second included the highly relevant information plus some moderately relevant information (*A small proportion of the captured CO2 can be used for the production of carbonated drinks*); the third included the highly relevant information plus some irrelevant information (*In English, CCS is referred to as "CO2 storage" or "CO2 sequestration." In French also two terms are used, namely "CO2 stockage" and "CO2 séquestration."*). Although all three messages included the same highly relevant points in favor of CCS, these pertinent arguments were less persuasive when diluted by irrelevant information (de Vries, Terwel, & Ellemers, 2014, p. 117).

Importantly, in the CCS study, the message was only weakened by irrelevant information when no source was identified; when participants were told that the source was an environmental nonprofit group or an industrial organization, dilution did not occur. This is a good example of how message and source characteristics can interact. Content of messages can also interact with audience characteristics to affect persuasion. For example, one study tested how people responded to messages about the potential risks of climate change that varied with regard to uncertainty; in the *low uncertainty* condition, the message stated risks in absolute terms (e.g., there is an 80% chance that a quarter of all species will go extinct), while in the *high uncertainty* condition, the message was more equivocal (e.g., there is a 70–90% chance that a quarter of all species will go extinct). Participants' reactions to these messages were influenced by their beliefs about the purpose of science. Reactions to the two different message types were not significantly different among participants who perceived science as *the search for absolute truth*; in contrast, participants who viewed science as *a process of ongoing debate* were significantly more persuaded by the message with high uncertainty than the one with low uncertainty (Rabinovich & Morton, 2012; see Figure 5.6). Similarly, in the previously described study about perceptions of scientists' motives (i.e., to inform versus to persuade), participants who believed that scientists *should* persuade were more receptive to persuasive than informative messages, while the opposite was true for participants who believed the purpose of science is merely to inform (Rabinovich & Morton, 2012). In general,

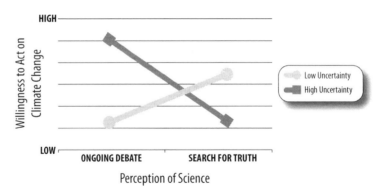

FIGURE 5.6 *Interaction between message content and audience.*
Adapted from Rabinovich & Morton (2012, p. 999).

persuasive messages are most influential when they are tailored to the intended audience.

While persuasive social influence indirectly affects behavior by changing thoughts, some social antecedents are direct behavioral requests. When people respond to these, they are exhibiting **compliance**. Whether we feel inclined to comply with a request can depend on factors such as who is asking and how they make their request. Some individuals (e.g., professional salespeople) are particularly skilled at getting people to do what they ask, but it isn't just natural giftedness that accounts for their proficiency. Social influence experts use strategic tactics that have been found effective at increasing the likelihood of compliance (Cialdini, 2006).

One method is the **foot-in-the-door** technique, which you have almost certainly experienced: Have you ever been asked by a door-to-door canvasser to make a financial contribution to a nonprofit organization? If you have, there is a good chance that before asking for money, the canvasser asked you to sign a petition or a mailing list. Why not just ask for the money up front? The reason is that people are more likely to comply with a request (that they might be tempted to refuse) when it is first preceded by a less costly request (that they almost certainly will *not* refuse; see Figure 5.7).

The foot-in-the-door technique is effective for encouraging a variety of behaviors, including energy conservation. In one field experiment, homeowners were randomly assigned to one of four conditions: In the first condition, they were asked outright to decrease their household electricity consumption; in the foot-in-the door condition, the conservation request was preceded by a request to fill out a brief energy conservation questionnaire; in the third condition, participants were asked to fill out the questionnaire, but were not directly asked to conserve energy; and, in the fourth (control) condition, participants were not subject to either request. After 12 weeks, participants in all three request groups had conserved more energy than

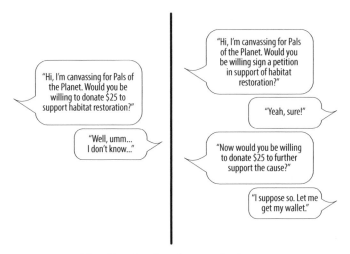

FIGURE 5.7 *The foot-in-the-door technique.*
People are more likely to comply with a request when it is preceded by a less costly one.

the control group, but the foot-in-the-door group conserved significantly more than the other two (Katzev & Johnson, 1983).

One reason that the foot-in-the-door technique works is that *people are inclined toward consistency* (Burger, 1999). Having complied with the first request (to fill out a questionnaire on energy conservation), a person is likely to comply with the second, related request (to conserve energy), because to do otherwise would be inconsistent. In general, people feel the desire to be, and appear, consistent (Gawronski & Strack, 2012). We like to think of ourselves as rational and reliable, rather than fickle and capricious, and we want others to perceive us accordingly. Some individuals value consistency more than others, and the tendency is stronger in individualistic cultures than in collectivist ones, but the desire for consistency is basically universal (Guadagno & Cialdini, 2010; Petrova, Cialdini, & Sills, 2007). Recall the evolutionary explanation for conformity described earlier in the chapter; our desire to be seen as stable and steadfast may be a closely related phenomenon.

What this means is that feeling hypocritical is aversive for most people. An unpleasant state of **cognitive dissonance** arises when we experience a contradiction between our behaviors and attitudes, or between our various behaviors (Festinger, 1957). Sometimes dissonance arises from discrepancies between thoughts about oneself. For example, if you consider yourself to be a good person, but then you do something unwise or unkind, you may experience dissonance (Aronson, 1968; Tavris & Aronson, 2007). Importantly, dissonance arises only when we feel like we behaved *voluntarily*; if we feel like we were forced to behave in a way that is inconsistent with our beliefs or our sense of self, we can justify it to ourselves as a circumstance that was out of our control. It is when we have **insufficient external justification** for acting in a discrepant way that we must resort to other tactics to relieve our dissonance. One common method is to adjust our behavior so as to bring the disparate elements back in line.

For example, imagine that you have just finished a workout at your campus gym and are headed to the locker room to shower. En route, a woman approaches you and identifies herself as a representative of the campus water conservation office. She asks you if you can spare a moment to answer a few questions. You comply. She asks if you are in favor of water conservation. Naturally, you say "yes." She then asks you a few questions about your showering habits: "When showering, do you ALWAYS turn off the water while soaping up or shampooing? When you take showers, do you ALWAYS make them as short as possible, or do you sometimes linger longer than necessary? In your view, about how long does it take an average person to shower and shampoo without wasting any water? About how long is *your* average shower in the locker room?" At this point, you are probably feeling a bit hypocritical, because you realize that you said you believe in water conservation but you don't always practice conservation yourself. Suppose she next shows you a flyer that reads: "Please conserve water. Take shorter showers. Turn showers off while soaping up. IF I CAN DO IT, SO CAN YOU!" She asks you for help with campus water conservation efforts by printing your name on the flyer with a bold black marker. She explains that copies will be posted all around campus. Your desire for consistency compels you to comply. She thanks you and sends you on your way to the showers. What do you suppose happens when you get there?

In a field experiment fitting the description above, a researcher staking out the shower room found that participants who had been induced to feel hypocritical took significantly shorter showers than did control subjects who were not subject to the intervention (Dickerson, Thibodeau, Aronson, & Miller, 1992). Similar experiments have yielded consistent results with regard to residential water use and energy conservation (Aitken, McMahon, Wearing, & Finlayson, 1994; Kantola, Syme, & Campbell, 1984).

By signing the flyer, participants in the shower study endorsed behaviors that they did not practice, and didn't strongly believe in. At the outset, they casually said they supported water conservation, but clearly their initial attitudes were not strong enough to guarantee that they would personally conserve water. When individuals find themselves in a situation like this, promoting a position that they do not hold, they are displaying **counter-attitudinal advocacy** (Festinger, 1957). To resolve the dissonance aroused by this inconsistency, advocates are likely to begin practicing what they preach. This is especially true when they have preached in a *public* way (Stone, 2012; Stone & Fernandez, 2008). For example, when residents in a community field experiment were induced to become "agents" for promoting grass mulching among their neighbors, the agents themselves ended up bagging fewer lawn clippings—and so did their neighbors (Cobern, Porter, Leeming, & Dwyer, 1995).

When behaviors are public, they are subject to scrutiny and evaluation. Therefore, public inconsistency not only arouses dissonance within the individual, but also activates concerns about self-presentation. People's desire to be perceived positively by others is a powerful social influence.

We Do What Makes Us Look Best

Other people's opinions of us matter. This is why we conform to social norms and feel pressured to appear consistent. We care so greatly about how others perceive us that we engage in **strategic self-presentation** in an attempt to control their impressions (Schlenker, 2012). We want others to like us and approve of us. Furthermore, we strive to be admired, desired, and envied. Perhaps it makes people sound shallow to suggest that we are driven to seek social approval and status, but these are likely innate behaviors with their roots in our evolutionary heritage.

People's concerns about keeping up appearances contribute to unsustainable behaviors in several ways. Perhaps the most obvious manifestation is the pervasive use of material goods, such as clothes, cars, and collections to express identity and display prestige. Certainly, some possessions give the owner personal satisfaction, and would be valued even if no one else knew of their existence, but many of the things we prize serve the purpose of communicating to other people who we are, where we stand, and how much we've got. Imagine your t-shirts without their brand labels, team insignias, band logos, or witty messages. If your t-shirts were no-name generics, would you be as eager to wear them?

Of course some individuals are more materialistic than others (more on this in Chapter 7), but to some degree, we all fall prey to the desire for more or different things when we engage in **social comparison**, evaluating ourselves against other people (Festinger, 1954). We compare ourselves to others on a multitude of dimensions, including wealth, status, and popularity. In consumer culture, where we are surrounded by images of attractive people enjoying luxury goods in expensive surroundings, our social comparisons can lead to a sense of **relative deprivation**, a feeling that we may have our needs met, but they could be met with more style or grandeur. Advertisements promise us that if we just have the *right* stuff, we will secure social approval—and to some extent this is true. What is not true, however, is the marketing message that possessions will bring us satisfaction and happiness. Research studies find that when it comes to emotional fulfillment, material goods do not deliver the goods (Dunn, Gilbert, & Wilson, 2011).

In industrialized cultures, **relative social status**, how we rank compared to others, is established and expressed through possessions; the key to cool is consumption. On the flip side, conscientiously *not* consuming may actually hurt social status in some circles. One study found that college students considered the use of clotheslines and public transportation to be low status behaviors (Sadalla & Krull, 1995). The researchers reasoned that because these behaviors might stem from economic necessity, they might be perceived as low status in a socioeconomic sense, but not necessarily in a general sense. Therefore, they replicated the study using recycling as the target behavior, but found the same thing. Together, the results seemed to indicate that conservation behaviors are perceived as low status in general. However, a study conducted 15 years later found no evidence that conservation behaviors were perceived as low status (Welte & Anastasio, 2010). These latter researchers concluded that public perceptions had changed. What no researchers have tested yet, however, is how participants would perceive a clearly high status person engaging in conspicuous conservation; picture a dumpster-diving debutante or a CEO with a ten-year-old cell phone. It seems likely that self-presentational concerns prevent some high status people from engaging in sustainable behavior.

Status concerns may suppress some sustainable behaviors, but they may actually facilitate others. Among the technological innovations developed to decrease resource use are some pretty pricey items, including solar panels and electric vehicles (the least expensive Tesla retails for $70,000). Their cost makes

'And getting rarer by the day, madam!'

these products potential status symbols. However, one study suggests even more moderately priced green goods, such as the Toyota Prius hybrid vehicle (starting retail about $20,000) may be attractive when people are feeling motivated by status concerns (Griskevicius, Tybur, & Van den Bergh, 2010). Evolutionary biologists posit that one of the ways members of many species convey status is by showing that they can afford to incur costs for the benefit of others; in this study, therefore, the researchers theorized that for humans, green products may serve as "costly signals." By using them, instead of more luxurious but less environmentally friendly products, a person is showing a willingness and ability to take one for the team. When the researchers experimentally prompted participants' status concerns, participants' desire for green products did indeed increase. This effect only occurred when participants were shopping *publicly* and when the green products *cost more* than nongreen alternatives, emphasizing the connection to status.

As the researchers suggested, "knowing that a desire for status can spur self-sacrifice . . .presents a powerful tool for motivating prosocial and proenvironmental action" (Griskevicius, et al., 2010, p. 402). In fact, all of the information presented in this section on social influence, and the prior section on behavioral modification, can serve as the basis for interventions to promote sustainable behavior. Such applications are the topic of the next section.

ENGINEERING SUSTAINABLE SITUATIONS

Given what psychologists know about the power of situational influences, it seems logical that changing behavior in a sustainable direction will require altering the situation. **Behavioral engineering** is the intentional manipulation of contingencies that currently create or maintain destructive actions to motivate more sustainable behaviors instead (Geller, 1987, 1992).

The idea of behavioral engineering is off-putting to many. As described earlier, we tend to attribute sinister motives to those who would dare to manipulate us, and we rebel against perceived limits to our behavioral choices (Brehm & Brehm, 1981). While these reactions might be understandable, they belie the fact that people feel quite comfortable with behavioral engineering in other contexts. People *seek* behavioral engineering when they hire consultants to help solve problems, or read books to improve test-taking skills. In fact, you (and/or your parents) *pay* professors to shape your behavior. All behavior is controlled, to some degree, by the situation. Recognizing this fact is empowering. Understanding the role of situational forces in human behavior can lead to the creative redesign of residential, commercial, and industrial environments and situations.

Behavioral engineering employs two main strategies: (1) those which focus on the stimuli that signal behavior, called **antecedent strategies**; and (2) those which focus on the outcomes of behavior, termed **consequence strategies**.

Antecedent Strategies: What to Do and When to Do It

Environments signal appropriate behavior. For example, access to recycling facilities tends to increase recycling behavior (Barr, 2007), and neighborhoods that offer farmer's markets increase the purchase of locally grown produce (Monroe, 2003). In these examples, opportunity and convenience are obviously increased, but physical access is not the only thing affecting behavior. In both of these situations, there are antecedent stimuli that further encourage behavior. These include information, modeling, and prompts.

Information. Providing information about environmental problems and behavioral solutions is the most frequently studied antecedent (Gifford, 2014). This approach includes providing justification for *why* certain behaviors are desirable. It can also provide instructional detail such as which packaging items are compostable (e.g., Sussman, Greeno, Gifford, & Scannell, 2013). A recent meta-analysis of 94 experiments found that information has a significant, albeit moderate, effect on sustainable behavior (Osbaldiston & Schott, 2012). Information seems best suited for increasing easy behaviors with few barriers to implementation (Steg & Vlek, 2009), and for gaining approval of organizational or civic policies (Gärling & Schuitema, 2007).

Normative information can serve as an especially effective antecedent. For example, perhaps you have stayed in a hotel room that had a sign in the bathroom encouraging the reuse of towels. Often these signs provide basic instructions and facts, such as "if you would like your towels replaced, please leave them on the floor. Reusing towels saves valuable resources." This message is informative, but not very compelling. With the addition of normative information, however, the signs are more effective. Both descriptive normative information (e.g., "The majority of the guests in this hotel reuse their towels") and injunctive normative information (e.g., "Many of our guests value conservation") increase towel reuse. The combination of the two is most effective, and their impact is increased when the message mentions a relevant reference group (e.g., "The majority of guests *in this room* reuse their towels" (Goldstein, Cialdini, & Griskevicius, 2008; Schultz, Khazian & Zaleski, 2008).

Prompts. You may have seen a sign on the door of a grocery store asking: "Did you remember your reusable bags?" This is an example of a **prompt**, or cue meant to trigger behavior. In this case, the prompt reminds people at a point when it is reasonable for them to retrieve forgotten bags. Prompts have earned some of the strongest empirical support for their impact on sustainable behaviors (Osbaldiston

& Schott, 2012). The more specific the prompt, the greater its efficacy. A sign saying "faculty and students—please turn off lights after 5 p.m." is more effective than one that reads "Conserve Electricity." Polite prompts are more effective than demanding ones (the word "please" can make a difference), and the closer the prompt to the behavior point the better (a sign over a light switch is more effective than a sign across the room). Polite, specific, and well-placed reminders can significantly change behavior (Lehman & Geller, 2004; Sussman & Gifford, 2012).

Information and prompts are not necessarily mutually exclusive. For example, backyard composting is usually unobservable to neighbors. But when householders were asked to post decals that advertised their participation in a composting program, backyard composting in the neighborhood increased (McKenzie-Mohr, 2000b). In this example, the stickers served as prompts that also communicated a descriptive norm.

Modeling. When attempting to influence others, demonstrating appropriate behavior often leads to better retention and increased use of a new behavior than simply describing when and how to perform desired behaviors (Muchinsky, 2012). The power of peer modeling for increasing sustainable behavior was cleverly demonstrated during a field experiment in a men's shower at the University of California-Santa Cruz. Although most students at this university would describe themselves as environmentalists, very few conserved water in the shower. Even when the researchers put up a sign asking users to "(1) Wet down, (2) Turn water off, (3) Soap up, (4) Rinse off," only 6% followed these water-conserving instructions. However, when the researchers had a confederate demonstrate the appropriate behavior whenever someone entered the shower room, compliance with the instructions rose to 49%. When two confederates were used, 67% imitated the models (Aronson & O'Leary, 1983).

Consequence Strategies: Was It Worth It?

Currently, most environmentally irresponsible behaviors are rewarded by highly prized outcomes such as convenience, social status, comfort, and pleasure, whereas environmentally responsible behaviors are not. In fact, some of the consequences of environmentally responsible behaviors may even be punishing. Understanding the punishers that discourage sustainable behavior and the reinforcers that keep unsustainable behavior in place can help us identify additional opportunities for engineering the situation. For example, Wales recently introduced a five-pence charge for using disposable bags when shopping. Taking advantage of this real-life shift in consequences, researchers conducted a quasi-experiment to evaluate its effects (Poortinga, Whitmarsh, & Suffolk, 2013). They contacted random samples in Wales and England via telephone both before and after the charge was introduced. As expected, researchers found a significantly larger drop in disposable bag use in Wales compared to the English comparison group. However, while punishment generally showed the expected reductions in behavior, attempts like this to curtail repetitive, habitual behavior are likely to fail if alternatives are not consistently prompted or reinforced. Also, knowing that people respond more favorably to rewards, an important priority for changing situational contingencies is to *reinforce* environmentally appropriate behaviors.

When we think of reinforcement, we often imagine **extrinsic** rewards, which are external benefits, such as money and social recognition. However, consequences can be **intrinsic** as well, meaning just doing the behavior can feel rewarding in and of itself. Intrinsic reinforcers are things such as feeling a sense of purpose, developing competence, and making progress toward a goal. Both intrinsic and extrinsic rewards increase behavior but in different ways (you'll learn more about this in Chapter 8). The short story is that people are more likely to maintain environmentally responsible behaviors when reinforcers are intrinsic; that is, when the activity is enjoyable or because it aligns with the person's values (e.g., Leiserowitz, Maibach, & Roser-Renouf, 2009). Of course, it is difficult to engineer intrinsic consequences. Changing the situation typically means changing the extrinsic rewards. Common forms of reinforcement can be divided into two basic categories: incentives and feedback.

Incentives. Financially rewarding behaviors through rebates, direct payments, and coupons was a popular methodology in the 1970s and was shown to be effective for increasing bus ridership (Everett, Hayward, & Meyers, 1974), increasing litter clean-up (e.g., Baltes & Hayward, 1976), and reducing energy consumption (reviewed in Abrahamse, Steg, Vlek, & Rothengatter, 2005). Use of extrinsic rewards like these has declined, as they are difficult and costly to implement, and changes do not typically last once the rewards are removed; that is, the new behaviors are easily extinguished (e.g., Geller, 2002). It is beneficial, however, to use extrinsic incentives like rebates and tax breaks to encourage *one-time actions* that produce significant increases in efficiency, such as purchasing a fuel-efficient vehicle and energy-efficient appliances, insulating a house, or investing in solar power (Gardner & Stern, 2008). A case example comes from Norway, where electric vehicle ownership doubled after

the introduction of numerous incentives. For a time, people could purchase an electric car tax free; this is *huge*, literally, since Norway's auto taxes generally double, and sometimes triple, the cost of purchasing a car. Additional perks include access to high occupancy vehicle lanes, free charging stations, and exemption from paying tolls, parking fees, and ferry fares.

Informational Feedback. Simply giving people feedback about what they have accomplished can be reinforcing, in which case, there is no need to give them additional material rewards. Informational feedback is effective because it communicates progress toward a goal (more on goal setting in Chapter 8). Many environmental behaviors can be shaped by informational feedback (reviewed in Abrahamse, et al., 2005; Osbaldiston & Schott, 2012). For instance, in one intervention, simply posting the number of aluminum cans retrieved from a recycling container increased the number of cans subsequently recycled by 65% (Larson, Houlihan, & Goernert, 1995). Informational feedback can effectively reduce energy use as well (Carrico & Riemer, 2011; reviewed in Vine, Buys, & Morris, 2013). Consumers prefer receiving information about their energy use via specific usage information (e.g., which appliances use the most energy) and graphs that demonstrate energy use patterns over time (Karjalainen, 2011). We are most likely to reduce our energy consumption when feedback is clear, timely, and customized (Vine, et al., 2013). Putting the feedback into context, such as by specifying the number of trees required to offset a person's carbon emissions, can also help (Jain, Taylor, & Culligan, 2013).

Social Feedback. Engineered social feedback typically includes information that encourages people to make social comparisons between their own environmental behavior and the behavior of others. In a classic field experiment, Schultz and colleagues (2007) provided comparison information on homeowners' energy bills. Each household was given descriptive normative information about average energy use in the neighborhood. This was presented alongside information about the individual household's use. Of course, some residences used more energy than average and others used less. As you read earlier, people tend to conform to social norms. In this case, higher users reduced their usage to be more consistent with the neighborhood average. However, the power of normative influence led many lower users to *increase* their usage!

The key to alleviating this "boomerang effect" was to include additional feedback in the form of social approval (i.e., injunctive normative feedback). People whose household energy consumption was below the neighborhood average received a ☺ with their feedback. The simple inclusion of a ☺ kept the below-average users at their low usage rate. Similar interventions have since been implemented across the United States, and have yielded sustained reductions in energy use (Allcott, 2011; Ayres, Raseman, & Shih, 2013).

Perhaps you wondered whether the higher users also received injunctive normative feedback. Indeed they did: The above-average group received a ☹. Although social disapproval did result in some reduction of energy use among these residents, it also provoked some resentment. In private sector applications, the ☹ emoticon's punishing effects infuriated paying customers, and have subsequently been dropped from use. As Skinner said, reinforcement generally works better than punishment.

Even when social comparison feedback is presented alone, without expressions of approval or disapproval, the realization that one's behaviors compare unfavorably to the behaviors of others can motivate behavior change (Schultz, et al., 2007). This may be especially true when those others represent a particularly relevant reference group. For example, college students' intentions to engage in proenvironmental behaviors increased after they received ecological footprint feedback indicating that their habits were less sustainable than those of fellow students (Toner, Gan, & Leary, 2014). Presumably, this feedback mattered more to them than more general footprint feedback would because it made them feel that they were not conforming to the standards of their close community.

Combining Contingencies

Behavioral engineering often involves the use of multiple strategies simultaneously. Sometimes a combined approach can be especially effective (Osbaldiston & Schott, 2012). For example, researchers conducted a quasi-experiment in which they studied the use of train station stairs vs. escalators during morning rush hour in London (Lewis & Eves, 2012). Over the course of several months, they measured baseline stair climbing, then added prompts that included a health justification for using the stairs, and after a few more weeks added congratulatory messages on the landing. The prompts and rewards were removed briefly, and then introduced again in reverse order. The researchers found that no matter the order of introduction, only a combination of prompt and congratulatory message significantly increased stair climbing.

In another example, researchers conducted a field experiment to test three different community interventions to promote recycling (DeLeon & Fuqua, 1995). Households were randomly assigned to one of three conditions: public commitment only (i.e., assenting to having their names published in a local newspaper article about the project), informational feedback only (i.e., reports on the amount of recyclable paper generated by households in this condition), or combined commitment and feedback. Public commitment alone had no effect. Feedback alone did yield a significant increase in recycling behavior, but the effect of the combined intervention was almost twice as large.

CONCLUSION

Behavioral engineering is an important strategy for sustainable behavior change. Currently, most contingencies do not support proenvironmental behaviors, so a proactive approach to altering those contingencies is warranted. Making simple changes can direct people toward more sustainable behaviors. For example, "choice architecture" involves setting up antecedents that make the better options easiest without limiting people's freedom to choose (Thaler & Sunstein, 2009; more about "green defaults" in Chapter 6). Take a standard gala invitation that asks people to RSVP about dinner options. Options might include delicious-sounding beef, chicken, and fish followed by a note to let the hosts know if a vegetarian option is needed. What do you think would happen if guests were instead offered a choice of delicious-sounding vegetarian plates, followed by a note to let the hosts know if a meat option is needed?

Single changes like this can lead to more changes. Baby steps can be a good way to create momentum in a new direction, though on the individual level, this is not always the case. For example, sometimes when people take a small first step, such as recycling in an effort to curb climate change, they stop there because just doing *something* is enough to alleviate the (punishing) worry that is motivating them; this is known as the **single action bias** (Weber, 1997). When the first behavioral change is a big one, people sometimes feel entitled to be less environmentally responsible in other domains, causing a **rebound effect** (Herring & Sorrell, 2008; Mazar & Zhong, 2010); for instance, people who buy an energy efficient car might feel less guilt about taking an extra road trip, therefore negating any savings in fuel use. Still, behavior change often begets more behavior change. And behavior change on the part of individuals can create broader *systemic* change.

Individuals can influence the larger system by starting with their own situational context. **Self-control** occurs when one changes antecedent and consequent stimuli in order to change one's own behavior (Skinner, 1953). If you have ever dieted, put yourself on a study schedule, or disciplined yourself in some other way to achieve a goal, then you have implemented a self-control project. This approach can be used to change environmentally relevant behaviors (see Appendix: Self-Change Project). In the process of changing their own context, people are likely to become aware of how their behaviors are limited by situational constraints, such as lack of access to environmentally friendly products (e.g., those made from recycled, durable, or less toxic materials). This can inspire individuals to give feedback to the decision makers who control infrastructure (e.g., companies and politicians). In other words, consumers and citizens have the power to alter the contingencies that shape the behavior of the powers that be.

In the process of making personal changes, individuals can inadvertently, or intentionally, model environmentally responsible behavior to others. This can lead to more widespread behavior change within a reference group (Rogers, 2003). When you reuse an envelope, carry your own to-go container into a restaurant, or show up at the board meeting wearing your biking helmet, you demonstrate behavior to others that they may begin to enact as well. All it takes is one influential individual to inspire the establishment of a new social norm. At first, people may provide punishing social feedback in response to behaviors that seem weird or uncouth, but norms about feedback can also be transformed. For example, currently, it is customary to keep quiet about other people's unsustainable behaviors; we might sometimes reinforce them for doing the green thing, but most of us are unlikely to overtly admonish someone for behaving in an ecologically irresponsible way. Yet, there is empirical evidence that shaming someone for unsustainable behavior can motivate that person to be more ecologically responsible the next time (Swim & Bloodhart, 2013). Shifting social feedback in a sustainable direction could lead to new injunctive and descriptive social norms.

All behaviors are subject to situational contingencies, but it is not just the situation that influences behavior; our behaviors can also influence the situation. And, while there is no doubt that we react, often reflexively, to the situation, we also have the capacity to *decide* how to behave when we set our minds to it. Whether we're thinking hard or not determines whether we can override the power of the (unsustainable) situation. To better understand the role of human cognition in the quest for sustainable behavior, turn to the next chapter.

CHAPTER 6

It's Not Easy Thinking Green

- **Two Systems for Thinking**
  - **The Analytic System**
  - **The Automatic System**
  - **Careful Reasoning versus Quick Intuition**
- **Cognitive Roots of Environmental Degradation**
  - **Perceptual Limits**
  - **Temporal Discounting**
  - **Affect**
  - **Availability Heuristic**
  - **Biases**
  - **Automaticity**
- **Cognition for a Sustainable World**
  - **Raise Awareness**
  - **Increase Personal Relevance**
  - **Elicit Emotion**
  - **Encourage Intention**
  - **Provide Appropriate Knowledge**
  - **Moving Toward a Greener World . . . Automatically**
- **Conclusion**

Before diving into the material of this chapter, take a moment to answer the following questions:

1. What is the first thought or image that comes to your mind when you think of global warming?

2. Which of the following are you most concerned about? The impacts of global warming on . . .

   (1) you and your family;

   (2) your local community;

   (3) the United States as a whole;

(4) people all over the world;

(5) nonhuman nature;

(6) not at all concerned.

These questions are from a U.S. mail-in survey published in 2005. When researchers asked 673 U.S. citizens about their global warming beliefs, 97% of the thoughts or images reported as "coming to people's mind" could be coded into one of eight different categories (see Figure 6.1). The top three categories, listed by about half of all participants, were images of melting glaciers and melting ice; general images of rising temperatures and heat; and images of impacts on plants and animals (Leiserowitz, 2005). Perhaps what you pictured falls into one of these categories?

Nearly 61% of the participants reported images that are distant from their daily lives. This may help explain responses to the second question; a majority of

FIGURE 6.1 *Types of images and thoughts people report when they think of global warming.*
Adapted from Leiserowitz, 2005.

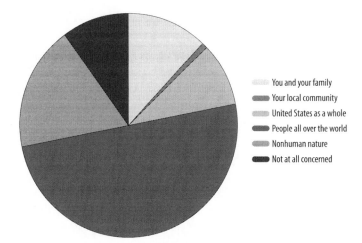

FIGURE 6.2 *Global warming concerns.*
The majority of people reported that they were most concerned about the impact of global warming on "people all over the world" (adapted from Leiserowitz, 2005).

respondents (68%) chose either "people all over the world" or "nonhuman nature" as their top concern. In comparison, only about 13% chose one of the first two categories ("you and your family" or "your local community") (see Figure 6.2). In other words, a large majority of people underestimates the personal impacts of climate change.

As noted in Chapter 1, most U.S. citizens are aware of climate change and about half express significant concern about it, however, only about one third believe it will pose a threat to their way of life (Gallup, 2014a). Even fewer are taking deliberate action in response to the issue. This is despite the verdict from climate scientists that climate change is real, it's already happening, and if immediate action is not taken, it will impose catastrophic impacts on our collective lifestyles.

Why do you think we see so little public attention and response to "the biggest crisis our civilization has ever faced" (McKibben, 2014)? Though awareness of the issue is high, concern is lower, and action lags far behind. If awareness of the growing potential for catastrophe is not enough to inspire widespread action, what *does* trigger a response? In the first chapter of this book, you learned that so-called *environmental* problems are actually *behavioral* problems. This chapter takes that idea one step further: Because all behavior begins in the brain, environmental problems are, at their root, *thinking* problems. This chapter explores how our thinking patterns contribute to environmental harm and our failure to respond to the hazards we have created, and how psychological research suggests we can learn to think differently.

TWO SYSTEMS FOR THINKING

Reflect for a moment on the process you went through to answer the questions above. You probably had at least one image in your mind right away. If the question prompted several images, you then had to consciously settle on one of them. To do this, you might have consulted your memory for something you heard on the news or discussed with a friend. Perhaps you even felt a twinge of sadness, fear, or frustration as the images emerged.

These phenomena are all examples of **cognition**, the thought processes of the human brain. Some types of thinking are conscious, deliberate, and analytic, such as consulting your memory for specific information, or choosing between two possible ideas. Other types of thinking happen automatically and without conscious effort, like seeing an image in the mind's eye or feeling emotion. Cognitive and social psychologists propose that these two types of thinking are performed by distinct brain processes—an **analytic system** and an **automatic system** (e.g., Chaiken, 1987; Sloman, 1996; Kahneman, 2011). The two systems were finely calibrated over the course of evolution and, working together, give us the creative and adaptive thinking power that has allowed us to flourish as a species. However, as noted in Chapters 1 and 2, our amazing capacity to think has also enabled our species to alter the planet so extensively that our ancestors would hardly recognize it. Meanwhile, our own physical evolution has not kept pace with the changes we have created in the world; our brains remain much the same as those of early humans, and are therefore poorly prepared to recognize and respond to self-imposed, modern risks (e.g., Ornstein & Ehrlich, 2000; more on this in Chapter 9). To help you better understand why, the following sections provide a deeper look into the two systems that underlie human cognition.

The Analytic System

Your analytic system is working every waking moment, composing your thoughts and shaping your conscious experience. It guides your attention by allocating mental energy to things it deems important. Slow, careful, and deliberate, the analytic system embodies sophisticated mental skills: the ability to think logically, to seek new facts and study data, to plan, and to make intentional choices (Kahneman, 2011).

Analytic thinking gives us the capacity to envision, invent, and think our way to creative solutions to solve problems and meet our needs. It is the system that holds our opinions and our conscious values, and has the ability to construct an argument to support them. But these advantages of analytic thinking come at a cost. Careful reasoning takes more time than trusting quick intuition. In addition, the analytic system, like the narrow beam of a flashlight, works best when focused on just a few things. That is why your logic suffers when you're multitasking or distracted. Finally, as any student who has studied for a test or written an argumentative paper knows, focused attention and deliberate thinking are *hard work*. Typical daily mental tasks require as much as 25% of the body's total calories, and focused thinking drives the demand up even more (Bor, 2012). Because it is strenuous and calorically expensive, the brain avoids hard thinking unless it's truly necessary. When it is not working at full throttle to solve a problem, the normal mode of the analytic system is a comfortable, easy awareness, such as a state of pleasant mind wandering (Kahneman, 2011).

Though we may feel like the analytic system is in control of our behavior, conscious mental experience reflects a mere fraction of what goes on in the brain. Like the tip of an iceberg bobbing above the surface of the ocean, the intentional thinking you do is dwarfed by what is hidden below the surface. In other words, much of the time the automatic system, doing its work beneath conscious awareness, is running the show.

The Automatic System

In contrast to the analytic system, the automatic system is speedy and nimble. It is emotional rather than rational, and intuitive instead of carefully reasoned. The analytic system is less interested in pesky numeric data and facts but is attuned instead to sensory cues in the environment. Its specialty is responding to clear and present dangers. In fact, it is the part of our brain most similar to that of other animals; hardwired to detect signs of threat, it rapidly mobilizes the body to respond.

It would probably surprise you how many of your daily decisions are made unconsciously by the automatic system, using cues that your conscious mind hardly detects. Of course, many mental duties of the automatic system are not what you would consider "thinking": rushing through the grocery store on autopilot, driving to school (and then wondering how you got there, or if you stopped at the red light, because your mind was elsewhere!), or feeling an emotion such as sadness, joy, or annoyance. However, the brain steers all of these actions. Your automatic system orchestrates these behaviors and many more, handling multiple tasks simultaneously with speed and ease. Because its work requires little effort, the automatic system can take charge of the easy and well-practiced mental tasks needed to get through daily life, leaving enough cognitive energy for the analytic system to do its important work when needed.

Careful Reasoning versus Quick Intuition

The two systems of the brain normally work together in harmony. Occasionally, however, the flow is disrupted, and what follows is like a scene from *Star Trek*. Picture Mr. Spock, your analytic system, calm but intent as he describes an elaborate plan he has developed to save the ship. It is logical and based on all the available evidence; it is sure to succeed. However, Captain Kirk, your automatic system, paces back and forth restlessly, betraying his vehement disagreement. He can't put it in words, but the idea just doesn't *feel* right (see Figure 6.3).

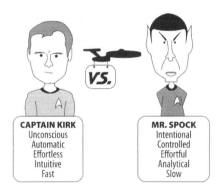

FIGURE 6.3 *The brain's two thinking systems.*
Adapted from Thaler & Sunstein, 2009.

When conflict arises between the sensory-driven automatic system and the logical analytic system, the unconscious and speedy impulses have the advantage. Therefore, despite our capacity for deliberate thinking, our behavior is often based on immediate gut feelings. To illustrate this dilemma, consider the following scenario. A package of ripe, red raspberries is on sale at the grocery store. The automatic system gleefully thinks, "Yeah! Delicious berries!" and your hand reaches out. However, you then hesitate as the analytic system, just a moment behind, thinks, "Wait a minute! It's January so these raspberries were probably flown in from Chile. Maybe I should buy something else—I want to reduce my carbon footprint."

Now imagine the same scenario, but after a long day of classes and your part-time job. Already tired, you face the prospect of a long night of studying ahead. A friend calls and you hurry through the grocery store listening to her tales of woe and giving her supportive advice. Simultaneously, you realize you can't find your wallet and begin searching through your bag. This time, as you spot the raspberries, what do you think happens?

If you guessed that dessert included raspberries and cream that evening, you are likely correct. As the scenario implies, when the analytic system is overloaded, when we are in a hurry, or when we lack clear information, our ability to make a careful and rational decision suffers. Most of the time, this works out just fine. Our brains do not need to carefully analyze every choice because the automatic system has built up a set of intuitive responses that serve us well in most situations. Shopping on autopilot, you procured a delicious dessert despite being exhausted and attending to an important relationship. However, because our industrialized society is built on an unsustainable food system, a choice based on quick intuition is likely to lead to environmental harm. In fact, most of today's environmental challenges require active attention and effort to detect, understand, and address. Our hectic, modern lifestyles often preclude such uses of the analytic system's careful thinking, leaving us to rely on automatic impulses. This strategy can get us in trouble if unconscious responses don't match the broader needs of the situation. The following sections describe six features of cognition that contribute to ecologically harmful behavior, as well as research-based strategies for better thinking about environmental issues—thinking that spurs action rather than complacency.

COGNITIVE ROOTS OF ENVIRONMENTAL DEGRADATION

Psychologists often point out that the human brain is more powerful than even the most advanced computer. However, whereas computers are consistently logical, the brain is prone to distortions. Ironically, the very features of the brain that do its work quickly and efficiently lead to cognitive errors that hamper our ability to notice and respond to ecological crises.

Perceptual Limits

Although you may not consider your five senses—sight, hearing, taste, smell, and touch—as part of thinking, the sensory organs transmit information to the brain in the process called **sensation**. Within the brain, **perception** involves interpreting sensory information by drawing on memories of past experience. Perception creates meaning from what the senses tell us. Both sensation and perception occur rapidly and below conscious awareness, driven by the automatic system.

At their most basic level, sensation and perception keep us alive by helping us avoid dangers in the world around us. As noted in Chapter 1, our ancient ancestors typically encountered threats in the form of sudden, dramatic events, obvious and unambiguous to the senses, such as attacks by wild animals, clashes with rival tribes, storms, and other acts of nature. Survival is a primary duty of the automatic system, which is always on the lookout for signs of trouble. Upon detecting danger, the automatic system puts the body into high alert and readies it for action with lightning speed.

Though attacks by wild animals are certainly still a possibility, they are no longer a daily threat for most of us. Some of the greatest dangers today are not nearly as clear and obvious. Human sensory organs are good at picking up abrupt changes, but poor at perceiving slow, incremental shifts. Stimuli that do not change, or that change very slowly, quickly lose their ability to activate a response from our sensory neurons; this is called **sensory adaptation**. Consequently, situational features that remain constant fade from awareness, while those that change too slowly never reach awareness at all. Just like the frog who will jump out of a pot of very hot water if suddenly thrown in, but will allow itself to be boiled to death if placed in a very slowly heating pot, people are not built to notice the slow time course and cumulative nature of most environmental changes.

Furthermore, some environmental hazards are truly undetectable by our senses; sensory receptors on the tongue or skin are not capable of sensing most pollutants in air, food, or water. And, perception is constrained by geographical location and distance, in effect making many problems invisible. For most U.S. citizens, Amazon deforestation, the submergence of low-lying islands due to rising seas, and melting permafrost in the Arctic are too far away to notice. This means that one of the first breakdowns in our ability to respond to environmental crises is a failure to detect that they even exist.

As an example of a mounting concern that is difficult to detect, consider plastic. Does plastic seem innocuous to you? Most likely you are surrounded by it as you read these words—the chair you're sitting in, the bag your sandwich was packed in, your eyeglasses, the casing around your cell phone. Colorful and ubiquitous, most of us hardly give a thought to the plastic in our lives. It certainly doesn't set off any "danger!" alarm bells. But might there be something sinister lurking there?

In fact, as you read in Chapter 1, plastic pollution is accumulating in the world's oceans at an alarming rate, and you'll learn in Chapter 9 that plastics contain chemicals that interfere with hormonal function. Does reading about these problems inspire you to remove all plastic from your life? Probably not. Though you might feel uncomfortable after hearing about a new set of worries, there are no apparent sensory cues in plastic items that indicate toxicity. Without such direct sensory warnings that the automatic system interprets as "danger!" even the most significant problems are easy to dismiss, especially if the problem is as pervasive as plastic.

Perceptual limitations also prevent those of us in industrial societies from noticing our own role in resource overconsumption. Unlike some places of the world, where people must carry their daily water, fuel, and food supplies, people living in industrialized countries lack a concrete sense of how much water, energy, or other resources they use. Flip a switch and the lights go on, turn on the tap and water simply comes out and flows down the drain. This leads people to severely underestimate their own impacts. How much water do you think you use for an average shower? For brushing your teeth? For washing your clothes? Most Americans guess that their typical household activities use about half as much water as they actually do (Attari, 2014). Perceptual limitations mask both the dangers of environmental degradation and the extent to which we each are personally complicit in the problems. It takes deliberate thinking and information seeking, the work of the analytic system, to overcome these limits.

Temporal Discounting

Are you ever compelled to do something despite knowing it has negative long-term consequences, such as buying take-out food in Styrofoam containers that end up sitting in a landfill for thousands of years? You are not alone. Immediate needs and wants frequently override concerns about the future because humans value short-term circumstances more than those far away in time. **Temporal discounting** makes it unlikely that we will give something up today to alleviate long-term problems (recall the discussion of contingency traps in Chapter 5). In the unpredictable environment of our early human ancestors, focusing on the needs of the moment was a successful strategy (e.g., van der Wal, Schade, Krabbendam, & van Vugt, 2013). In today's world, however, behaviors that offer immediate benefits, such as the convenience of packaged take-out food or the pleasure of going for a scenic drive, have long-term, negative consequences that are already taking their toll on our health and the environment.

In addition to making us less likely to protect ourselves from future dangers, temporal discounting causes us to avoid costly investments in the present despite later payoffs. Future benefits are simply less compelling than short-term costs. It is difficult to spend money now to purchase a new, more fuel-efficient car even if it would save significant amounts of energy and money in the long run, and also reduce greenhouse gas emissions (Bazerman, 2006). Behaviors that reduce energy waste (e.g., more efficient appliances, home insulation) result in a return on investment of 30% to 50% per year, much greater than the performance of most stocks, bonds, and money market funds, even in a strong economy (Bazerman & Hoffman,

1999). Yet, most people do not make these efficiency purchases because they are put off by the upfront expense.

Affect

Affect refers to the automatic system's spontaneous emotional response to every situation we encounter. You probably have experienced how commanding a motivator emotion can be (e.g., Slovic, 1987; Slovic & Peters, 2006). The sight of things you love evokes a tiny but sure rush of positive affect, while things you dislike result in an upward tick of negative affect. These unconscious signals guide behavior; we are drawn toward things that the automatic system deems positive and repelled by things felt as negative. This instantaneous sense of "I like this" versus "Get me out of here!" cuts down on the conscious thinking we need to do to understand and navigate the world around us.

Affect is most powerful when the automatic system encounters a sign that something potentially threatening is happening. Our unconscious fear response is swift and strong, drawing cognitive resources away from the analytic system while instinct takes over. But, invisible risks inspire little emotion. Because we don't perceive the harms in plastics, pesticide residues, and distant deforestation, they don't set off our internal alarm bells. Paradoxically, analytic thinking, which helps us grasp the scope and danger of environmental risks, does not give people the emotional kick-in-the-pants provided by the automatic system (e.g., Weber, 2006).

The invisibility of environmental threats, coupled with their typically delayed impacts, contributes to **psychological distance**. This is the sense that a problem is far removed from the personal here and now (Trope & Liberman, 2010). The closer something is to your direct personal experience, in terms of time, physical space, social relevance, and how certain it is to happen, the more important it feels to you psychologically. To motivate emotion and action, an issue must feel psychologically close and real, rather than distant and hypothetical (e.g., Milfont, 2010; Spence, Poortinga, & Pidgeon, 2012).

Climate change may be the most important illustration of how perceptual limits, affect, temporal discounting, and psychological distance all impact our reactions— or lack thereof. Climate change requires deliberate thinking to comprehend and respond appropriately. For instance, because people are accustomed to fluctuations in daily and seasonal weather, unusual conditions may still be perceived as part of "normal" weather variability. The more serious impacts of climate change continue to feel far in the future. Thus, most Americans think of climate change as a real but distant, and therefore not personally threatening or urgent, problem (Leiserowitz, et al., 2014; Gifford, 2011).

So should we try to scare people into environmentally friendly behavior? Beware! Such attempts usually backfire (Moser, 2007). Anxiety and fear are extremely uncomfortable emotions; that is why they motivate action so powerfully— to reduce that discomfort. However, it is sometimes easier to pretend a problem doesn't exist than to take action. Freud (1938/1964) was on to something in describing **defense**

mechanisms such as denial and rationalization; though his specific ideas are not empirically validated, we do use mental strategies to minimize negative emotions. This is generally called **emotion-focused coping**, and while it may alleviate anxiety for a while, it leads to greater suffering in the long-term if it allows the underlying cause of discomfort to worsen. **Problem-focused coping**, on the other hand, is a concerted effort to eliminate the source of discomfort (Lazarus & Folkman, 1984). Problem-focused coping might involve buying a home water filtration system in response to worries about pesticides in the water, or better yet, joining a local campaign working to reduce pesticide use in the first place.

What determines people's coping response to an anxiety-producing stressor? Much depends on the situation. If the problem is completely outside of a person's control, then emotion-focused coping is an appropriate response. Why bother adding to the stress by making an effort bound to end in failure? Similarly, if people have no idea how to tackle the anxiety-producing problem, emotion-focused coping may feel like the only option (Moser, 2007). Upon learning about pesticide contamination in the local watershed, the choice between running out to join a clean water campaign versus hiding under the covers may depend upon people's feelings about whether their efforts will make a difference. If the problems feel out

of control, if no effective solutions are available, or if people feel incapable of taking part in solutions, they give up and do their best to feel better by ignoring the problem or distracting themselves with other concerns (Moser, 2007).

As if ignoring the problem was not bad enough, frightening environmental messages may also *encourage* environmentally damaging behavior, particularly if they imply one's death. **Mortality salience** means awareness of the inevitability of one's own demise, and not surprisingly, it creates anxiety. The brain rushes to protect us by seeking reassurance that our lives are meaningful and important (i.e., *terror management theory*; Solomon, Greenberg, & Pyszczynski, 2004). In consumer culture, importance is signaled through wealth and possessions. Therefore, when faced with the idea of mortality, many people turn to buying more, and fancier, stuff to dampen their death anxiety (Arndt, Solomon, Kasser, & Sheldon, 2004; Dickinson, 2009), further contributing to environmental degradation.

Unfortunately, "OMG, we're all going to die!" feelings aren't the only way affect steers behavior away from sustainability. Do you feel a flash of revulsion at the thought of having a bin of worms in your basement? Vermicomposting (Chapter 3) is an effective environmentally friendly behavior, but hardly widespread, likely because of the visceral "yuck!" response it evokes from many people. Similarly, people may initially balk at the thought of taking shorter showers, turning down the heat in the winter, or eating less meat. To the automatic system, these lifestyle adjustments sound like unappealing sacrifices. Though the analytic system might offer perfectly sound reasons ("A worm bin would be a great composting option for the winter!"), logic is often not powerful enough to overcome a strong aversion.

Affect frequently drives decisions that we believe we are making on purely rational grounds (Haidt, 2001). We think we are weighing the pros and cons so as to make a logical choice, but really, our judgment is being distorted by our feelings. When people have a positive feeling about something, they often conclude its risks are low and its benefits high; if they have negative feelings, they do the opposite, perceiving the risks high and the benefits low (Slovic, Finucane, Peters, & MacGregor, 2002). Thus, if you are enjoying a colorful, new, plastic iPhone case, you will tend to downplay the health and environmental risks attributed to plastics. Or, if your gut feeling toward vermicomposting is intensely negative, you may find yourself thinking, "I'm sure my basement is too cold for a worm bin."

Availability Heuristic

To save mental effort, the brain has developed many shortcuts, called **heuristics**, which allow it to come to a decision by ignoring all but a few pieces of information. For example, two of the foremost experts on cognitive errors, Amos Tversky and Nobel Laureate Daniel Kahneman, famously noted that if people can think of an event quickly, they believe it is more likely to happen. This tendency to make a judgment based on what is effortlessly available in memory is called the **availability heuristic** (Tversky & Kahneman, 1973). More recent events are easier to recall, as are vivid events and those with high and recurrent media coverage, and they therefore feel more threatening. For example, disastrous oil spills get a lot of attention, but people rarely worry about gradually leaking oil pipelines, despite the fact that the annual volume of oil seeping from pipelines is more than that of many big spills.

Perhaps you experienced the availability heuristic when you answered the first question of this chapter: "What is the first thought or image that comes to your mind when you think of global warming?" This heuristic was at work if the scenario you imagined was based on events you've witnessed with your own senses or experienced vicariously. Seeing your neighborhood struck by a flash flood, or hearing your neighbor's emotional description of the event, creates a vivid understanding far more powerful than being told by a scientist that floods are likely to increase in frequency or severity (Weber, 2006). Mental availability of a risk spurs action. Indeed, people with firsthand exposure to a hazard are more motivated to take action to prevent future hardship (Kahneman, 2011; Whitmarsh, 2009).

Of course, there are limits to the power of personal experience. Availability is stronger for more recent events. As memory fades, so does the ease of recall, and with it, motivation to do something (Kahneman, 2011). In addition, new worries pop up all of the time, and more recent concerns push aside older ones. People have a limited capacity to feel concern, and this *finite pool of worry* means that the more emotion we direct toward one problem, the less attention and concern is available for other issues (Weber, 2006).

In addition, it is difficult for human imagination to picture dangers beyond the scope of our own and others' experience (Kates, 1976; Kahneman, 2011). This means we underestimate the risk of unprecedented and unknown threats. Predictions of societal collapse and extreme human suffering are simply too far outside of our personal and cultural realities to be available to the mind.

Our perception of risks is influenced not only by what most easily comes to mind, but also by our current sensory experience. Physical conditions sometimes heighten perceptions of risk. Take a moment to notice the temperature in the room. Is it cool, comfortably warm, perhaps even too warm? Whatever the temperature, it is subtly altering your response to climate change. Current sensory experience is a form of **priming**, an unconscious activation of related concepts. Studies have shown that warm temperatures prime climate change: People in a warm room express higher belief in climate change than people in a cool room, and people also more readily agree that climate change is real on warmer days than colder days (Risen & Critcher, 2011; Joireman, Truelove, & Duell, 2010). The sensory feeling of warmth makes it easier to imagine higher global temperatures.

Visual cues prime environmental concern, too. Seeing dead plants in the room increases people's belief that climate change is real, probably because it activates associations with drought and heat waves (Guéguen, 2012). Climate change concerns can even be primed by language. For example, people presented with words connoting heat, such as "boil" or "Equator," expressed higher belief in climate change compared to a control group (Joireman, Truelove, & Duell, 2010).

Finally, what is available in the mind depends on our **mental models**, the simplified memory representations the brain holds for the things we think about.

FIGURE 6.4 *A mental model of the world.*

An example of a mental model is shown in Figure 6.4. Though it is filled with inaccuracy, and retains only a few general features of the thing it represents, it is still easily recognizable as a map of the world. As we learn more about a topic, its mental model improves; experts' mental models are significantly more accurate and sophisticated than nonexperts'. For most easy tasks, our simple, inaccurate, and incomplete mental models are good enough. However, for complex issues with high stakes, like climate change, faulty mental models can have damaging consequences (Bostrom & Lashof, 2007). For example, 11% of people in the United States hold a completely false mental model of climate change, equating it with the hole in the ozone layer (Leiserowitz, 2007). The solution most relevant to this incorrect mental model—stop using aerosol sprays—does nothing to address climate change.

Biases

Like affect, preexisting beliefs, things we "know" to be true, help us make sense of the chaos in the world. To paraphrase Bloom (1987), too much open-mindedness would make our brains spill out! Our beliefs provide stability and order for our thinking. They give us confidence that we've got things figured out. We therefore have an unconscious desire to defend them. No one is immune from **belief perseverance**, the irrational drive to maintain cherished notions, even in the face of contradictory evidence. Motivation to hold onto our beliefs shapes our interpretation of every piece of information we encounter and every experience we have.

While preexisting beliefs are necessary for the sake of sanity, they sometimes so potently color perception that they can get in the way of logically interpreting events. For example, the **confirmation bias** causes us to ignore new ideas. When testing hunches against incoming data, people tend to make the mistake of looking for confirming rather than disconfirming information (Wason, 1960). Information is easier to process and store when it conforms to ideas we've already established in our minds. Plus, people generally do not like the disappointment or embarrassment of being wrong, so they do not seek disconfirming evidence. Have you ever wondered why some people strongly believe in *global warming*, while others vehemently deny that such a problem exists? Believers see the record high temperatures every year of the last decade as persuasive evidence, while the deniers focus on the record low temperatures and heavy snowfalls. This is confirmation bias at work: Both sides are motivated to find what they want to see in the data. Of course, scientists and other well-informed folks know that *both* of these weather extremes are evidence of climate change.

Though biases are universal, they can manifest in culturally specific ways. One of the most powerful and pervasive biases in the United States is motivated by political partisanship Attachment to a party can be deep and emotional. For some people, political affiliation is an important part of identity (Leeper & Slothuus, 2014). Given the emotional bond people have with their political party, they tend to automatically agree with people who share the party affiliation, and avoid looking too deeply at what those opinions really mean (Leeper & Slothuus, 2014). Little factual information is necessary if we simply follow the rule of "if my party likes it, then it must be ok." This is especially true for a complex issue such as climate change, which is difficult for nonscientists to assess on their own (Weber & Stern, 2011). Thus, as the divide between U.S. Republican and Democratic leaders has widened, with Republicans expressing skepticism or resistance to acting on climate change, and Democrats acknowledging the issue's significance, a similar divide has occurred among Democratic and Republican citizens (Dunlap & McCright, 2008a; see Figure 6.5).

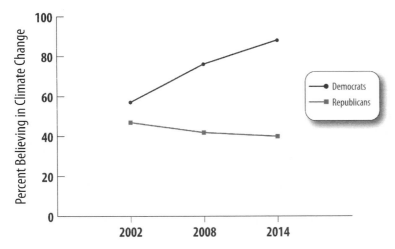

FIGURE 6.5 *The growing political divide between Democrats and Republicans on belief that climate change is happening.*

This figure is based on data from Dunlap & McCright (2008a) and Leiserowitz, et al., Politics & Global Warming (2014).

In addition to creating a political stalemate on climate change, the partisan divide has negative social consequences: We attribute negative motives to those on the other side of a conflict. This unnecessarily exacerbates people's bias against those who disagree with them, especially given that we tend to perceive those on the opposing side of a partisan debate as having more extreme views than they really do (e. g., Kennedy & Pronin, 2008; Reeder, Pryor, Wohl, & Griswell, 2005).

Research shows that people adapt their behavior to be in line with how they believe other members of their favored party act. In the United States, Democrats have a "greener" reputation and, consistent with others in their party, people who think of themselves as Democrats also report more environmentally friendly behavior than people who think of themselves as Republicans. Not only that, the level of environmentally friendly behavior differs depending on the strength of party attachment. Democrats with stronger party affiliation report more green behavior than their peers with weaker party affiliation, whereas Republicans with *weaker* party affiliations report more green behavior than their peers with stronger party affiliations (Coffey & Joseph, 2013).

Perhaps you are not (yet) politically active and have little interest in party labels. But note that other labels create unconscious bias as well. For instance, nearly every news story about an environmental issue mentions people who work to protect our communities and our natural resources from the effects of pollution, otherwise known as **environmentalists**. What comes to mind when you see this label? The word environmentalist may invoke a pretty negative stereotype that is fairly difficult to dislodge from people's minds. **Stereotypes** help the brain quickly categorize people, a skill that was probably a useful survival tactic in early human history when it was a matter of life and death to quickly decide if someone was friend or foe. However, the bias it creates can lead to faulty judgments. In one study, research participants read a short fictitious scenario: "Senator Johnson, under pressure from *concerned citizens,* voted against the Omnibus Budget Bill because it contained a provision to drill in the Arctic National Wildlife Refuge." For a second group, the words "concerned citizens" were replaced with the term "environmentalists" (i.e., "Senator Johnson, under pressure from *environmentalists* . . ."). Participants in the first condition were significantly more likely to say that Senator Johnson's decision was appropriate than those in the second condition. In fact, people reading the second version so vehemently disliked the influence of environmentalists that they disagreed with Senator Johnson's decision even when, on the very same survey, they themselves stated there should be no drilling in the Alaska National Wildlife Refuge (Amel, Scott, Manning, & Stinson, 2007).

A final example of biased cognition can be experienced at your local grocery store. Does having to pay five cents for a grocery bag make you more likely to carry along a canvas tote? Surprisingly, this small amount of money significantly impacts people's bag use. This illustrates **loss aversion**, a heightened negative reaction to giving something up, even something as small as a nickel. Many grocery stores offer an incentive, such as a five-cent discount (a gain), if people bring a reusable bag for their purchases. At other stores you must pay a five-cent penalty (a loss) if you take a disposable bag. An observational study looked at the grocery store sales records at stores with different bag incentive policies. The transaction records showed clearly that when faced with the prospect of having to pay an extra five cents for a bag from the store, customers took significantly fewer disposable bags. In contrast, customers shopping at stores with a five-cent bonus for bringing a reusable bag took just as many disposable bags as stores that gave no bonus at all (Homonoff, 2013). As loss aversion predicts, avoiding a five-cent loss was motivating in a way that earning a five-cent bonus was not.

When the stakes are higher, loss aversion can have significant real-life consequences. A more diffuse type of loss aversion may be at work in people's reticence to embrace significant change. Decreasing our energy use, changing personal transportation habits, lowering our consumption; these actions all imply "giving up" things we are accustomed to. Indeed, one study found that people exposed to an explicit message that climate change would require personal sacrifice felt less engaged with the issue (Gifford & Comeau, 2011). It is likely that loss aversion makes us reluctant to change the way we live, despite research that consistently shows our current lifestyles do not make us happy, and that, in contrast, sustainable living tends to increase people's wellbeing (e.g., Jackson, 2005).

Automaticity

Many daily unsustainable behaviors are firmly programmed into the automatic system and, as such, are difficult to alter. As you read in Chapter 5, *habits* are sets of behaviors performed so frequently that they become overlearned. Behaviors that were initially controlled by the analytic system become automatic through practice; this means that they are outside of conscious control. People go through much of their daily routines on autopilot; their automatic system is completely in charge while their conscious mind is otherwise occupied.

Habits are jump-started by context; at the grocery store, for example, you may find yourself reaching for the things you usually buy, even if you've formed a new conscious intention to avoid items with too much packaging. This is because the familiar cues of the placement in the store and the product label prompt a habitual response. Unfortunately, the world we have created encourages and supports unsustainable habits, such as using disposable containers, taking plastic bags at the store, using more energy than we really need, and many others. Without significantly changing the context and removing the cues that launch us into them, these automatic tendencies are notoriously difficult to change (Verplanken & Wood, 2006).

You now have a sense of the features of cognition that get in the way of environmentally responsible behavior: The senses don't detect most environmental hazards and, without the notice of the automatic system, we don't look any further. Our emotional responses lack urgency and often guide us toward the wrong behaviors. We place less value on the future and therefore put off taking action to prevent potentially distant difficulties. Neither the most serious problems nor the important solutions are easily called to mind. Our biases cause us to misinterpret the evidence in front of us. And finally, we become stuck in unsustainable behavior patterns programmed in the automatic system and cued by the situation. Though these cognitive failures may seem intractable, research findings suggest that each can be overcome. The human brain is amazingly flexible and capable of learning new patterns of thinking; it just needs a little help to get there. The next section reviews findings from a broad array of studies that detail how new patterns of environmental thinking may be fostered.

COGNITION FOR A SUSTAINABLE WORLD

Sustainability is a new idea to many people, and many find it hard to understand. But all over the world there are people who have entered into the exercise of imagining and bringing into being a sustainable world. They see it as a world to move toward not reluctantly, but joyfully, not with a sense of sacrifice, but a sense of adventure. A sustainable world could be very much better than the one we live in today.

Meadows, Randers, & Meadows (2004, p.253)

The late Donella Meadows, scholar, teacher, farmer, and scientist, was known for her optimism and her faith in human ability to move society toward a more equitable and sustainable way of living. The demands of a sustainable society are unfamiliar to many, and for some, it represents uncomfortable change, an unknown future, or sacrifice. However, the essence of sustainability is something that most people wholeheartedly endorse: high quality of life for all, today and into the future. But, because we tend to overlook, dismiss, and misinterpret the signs of dangerous environmental degradation and our role in it, our cognitive failures are putting us at risk, both as individuals and as a society. Thus, we will need to acknowledge the quirks of the automatic system and harness the tools of the analytic system to collectively engage in thoughtful and deliberate efforts toward environmental sustainability.

Efforts to change environmental thinking abound. The problem is that they typically rest on the assumption that human beings are rational decision makers. Yet, as you have just read, human cognition is not purely analytic and rational. Much of the time, our thinking is prone to bias and distortions stemming from the limited capacity of the analytic system and the quick and unconscious responses of the automatic system. Because so much environmentally harmful behavior likely stems from irrational thinking, purely rational approaches to change it, such as informational campaigns, are likely to fail.

Luckily, researchers in conservation and environmental psychology are an innovative and persistent bunch. Their studies explore new methods of presenting, teaching, and talking about environmental risks that overcome the counterproductive tendencies of the automatic system, while also awakening the interest of the analytic system. Both are necessary. Without the analytic system's conscious and deliberate attention, complex threats cannot be properly understood or responded to. But at the same time, without the affective support of the automatic system, the analytic system has little chance to engage in the first place. Some of the most intriguing ideas about changing environmental thinking are described below.

Raise Awareness

If you are unaware of a problem, you certainly cannot *intentionally* act to solve it. Awareness is therefore a critical first step for thoughtful engagement with any problem that requires an active response (Frantz & Mayer, 2009; Latane & Darley, 1970).

As mentioned before, the information campaigns used in many large-scale environmental efforts are rarely successful in their goals to increase concern and action (e.g., Crompton & Kasser, 2009; Latane & Darley, 1970). This failure is because information alone does little to engage the automatic system, the brain's gatekeeper, which is in charge of sifting through the deluge of data coming in through the senses and throwing out all but what is most important. This is a tough job these days: We live in an age of information overload and are bombarded with more than the brain can possibly handle, and certainly more than our ancestors contended with. How many environmental solicitations have you received in the mail, only to discard them after (or without) giving them a passing glance? Your automatic system likely judged them as low priority. Without a blip of curiosity or concern from the automatic system gatekeeper, the conscious analytic system is unlikely to look more deeply, particularly if it is already occupied with other pressing thoughts.

Another vexing problem for people hoping to raise awareness is the aforementioned lack of visual cues associated with many environmental risks, which renders them invisible. Humans tend to rely more heavily on visual information than on any other sense (hearing, smell, touch, or taste), assuming they are not visually impaired. Although it can be a serious limitation, this visual dependency also provides opportunity: Creating tangible and visible representations of an otherwise invisible problem is likely to capture the notice of the automatic system and spark an emotional response. Simple but thought-provoking images can effectively raise awareness of otherwise invisible issues (see Figure 6.6).

Researchers have also developed a range of visualization tools, including video games, interactive displays, virtual reality worlds, installations on city streets, and more, that appear promising for helping people understand otherwise intangible problems (Mulkern & ClimateWire, 2013; Sheppard, 2012). For example, a 3D visualization of rising sea levels shows how well-known urban scenes from coastal cities such as Vancouver, Miami, or Boston would look after various degrees of sea level rise (see Figure 6.7). In addition, one research study used immersion in a virtual world to allow people to make a sensory connection between their own use of

FIGURE 6.6 *Making water pollution visible, tangible, and personal.*
With permission from Wesley/Wespionage

FIGURE 6.7 *Making climate change impacts visible with 3D representations of sea level rise in coastal cities.*
Go to http://www.boston.com/yourtown/specials/boston_under_water/ for a vivid interactive display of how rising sea levels will affect Boston.
With permission from Canstock Photo

paper products and deforestation. Participants were given a short tutorial on how nonrecycled paper products contribute to deforestation. They then put on a head-mounted virtual reality device through which they entered a virtual forest complete with trees, twittering birds, and the peaceful sound of a breeze. They were told to grip a handset shaped like a chainsaw handle and chop down a nearby tree. They heard the chainsaw roar to life. The display showed the chainsaw responding to their motion while the handset gave off realistic tremors and resistance, as if they were cutting through real wood. Participants continued cutting until the virtual tree fell to the floor of the forest with a crash. Upon leaving the lab after completing this destructive deed, participants encountered a confederate who asked for help mopping up a spilled glass of water. Those who had been in the virtual world used significantly fewer paper towels for the mop-up than a second group of participants who had instead read a vivid description and imagined the act of cutting down a tree (Ahn, Bailenson, & Park, 2014). Personally cutting down the (virtual) tree not only made participants more paper-conscious, it also left such an emotional impact on them that some reported thinking about it months later when buying paper products (Mulkern & ClimateWire, 2013).

Once the automatic system has taken notice, the conscious mind has the opportunity to examine an issue further. However, whether the analytic system directs attention toward new information is highly dependent on a second factor: personal relevance.

Increase Personal Relevance

In the everyday flood of information, how does the automatic system decide what is important versus what should be filtered out? Because this system evolved to keep us alive above all else, it is most attuned to the directly personal and immediately meaningful. Ecological problems, when perceived as something "out there," are unlikely to pass the filter test of the automatic system. "Save the environment? What has that got to do with me?!" The issue is quickly dismissed and never even considered for a deeper, conscious examination.

What can be done to make environmental information more personally meaningful? One answer lies not so much in the content but rather in how a message is crafted. **Framing** involves presenting information in ways that illuminate its relevance. Different frames make certain pieces of information come forward in the mind, while other ideas are pushed aside. Slight changes in wording, emphasis, perspective, or accompanying images can result in very different responses, both emotional and analytic, despite the factual content remaining the same.

Certain frames successfully increase the personal relevance of environmental degradation. This can spark deeper thinking and bring conscious attention to an issue, the necessary step to building a stronger understanding and better mental models (Rogers, Kuiper, & Kirker, 1977). For example, water pollution is often referred to as "an environmental problem," emphasizing impacts on ecological systems and wildlife. In contrast, speaking about "the water our children drink" reminds people of the risk posed to people and shifts attention away from the consequences "out there." Similarly, describing the impacts of climate change in terms of familiar human health risks, such as increases in asthma, heat stroke, allergies, and pest-borne diseases, appears to help people connect it their personal lives more than when it is understood in terms of melting ice caps and polar bears (Maibach, Nisbet, Baldwin, Akerlof, & Diao, 2010). Information can also be designed for a particular audience and, in fact, tends to be more powerful than generally framed information. In a study involving bird-watchers, a frame emphasizing climate change's impacts on birds was more effective for increasing intentions to take action than a frame describing similar impacts on people (Dickinson, Crain, Yalowitz, & Cherry, 2013).

Framing can also tap into people's beliefs and biases. For example, climate change deniers are willing to engage in many proenvironmental behaviors when they see those activities framed as contributing to a better society, public health, and technological progress, but *not* when framed as solving climate change (Bain, Hornsey, Bongiorno, & Jeffries, 2012). Furthermore, in the United States, describing how environmentally friendly acts are patriotic or important for national security enlists the support of people who are otherwise committed to the "American way of life" and deeply skeptical of environmentalism (Feygina, Jost, & Goldsmith, 2010). People want to live life aligned with their deeply held personal values; they behave more sustainably when consciously reminded that the environment is important to them (Verplanken and Holland, 2002) and express greater support for environmentally friendly policy when other core values, such as family, are highlighted (Sheldon, Nichols, & Kasser, 2011).

Elicit Emotion

In addition to increasing personal relevance, framing can activate important emotions. Because affect is such a powerful force in shaping behavior, the right emotional response is critical. However, psychological research is unclear as to what, exactly, is the most effective and motivating emotional response. Without a sense of personal and urgent risk, people are less motivated to become engaged and take action (Weber, 2006). On the other hand, fear-based appeals can overwhelm and disempower (O'Neill & Nicholson-Cole, 2009). Thus, fear inducing messages should be used thoughtfully and with caution, and *never* without linking the problem to compelling actions individuals can take to alleviate the prompted worry. It is not enough that actions exist; to enable people to respond with problem-focused rather than emotion-focused coping, those actions must be specifically identified when the fear is activated (e.g., Moser, 2007; see Figure 6.8).

Of course fear is not the only emotion people feel, and it certainly is not the only emotion that motivates action. People are also moved by caring and empathy. In studies of perspective taking, the act of imagining oneself in another person's

FIGURE 6.8 *Messages that evoke fear must include concrete actions.*

shoes creates an empathic response (Batson, Early, & Salvarani, 1997). For instance, taking the perspective of a future human being suffering from the impacts of today's irresponsible environmental behavior increased people's motivation to take action, and inspired them to gather more information about environmental issues (Pahl & Bauer, 2013). People's empathy toward nonhuman nature can be heightened by perspective taking as well. A research study demonstrated this by asking college students to view a dead bird covered in oil from an oil spill. Participants who had been told to imagine how the bird might feel about what had happened reported significantly more empathy than those who had been told to view the image objectively. Furthermore, participants who had imagined the harmed bird's feelings also assigned more money to a fictitious environmental group on a subsequent task in which they proposed a budget for campus funds. Similar but slightly weaker results were found for participants who were told to take the perspective of a group of eight trees that had been cut down and lay by the side of a road (Berenguer, 2007).

People's emotional connection to other animals is another affective source of motivation to protect the environment. For example, many zoos have developed innovative programs to encourage conservation behaviors by linking them with the caring feelings people have toward other species, especially **charismatic mega-fauna**, the large, best-known and most beloved wild animals such as pandas and elephants. Tapping into people's feelings is a good way to inspire concern about habitat loss and other challenges these animals face in the wild (Skibins & Powell, 2013).

Though caring and empathy can effectively motivate sustainable behavior, they are often inspired by a negative event, such as destruction or loss. Psychological research suggests that eliciting positive emotions has a few advantages over provoking negative emotions when it comes to moving people toward change, especially if that change requires innovation. Recall that Skinner made the same point about reinforcement's advantage relative to punishment (Chapter 5). Positive emotions lead to stronger intentions to act upon new knowledge (Fröhlich, Sellman, & Bogner, 2013); broadened attention, making a wider range of options visible for consideration (Friedman & Förster, 2010); and enhanced creative thinking (Fredrickson, 1998).

Is it possible to motivate sustainability through purely positive emotions? To date, no studies have directly examined this question; however, related results are promising. For example, researchers presented participants with either a positive solutions-oriented framing of climate change action (e.g., *My neighborhood will be a healthier place to live if we walk more to cut greenhouse gases*) or a framing emphasizing giving things up (e.g., *To stop climate change, I have to make sacrifices*). The positive framing made people feel more competent to take proenvironmental actions, increased their level of emotional engagement with climate change, and bolstered their intentions to behave more sustainably. Also, pointing out the collective positive impacts of small, individual behaviors increases people's willingness to take personal action (Dickinson, et al., 2013).

Encourage Intention

After becoming aware of a problem, forming an intention is the next step on the way to deliberate action (more on this in Chapter 8). Intention is conscious, requiring the attention and effort of the analytic system. Forming an intention is not a perfect predictor of carrying out a behavior, as even the strongest intention can be thwarted by unexpected barriers. However, it is a necessary step on the path to conscious action.

One of the most intractable barriers to intentional sustainable behavior is the busy-ness of modern existence. Our stressful and hurried lives can push the conscious analytic system to its limits and leave little capacity to think deeply through all the issues we care about. Even when we are aware of ecological problems, see their personal relevance, and feel the necessary balance of urgency and hope, we may simply not have the time or the mental space to focus on anything more than getting through our daily routines.

When life becomes too busy, stressful, and overloaded by information, the automatic system takes control to give the conscious analytic system a break. Unfortunately, while on autopilot, we easily lose track of what makes life meaningful

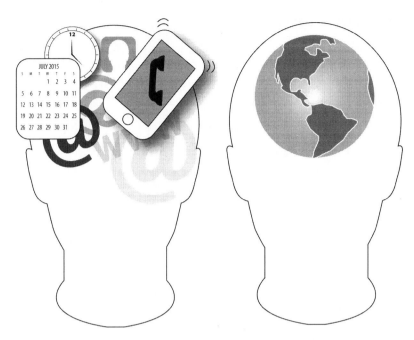

Mind Full, or Mindful?

and significant, such as our connections to friends and family, or a desire to make a difference in the world. Without the analytic system to help think through the consequences of our decisions, we may do things we later realize are incompatible with our values and long-term goals. Research suggests that an antidote may be **mindfulness**, the awareness of where one's mind is in the moment. One study found that higher scores on a measure of mindfulness were correlated with people's tendency to behave more sustainably (Amel, Manning, & Scott, 2009). In addition, people who take a moment to mindfully reaffirm their core values are better able to exert willpower (Schmeichel & Vohs, 2009). Other studies suggest that teaching people to practice mindfulness may be a good way to reduce overconsumption (De Wit, 2008; Goleman, 2009) and lower ecological footprints (Brown & Kasser, 2005), as well as promote subjective wellbeing and overall health (e.g., Kabat-Zinn, 2013).

Mindfulness is necessary to set an intention and to act upon it. Fortunately, over time, it's possible to live sustainably without being continuously mindful. Recall that old habits are cued by context, and new habits likely need different supporting circumstances and prompts. However, once a new context is created (e.g., you have bought a bike, a helmet, and a sturdy pannier), and you've practiced the new behavior a few times (i.e., you've ridden to and from work), new habits will form and you will no longer need to be as mindful for sustainable behavior to occur.

Provide Appropriate Knowledge

When asked, most people agree that environmental sustainability is important (Gallup, 2014b). Yet every day, you see people driving alone, throwing recyclables into the trash bin, drinking out of single-use plastic bottles, and leaving lights and computers on when they exit the room. One explanation for this inconsistency is that people may not understand the connection between their behavior and the environmental degradation they hear about on the news, or they may not know how to make their actions more sustainable.

As you will read in Chapter 7, levels of environmental knowledge in the United States are inconsistent across issues, and for some issues, are quite low. While it is true that people with higher levels of environmental knowledge tend to be more engaged in environmental action (Frick, Kaiser, and Wilson, 2004), this correlation does not necessarily indicate a causal relationship (Chapter 4). In fact, empirical evidence suggests that greater knowledge does not lead directly to greater environmental action, though it probably helps (Kaiser & Fuhrer, 2003; Gifford & Nilsson, 2014). Sometimes, instead of knowledge leading to action, it may be that the desire to act leads people to seek greater knowledge *and* change their behavior (e.g., Chess & Johnson, 2007). Still, despite the ambiguous link between knowledge and action, environmental organizations continue to spend their limited resources producing informational brochures and campaigns that are often, by themselves, ineffective at inducing behavior change.

Perhaps at this point you are saying to yourself, "But wait! Behavior can't change without some knowledge." Your instinct is correct. Though the role of knowledge in forming an intention is weak, once people have decided to act, certain forms of knowledge are helpful: effectiveness knowledge, how-to knowledge, and systems knowledge.

What behaviors are likely to contribute most to sustainability? Many people have no idea. In fact, when asked what actions they took to address climate change, people in southern England most often reported recycling (Whitmarsh, 2009). Given what you learned about ecological principles in Chapter 3, you can understand why recycling is nowhere on the list of the seventeen most effective household-level actions (Gardner & Stern, 2008). **Effectiveness knowledge** helps people aim for the most impactful behaviors, so that their conscious efforts are not misdirected.

However, effectiveness knowledge comes with at least one caveat: It sometimes has a counterproductive outcome due to the cognitive dissonance that people experience when new information is inconsistent with preexisting beliefs or behaviors. This dissonance may occur when people learn that difficult or unappealing lifestyle changes are most effective at reducing personal ecological impact. Few people are willing to take what they consider extreme measures, such as giving up air travel or becoming a strict vegetarian. To avoid cognitive dissonance, they may downplay the impact of those activities, rationalizing, "Oh, that doesn't make a difference." Thus, they can continue to enjoy current behaviors without guilt (Tobler, Visschers, & Siegrist, 2012).

Another form of knowledge essential for taking action is **how-to knowledge**. Knowing *how* to do something, and perhaps having a chance to practice, may be equally as important as the knowledge of *what* needs to be done. These two forms of knowledge, effectiveness and how-to, help ward off emotion-focused coping and allow people to engage in productive problem-focused efforts, concentrating on the high impact actions that make a real difference.

Finally, an understanding of the systems we are embedded in, both ecological and social, helps people identify how and where they can intervene. **Systems knowledge** provides the larger picture behind our problems and reveals how seemingly separate issues are interconnected (Chapter 3). Ecological systems knowledge includes a basic understanding of the larger earth systems that sustain life on the planet, such as the atmosphere, hydrosphere, and biosphere. Also important is understanding how human systems connect to the larger picture (i.e., where does my water come from, and where does it go when it leaves my drain?).

Systems thinking does not come easily to people, so researchers suggest using metaphors or analogies to concrete and familiar things. This way, people can form a

mental model of a complex system that captures its essential features. For example, the greenhouse effect can be understood as a "heat trapping blanket" covering the planet and trapping carbon in the atmosphere. This mental model helps people correctly identify lowering carbon emissions as part of the solution (Aubrun, Brown, & Grady, 2006).

An understanding of how our political and economic systems function can also point the way toward effective actions. Many people have no idea how to engage their political systems to encourage better policy or voice a preference for new infrastructure. This hinders them from imagining any form of action beyond simple personal acts like changing light bulbs, despite knowing that this does not address larger structural problems (Kenis & Mathijs, 2012).

Engaging these cognitive activities encourages new thinking patterns, exactly what is needed for intentional sustainable behavior. Though it would be ideal to inspire everyone to thoughtfully engage the analytic system and consider the environmental consequences of every decision, it simply isn't realistic. That is not to say that we can't make significant progress in that direction—this we can and must do. However, in the meantime, we need other tools to help us along. One strand of psychological research suggests that subtle changes in context can cue sustainable behavior through the unconscious automatic system, without requiring effortful thinking.

Moving Toward a Greener World . . . Automatically

Sustainable choices do not require conscious effort if the situation is constructed in the right way. Consider a home or apartment with a preset programmable thermostat that turns the heat down at night, or an office building that turns lights off unless a motion detector senses someone present. These are examples of "**green defaults**" that make sustainability the preset option unless people deliberately opt out (Sunstein & Reisch, 2013). They save people the burden of having to effortfully consider the sustainability of every action; if no choice is made, the most sustainable option is the one that occurs. Of course, changes in policy can have the same effect. For example, many countries, including the United States and Canada, have passed legislation to phase out inefficient incandescent light bulbs and make energy efficient alternatives the only ones available to consumers. Green defaults work differently than policy, though. Rather than taking away options, green defaults instead allow people to choose the less sustainable option if it is their conscious preference. For example, setting printers to the double-sided option is an effective way to encourage automatic paper saving, but individuals *can* print on one side if they choose. Institutions implementing this change have saved significant amounts of paper because most people simply accept the preselected default (reported in Sunstein & Reisch, 2013). Altering the context surrounding behavior allows the automatic system to respond as it always has, quickly and intuitively, but now with a more sustainable result.

CONCLUSION

There are common patterns of thinking among humans: Our automatic systems search for danger and sift through the minutia of daily life, and our analytic systems hang out in the background saving themselves for the heavy lifting of making significant decisions. Though our thinking patterns may tend to steer us toward unsustainable behavior in the contemporary, industrialized, context, there is abundant reason to believe that in spite of our circumstances, we can overcome our cognitive biases and harness our cognitive strengths to move in a more sustainable direction. Today's conscious choices will ultimately move us toward a future world in which policy and infrastructure help people develop more sustainable daily habits, and green defaults make ecologically responsible living even more widespread.

Though the cognitive tendencies described in this chapter are presumably universal among humans, there is some variability in how, and to what extent, people exhibit them. Humans are a diverse lot. Individual differences in thinking, along with differences stemming from personality, identity, and gender, are discussed in the next chapter.

CHAPTER 7

Putting the "I" in Environment

- **How I Think**
  - **Knowledge**
  - **Beliefs**
  - **Attitudes and Values**
  - **Thinking Style**
- **Who I Am**
  - **Personality**
  - **Identity**
  - **Gender**
- **Conclusion**

Picture the following situation: A student is walking down the hall with an empty plastic bottle in hand. As he approaches the trash can, clearly intending to toss the bottle into it, a well-meaning professor who happens to be passing by prompts him with a friendly smile, "Hey, there's a recycling bin over there!" The student responds by fixing the professor with a glare as he defiantly chucks the bottle into the trash can and vehemently declares, "I'm not a campus liberal!"

Whoa! As this anecdote (which really happened to one of us authors) illustrates, even when the situation encourages sustainable behavior and the behaving person is thinking deliberately, it is not a given that the behavior displayed will be proenvironmental. To fully understand people's behavior, it is necessary to consider more than external influences, such as those described in Chapter 5, and cognitive tendencies, reviewed in Chapter 6. The missing ingredient in the recipe thus far is **individual differences**. In this chapter, you will read about how people's environmentally relevant behaviors are related to variability in knowledge and beliefs, attitudes and values, thinking styles, personality traits, identity, and relationship with nature. Although these factors will be discussed one at a time, keep in mind that they are intimately interrelated and together comprise a cluster of characteristics that differentiate individuals from one another. Also keep in mind that some individual differences (e.g., attitudes and values) are changeable, and thus susceptible to intervention, while others (e.g., personality traits) are likely to be more stable and resistant to change.

HOW I THINK

How people think affects how they behave. As you learned in the last chapter, there are numerous universal cognitive tendencies that play a significant role in influencing environmentally relevant behavior. In addition to these, there are individual differences in thinking that are relevant for understanding why some individuals behave more sustainably than others. This section will describe pertinent differences in knowledge, beliefs, attitudes, values, and thinking styles.

Knowledge

By this point in the book, you understand that just providing information to people is not a sufficient strategy for inspiring behavioral change in a sustainable direction, but this does not mean that the amount of information an individual has is irrelevant. A person who doesn't know anything about ecological problems is not likely to act on them.

Depending upon how much environmental education individuals have experienced, either formally or informally, they differ in **ecological literacy**, an understanding of key ecological concepts and the reciprocal relationship between natural systems and human systems (Orr, 1992; Puk & Stibbards, 2012). Reading Chapter 3 may have increased your ecological literacy.

Of course, one does not need to understand ecological systems to recognize that there are problems. **Environmental knowledge**, the extent to which people are *aware* of environmental issues, whether they grasp the underlying processes or not, varies depending on the topic, and depending on demographics. For instance, a review of 15 surveys of environmental knowledge among U.S. citizens found that the majority of people knew something about household hazardous waste and species extinctions, but public knowledge was generally lacking when it came to energy production, water quality, and climate change (Robelia & Murphy, 2012). These were the overall trends, but individuals varied in their mastery of the topics. What stood out in several of the surveys as the single best predictor of an individual's environmental knowledge was *level of education*; people with more schooling under their belts knew more about the various topics than those with less education. Furthermore, the survey data revealed that middle-aged adults, especially men, demonstrated the highest levels of environmental knowledge. The authors speculate that this may be because age and gender are confounded with education (i.e., older people and men are likely to be more educated than younger people and women).

As you might expect, people with more environmental knowledge are considerably more likely to act in proenvironmental ways than people uneducated about environmental issues (Coyle, 2005). Environmental knowledge and ecological literacy not only affect everyday decisions and behaviors, they also affect political participation. To have meaningful exchanges about policy, people must be well-informed about topics, yet a significant portion of the voting public possesses relatively low levels of environmental knowledge (Robelia & Murphy, 2012). This is not to say that policy makers themselves necessarily have high levels of environmental knowledge or ecological literacy. Indeed,

> There is little difference in environmental knowledge levels between the average American and those who sit on governing bodies, town councils, and in corporate board rooms. . . . There is encouraging evidence that the public can learn about the environment and complex ecological relationships. That we are far from succeeding in making this a reality is due to the absence of a comprehensive coordinated approach to environmental education. We also consider low levels of knowledge about the environment as a signal that members of the public will be unprepared for increasing environmental responsibilities in the coming years. As environmental topics and problems become more complex and pervasive, our decades of reliance on trained experts within the private and public sectors to handle our needs are nearing an end. In the future, many leading environmental problems, ranging from water quality to ecosystem management, will require the efforts of more skilled non-experts acting as individuals, through small business, or as community leaders.
> (National Environmental Education and Training Foundation, 2005, p. 11)

Most formal **environmental education (EE)** is aimed at children. The best programs recognize that children of different developmental phases need different types of activities and foci. In fact, depending upon the age of the children, EE may not increase their environmental knowledge or ecological literacy so much as it piques their curiosity and inspires feelings of care for nature. For example, one developmentally appropriate EE model differentiates between early and middle childhood and early adolescence (Sobel, 1996). Young children (ages four to seven) benefit from opportunities to *empathize* with creatures in the natural world through play activities where they are encouraged to mimic or "become" a nonhuman animal. Middle childhood (ages eight to eleven) is a time for *exploration* of their environment, beginning with the areas surrounding their houses and school, and expanding to nearby forests and streams. Access to local green spaces is essential for exploration, and is particularly important for urban children of lower income families who may not have many opportunities to visit nature reserves and national parks (Fisman, 2005). Finally, *social action* is key to early adolescence (ages 12 to 15), so educational activities for them occur in groups that take responsibility for particular projects such as energy conservation or recycling programs (Sobel, 1996; see also Hart, 1997). For now, EE is not universally present in childhood education, resulting in young adults with differing levels of environmental knowledge and ecological literacy.

Given that most adults are not in school, their environmental education occurs in other ways. Some adults become educated by actively seeking the experience. They read books or browse websites; visit zoos, aquariums, botanical gardens, and national parks; join environmental organizations; and participate in activities such as wilderness treks and conservation campaigns (Walter, 2009). For the majority of adults, however, the primary source of environmental information is the media. Media coverage can be misleading, perpetuating myths and misunderstandings about environmental issues. For example, in an effort to be "balanced," some media outlets continue to present both sides of the climate change debate; but, giving equal time to the arguments for and against humans as the primary cause misrepresents the science (which, as you read in Chapter 1, overwhelmingly supports the anthropogenic explanation). Many adults are likely not being exposed to accurate information about environmental issues (National Environmental Education and Training Foundation, 2005).

Although people who are better informed are more prepared to actively participate in promoting a sustainable future, information is not always received the same way by different individuals. Once people have knowledge, what they do with that knowledge depends on their beliefs.

Beliefs

You may recall the "Readiness to Change" item presented in Chapter 1 (Figure 1.1). Two of the response options are "I have heard of environmental problems but I don't believe they are true" and "I have heard of environmental problems, but I am uncertain whether they are true." Some people are aware of concerns about climate change, for example, and may even know quite a bit about the evidence (e.g., that weather patterns are getting more extreme and average global temperature has been at record highs for the past dozen years), but this does not mean that they *believe* action is warranted. In the case of climate change, some of the so-called climate change skeptics or deniers don't actually doubt climate change itself; rather, they doubt that it is caused by humans (e.g., Sibley & Kurz, 2013). Furthermore, there are some individuals who are labeled skeptics/deniers, who don't doubt the validity of anthropogenic climate change, but believe the doom-and-gloom scenarios of dire future consequences for humans are overblown, so there is really nothing to worry about (Murray, 2011). In other words, individuals differ in terms of their beliefs about the same sets of information.

A study of more than 1,500 adults in the United States tested two competing explanations for why the public seems to be less concerned about climate change than is warranted, given the scientific evidence (Kahan, et al., 2012). One explanation is that the public lacks knowledge, that laypeople don't take climate change as seriously as scientists think they should because they don't understand the science. The other explanation, labeled the **cultural cognition thesis** is that people's perceptions of risks are influenced not just by knowledge, but also by the values of the social groups with which they affiliate and identify. This study found no support for the first explanation; there was not a clear association between scientific literacy and concern about climate change. For some individuals, higher scientific literacy was indeed positively correlated with concern, but for others, higher scientific literacy was *negatively* correlated with concern. In contrast, the cultural cognition explanation does account for this seemingly perplexing finding. Specifically, whether the correlation was positive or negative depended upon whether the individuals held a hierarchical, individualistic worldview or an egalitarian, communitarian one (see Figure 7.1). The researchers suggested that those who subscribe to the former

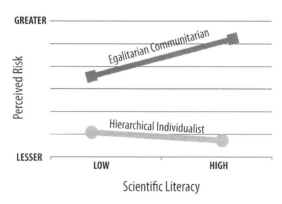

FIGURE 7.1 *Cultural cognition thesis.*
The same information can be interpreted differently depending upon people's beliefs. Figure adapted from Kahan, et al., (2012).

worldview react negatively to policies that would restrict the business activities they value, whereas the latter worldview includes the perception that unchecked commerce and industry contributes to social inequity. Therefore, two individuals with these opposing perspectives can look at the same set of information about the changing climate, but interpret it differently. As you learned in the previous chapter, people are biased thinkers.

One thing many individuals are motivated to believe is that the world is a *just* place, that people get what they deserve and deserve what they get (Lerner, 1980). Believing in a just world offers a false sense of security that if one is a good person, one is assured good outcomes. Perceiving social systems as fair also helps people to feel less guilty or upset when witnessing injustice experienced by others (Jost, et al., 2010). Recently, researchers investigated whether there may be individual differences in the tendency to perceive the world as *ecologically* fair, as a place in which "everyone gets what he/she deserves concerning ecological resources" (Baier, Kals, & Müller, 2013, p. 274). Their **Ecological Belief in a Just World (EBJW)** scale measures the extent to which respondents believe (falsely) that all people generally have equal access to healthy, unpolluted natural resources and the same opportunity to participate in decision making about where industrial hazards will be located.

When the researchers administered the scale to a sample of more than 300 German adults, they found that EBJW was significantly positively correlated with outrage about conservation-related limits to personal freedom and economic growth. It was negatively correlated with feeling a sense of responsibility for energy conservation and with being willing to purchase energy-saving devices. In other words, the more strongly one believes the world is ecologically fair for all, the more one feels entitled to resources and angry that access might be restricted. This individual difference variable may help explain why the reality of **environmental injustice** does not motivate everyone similarly. In the words of the researchers: "Even though perceptions of ecological injustice motivate proenvironmental behavior, not all people perceive the environmental inequalities as unjust" (Baier, et al., 2013, p. 274; more on (in)justice in chapter 8).

Two more response options on the "Readiness to Change" item presented in Chapter 1 are "I believe there are environmental problems, but I am not ready to make changes because of them" and "I believe there are environmental problems, and I would like to make changes because of them." Notice that both of these options indicate that the individual believes in the problems, but has not yet made any behavioral changes. Certain types of beliefs can affect whether informed and concerned individuals actually act on what they know. These include beliefs about the *effectiveness* of acting as well as beliefs about one's *ability* to act.

Some people believe that they have little influence over personal outcomes and world events; they feel like things happen because of chance, fate, luck, divine will, or other external influences. Individuals with this belief orientation are said to have an **external locus of control (LOC)** (Rotter, 1966). In contrast, individuals who believe that outcomes depend upon planning, effort, and personal commitment have an **internal locus of control**. Whom do you think would be more likely to behave in an ecologically conscientious way: people with internal or external locus of control? Since the 1980s, several studies have investigated whether locus of control is related to environmental values and behaviors, but the results are mixed and inconclusive. More recently, researchers using a specialized measure of **internal environmental locus of control (INELOC)** found support for their hypothesis that it would significantly predict a variety of proenvironmental behaviors, including relatively easy behaviors such as recycling and choosing green products, as well as more costly behaviors such as purchasing energy-efficient appliances and volunteering with environmental groups (Cleveland, Kalamus, & Laroche, 2012; see Figure 7.2).

Knowing that individuals vary in terms of environmental locus of control is useful for creating messages to inspire sustainable behavior. For example, marketers may be able to reach individuals with an internal environmental locus of control by simply reminding them that their behavior does indeed make a difference (Cleveland, et al., 2012). This type of message will not be effective, however, for individuals with an external environmental locus of control; they need to be persuaded of the connection between their effort and the desired outcome.

Still, it may not be enough to convince people that their behavior makes a difference. They also must believe that they have the ability to do the behavior in the first place. Many variables, including the challenge posed by the situation and one's skill set, combine to form **self-efficacy**, the belief that one has the ability to successfully pursue a course of action required to achieve a goal. Research around the globe indicates that self-efficacy perceptions significantly predict a variety of actions including sustainable consumer behavior, energy conservation, environmental activism, and recycling (summarized in Gifford & Nilsson, 2014). Individuals low in self-efficacy may benefit from training and practice to build confidence as they pursue sustainable behavior.

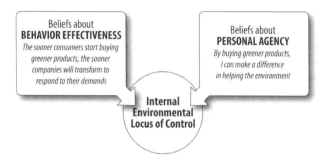

FIGURE 7.2 *Internal environmental locus of control.*

Attitudes and Values

An **attitude** is a positive or negative evaluation of something (Eagly & Chaiken, 1998). Attitudes are related to beliefs, but a belief is what the individual *thinks about* something, whereas an attitude is how the individual *feels toward* it. In the realm of sustainability, researchers have studied people's attitudes toward nature, environmental protection, and environmental policies. As described in Chapter 4, the structure of environmental attitudes is multifaceted.

Formation of an attitude can happen unconsciously, but sometimes it requires input from the analytic system you read about in Chapter 6. People don't tend to exert the cognitive effort required to develop elaborate attitudes unless things are particularly relevant or important to them. Which environmental issues seem most relevant can depend on geographical and economic factors. People living in wealthy countries, where their basic needs are more than met, can afford to give careful consideration to global environmental issues, while residents of less wealthy countries sometimes have a narrower, regional focus (Gifford & Sussman, 2012). Despite some variations, however, researchers generally find that environmental attitudes are a valid construct across cultures (e.g., Xiao, Dunlap, & Hong, 2013).

Like beliefs, attitudes do not always predict behavior. You might think that a person who holds proenvironmental attitudes would behave accordingly, but research suggests that this is not necessarily the case. Meta-analyses of studies on the relationship between environmental attitudes and behaviors have found modest correlations of around .40 (Hines, Hungerford, & Tomera, 1986; Bamberg & Möser, 2007). In fact, the **attitude-behavior gap** appears across most domains researchers have studied, not just in the realm of environmental attitudes. Inconsistencies between people's attitudes and their behaviors are due to the fact that many things besides attitudes influence our behavior. What factors might prevent a person with proenvironmental attitudes from behaving sustainably? In the next chapter, you will learn about theories that highlight how other factors such as norms and perceived self-efficacy interact with attitudes to influence behavior.

Another reason many studies fail to find a strong relationship between attitudes and behaviors is that accurate assessment is challenging. Researchers cannot follow participants around for an extended period of time across a wide variety of situations, so they may not be able to obtain an adequate sample of behavior to detect a proenvironmental or antienvironmental pattern. Also, measures tend to rely on self-report,

and as you learned in Chapter 4, people are not always honest when it comes to reporting things that they believe may be perceived as socially undesirable. In the case of environmental attitudes, people may misrepresent their actual attitudes depending upon what they think will make them look best in the eyes of the researcher. Self-report measures of attitudes, such as the New Ecological Paradigm Scale described in Chapter 4, are also limited in that they only tap into **explicit attitudes** that people are conscious of holding. Yet people also hold **implicit** or nonconscious attitudes (Greenwald & Banaji, 1995), which are important to measure as well.

Implicit attitudes are assessed with subtle measures that are not as vulnerable to social desirability bias, such as the **Implicit Associations Test (IAT)** (Greenwald, McGhee, & Schwartz, 1998). The IAT is administered on a computer that records the time (in milliseconds) it takes participants to sort words or pictures representing two categories about which they may hold contrasting implicit attitudes (such as *Nature* and *Industry*). When a stimulus (e.g., a picture of a tree) is presented on the screen, the participant presses one key if it belongs in the category on the left side of the screen (e.g., Nature) and another key if it belongs in the category on the right side of the screen (e.g., Industry). After a few of these straightforward trials, the task becomes more complicated, and that is when researchers are able to tap into implicit attitudes. See Box 7.1 for an explanation of how this test works.

BOX 7.1 *The Implicit Association Test*

After some easy trials sorting words or pictures into two categories (e.g., Nature and Industry), the participant completes some equally simple trials sorting positive and negative words, such as "wonderful" and "awful," into the categories Good and Bad. But then things get more complicated. On subsequent trials, the participant must sort both types of stimuli, presented in random order one at a time, into the left or right categories, which are now combinations, such as "Nature or Good" and "Industry or Bad."

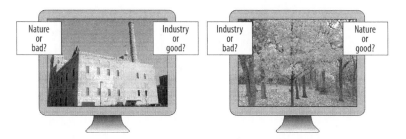

The participant completes multiple trials with the combinations reversed half of the time (i.e., "Nature or Bad" and "Industry or Good"). The logic of the test is this: Participants will sort more quickly and accurately when the category they favor is paired with "good" than when that category is paired with "bad." For example, if the participant feels more positively about Nature than about Industry, the participant will do better on the trials where the headers are "Nature or Good" and "Industry or Bad." This is because the task is easier when both of the "good" things go on one side (i.e., require the same key press) and both of the "bad" things go on the other.

Research employing the IAT has yielded mixed results. For example, one study found that implicit attitudes were better predictors than explicit attitudes of intentions to buy eco-friendly products (Vantomme, Geuens, De Houwer, & De Pelsmacker, 2005), but a later study found that implicit environmental attitudes did not predict environmental intentions or behaviors at all (Levine & Strube, 2012). Consequently, the relationships between explicit and implicit environmental attitudes, and between implicit environmental attitudes and behavior, remain unclear. However, research on other attitudes subject to social desirability bias shows a predictive advantage of IAT scores over explicit measures (Greenwald, Poehlman, Ulmann, & Banaji, 2009). Also, some research suggests that explicit attitudes may be more predictive of deliberate, intentional behaviors, while implicit attitudes may be more predictive of spontaneous behaviors (Perugini, 2005). As discussed in previous chapters, many environmental behaviors require deliberate intention until they become habits, so both types of attitudes are likely relevant in the psychology of sustainability. Implicit environmental attitudes seem to warrant further empirical investigation.

Importantly, attitudes are *changeable* through persuasion and other influences. Attitudes formed via conscious deliberation are more stable and resistant to change than attitudes formed heuristically (Bodenhausen & Gawronski, 2013), but in general, attitudes can be considered temporary and less enduring than **values**, which are "guiding principles" that inform both beliefs and attitudes (Schwartz, 1992). Values influence one's priorities and decisions, and predict one's **personal norms**, idiosyncratic feelings of moral obligation to act in a particular way (Schwartz, 1977; Steg, Dreijerink, & Abrahamse, 2005). Like social norms, personal norms are potent influences on behavior, but unlike social norms, the consequences of violating them are all self-inflicted, usually in the form of guilt and remorse. Personal norms operate independently of social norms; for example, a person may feel guilty when he or she forgets to take his or her reusable cup to the coffee shop because of his or her personal norm about not generating unnecessary waste, even if everyone else in the place is using single-use paper cups.

Three types of values are pertinent to proenvironmental beliefs, intentions, and actions: egoistic, altruistic, and biospheric (Stern, Dietz, & Kalof, 1993; Steg & De Groot, 2012; see Figure 7.3). **Egoistic values** lead to people's concern about the environment because of direct impacts to them personally, such as a waste dump in their neighborhood that could poison their water supply, or the construction of wind farms that might ruin their view. In contrast, people with **altruistic values** care about the environment because of its relevance for *other* humans, including children, community members, other cultures, future generations, and society in general. People with altruistic values would find air pollution in Mexico City a problem because of its impact on residents' health, even if they personally don't live or travel there; climate change is a concern because it will lead to food shortages and starvation in many parts of the world, and flooding of communities on far-away islands. Finally, **biospheric values** lead people to find environmental degradation problematic because it impacts ecological systems (the biosphere), including

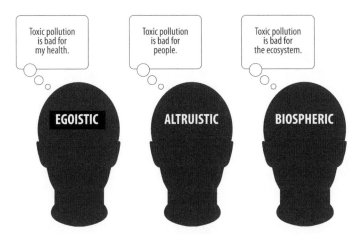

FIGURE 7.3 *Values related to environmental concern.*

nonhuman animals, plants, oceans, and forests. People with biospheric values regard ecological systems as important in themselves, beyond what they mean for human survival (altruistic) or personal comforts (egoistic).

Both biospheric and altruistic values directly predict environmentally friendly behaviors such as buying organic food, contributing to environmental groups, and recycling (Karp, 1996). This does not mean, however, that they are always overlapping. In some situations they can actually conflict. For example, imagine a person shopping for coffee who has to decide between organically grown and fair-trade; biospheric values will support the first choice while altruistic values will support the second (Steg & De Groot, 2012). Perhaps you can think of some more significant situations in which these two values orientations might clash.

As one might expect, values differ across cultures. Although altruistic values are most commonly observed around the world, United States and European samples show a priority of egoistic over biospheric concerns, whereas Latin American countries tend to show higher biospheric than egoistic concerns (Schultz, 2002a; Schultz & Zelezny, 2003). In a cross-continental study of six diverse countries (Brazil, India, Germany, New Zealand, the Czech Republic, and Russia), university students living in the first five scored higher on biospheric than egoistic values. Although the pattern was reversed in the Russian sample, Russian students showed the same tendencies as other students on a separate measure of concern for universal wellbeing (Schultz, et al., 2005).

Individual differences in how the environment itself is valued obviously are relevant for environmental behaviors, but there are other types of values that also influence people's behavior toward the environment. For example, some individuals with strong religious values find themselves compelled to behave sustainably as stewards of a divine creation (e.g., Smith & Leiserowitz, 2013). On the other hand, religious values inspire some others to resist conservation efforts in the interest of hastening the "end times," or apocalypse described in the Christian bible (e.g., Hendricks, 2005).

What other values do you think may be relevant for predicting environmental behavior? Researchers have devoted considerable attention to individual variation in the extent to which people prize and prioritize wealth and material possessions. As mentioned in Chapter 4, a meta-analysis of 17 published and unpublished studies on **materialistic values** found that they are moderately predictive of both anti-environmental attitudes and behaviors (Hurst, Dittmar, Bond, & Kasser, 2013). People with more materialistic values have larger ecological footprints, consuming more of the earth's resources due to their lifestyle choices regarding food, transportation, and housing, and highly materialistic individuals report fewer environmentally friendly behaviors such as riding a bike, reusing paper, and buying used products from secondhand stores (Brown & Kasser, 2005; Kasser, 2011).

There has been a surge in materialistic values in the United States over the last few decades. Thirty years of data on more than 350,000 high school students show that materialism among teenagers peaked in the 1980s and has remained fairly steady since (Twenge & Kasser, 2013). The researchers speculate that the trend is due to an overall, society-wide increase in the factors that cause some individuals to be more materialistic than others: unmet emotional needs and materialistic modeling. Children who feel insecure because their parents are unemployed, unavailable, or in an unstable relationship are vulnerable to becoming more materialistic, as are children who are exposed to materialistic role models and messages on television or in their own social networks. Studies of twins suggest that variability in materialism among individuals is determined almost entirely by life experiences like these rather than genetic inheritance (Giddens, Schermer, & Vernon, 2009).

Thinking Style

A primary theme of the previous chapter is that there are two general modes of cognition: effortful and analytic, and quick and automatic. As you learned, our tendency is to rely on the latter much of the time; it is only when people are sufficiently motivated and have enough cognitive resources available that they engage in the type of deliberate thinking that requires conscious attention. When in the effortful mode, individuals differ in the *complexity* of their thought. Some people are inclined to think harder than others. A person who finds it fun to do brainteasers, solve puzzles, critique evidence, and come up with solutions to perplexing problems is high in **need for cognition**; for people like this, thinking hard is rewarding. Someone who feels comfortable mulling over large amounts of information and is able to make connections and see patterns is high in **conceptual complexity**; people like this are good at integrating ideas. Finally, an individual who is uncomfortable with nuance or uncertainty, and is eager to finalize decisions, is high in **need for closure**; people like this have a limited tolerance for new information. Together, high need for cognition, high conceptual complexity, and low need for closure comprise **integrative complexity** (Suedfeld, 2009; see Figure 7.4).

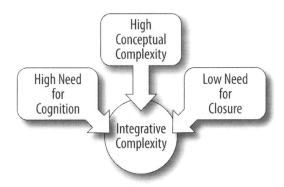

FIGURE 7.4 *Integrative complexity.*
Some individuals think harder than others; they enjoy it, are good at integrating ideas, and can tolerate ambiguity.

Research on how complexity of thinking relates to environmental attitudes and behaviors is limited. One study on college students' opinions of the Endangered Species Act (ESA) measured integrative complexity with writing samples. Researchers found that the students with *moderate* attitudes toward the ESA wrote more complex essays about it than did students who strongly favored or opposed the ESA (Bright & Tarrant, 2002). This correlational finding suggests that a complex thinking style is associated with seeing shades of gray in the realm of environmental policy.

Complex thinking is challenging and so is thinking ahead. As you learned in the last chapter, it is generally difficult for people to mentally make the connection between behaviors now and consequences that may occur far into the future, which is one reason why so many people are unmotivated by problems that to them feel temporally distant (e.g., climate change). Even when people can envision possible negative consequences in the future, immediate benefits are usually more salient. There are, however, some people who *habitually* think about the future when making decisions in the moment. These individuals are said to have a **future time perspective**; they routinely visualize upcoming events and outcomes, planning and goal-setting as they go, rather than basing their decisions on past experiences or things happening in the here and now (Karniol & Ross, 1996; Zimbardo & Boyd, 1999).

Because sustainable decisions sometimes require favoring long-term benefits over immediate ones, researchers tested the hypothesis that a future time perspective is associated with proenvironmental behaviors. A meta-analysis of 19 samples, including participants from seven different countries (Australia, Brazil, Germany, Mexico, New Zealand, Norway, and the United States) found that past and present time perspective did not predict proenvironmental behaviors, but future time perspective did (Milfont, Wilson, & Diniz, 2012). Individuals inclined to focus on immediate concerns are less likely to behave sustainably, but even for these individuals, proenvironmental concern can be temporarily increased by experimentally inducing a future-oriented mindset (Arnocky, Milfont, & Nicol, 2014).

When you think about the future, are you a "glass half full" kind of person, or a "glass half empty" one? Some people tend to anticipate good outcomes while others worry about bad ones. Which style of thinking, **optimism** or **pessimism**, do you think would more likely predict proenvironmental behavior? It's not an easy question to answer. On the one hand, optimistic individuals might be inspired to move their behavior in a sustainable direction because they feel hopeful about people's ability to mitigate the ecological crisis; on the other hand, maybe it's the pessimists who will feel most compelled because of their anxieties about what will happen if we do not act. An optimistic outlook can also lead individuals to underestimate future negative consequences of ecological degradation, thus undermining the motivation to behave more sustainably (e.g., Ojala, 2012). Similarly, if one is thinking too pessimistically, one may feel apathetic, hopeless, indifferent, or paralyzed by fear, and do nothing (Lomborg, 2001).

In the context of environmental economics, optimists and pessimists have been called **Boomsters** and **Doomsters**, respectively (Bailey, 1993). The Boomster view was exemplified by economist Julian Simon (1981), who argued that human beings are not limited by the carrying capacity of an ecosystem because, unlike other animals, humans have the intelligence to redesign their habitat by inventing technology. Therefore, human ingenuity is likely to produce technological solutions that will solve problems in ways unimaginable at present. Human beings are (as in the title of Simon's book) "the ultimate resource."

From a Boomster perspective, Doomsters are like fire and brimstone preachers; describing the coming environmental hell in graphic detail, they scare their

audience with dreadful prophecies, and then promise salvation through conversion to a new ecological worldview. Doomsters do not always agree with each other with respect to solutions. Some suggest that the problems will be solved by governmental regulations, while others argue that only transformation of people's deepest spiritual values will provide deliverance. For some, the complete "gloom and doom" perspective can lead to the conclusion that there is no hope for warding off environmental catastrophes. The problems can seem too huge, too complex, and too expensive for human beings to manage. It is likely that there are elements of truth in both the Boomster and Doomster positions. Although the ecological crisis is grave, some problems are being addressed by human ingenuity. The point here is that individuals' outlooks differ, and both optimism and pessimism can motivate or discourage efforts to behave more sustainably.

Outlook, values, attitudes, beliefs, and knowledge are all aspects of thinking that vary among individuals. As you have read, this variability helps explain differences in people's tendencies to behave in ecologically responsible ways (or not). With apologies to Descartes, thinking may let us know *that* we are, but *who* we are is a matter of individuality, as described in the next section.

WHO I AM

Try this exercise: Label the top of a blank page with the question "Who am I?" Number from one to twenty down the side. Now take a few minutes to fill in each of the 20 lines by completing the phrase "I am . . ." with the first single word or short phrase that comes to mind to describe yourself. If you can't think of twenty responses, that's ok. Once you have exhausted your ideas, take a look at what's on your list.

If you are like most individuals who complete the Twenty Statements Test (Kuhn & McPartland, 1954), your list includes a variety of descriptors. Some are **personality traits**, aspects of your general character that tend to remain consistent across situations, such as "shy" or "detail-oriented." Others are facets of your **identity**, such as roles (e.g., college student), ideological beliefs (e.g., liberal), interests (e.g., psychology major), and self-evaluations (e.g., good at math) (Kuhn, 1960). You probably feel that your personality and identity are uniquely *you*. It is true that each individual is unique, yet there are aspects of both personality and identity that people share. Researchers can, therefore, look for trends that reveal how particular types of personalities or identities relate to specific variations in behavior.

Personality

Given the diversity of people in the world, it may surprise you to learn that many psychologists believe all variation in people's personalities can be captured by just five dimensions. The **Big Five**, as they are called, include Extraversion,

FIGURE 7.5 *The Big Five Personality Traits.*
Each of these predicts proenvironmental behavior, particularly Openness and Agreeableness.

Neuroticism, Conscientiousness, Agreeableness, and Openness to Experience (McCrae & Costa, 1987; see Figure 7.5). People high in **Extraversion** are outgoing, sociable, talkative, and like being at the center of things. In contrast, introverts are socially withdrawn, quiet, and reserved. Individuals high in **Neuroticism** are prone to psychological distress such as worry, anxiety, depression, and feelings of inadequacy, compared to those with low neuroticism who are more happy-go-lucky. People high in **Conscientiousness** are detail-oriented and dependable, as opposed to disorganized and irresponsible. People high in **Agreeableness** are cooperative, trusting, and sympathetic, rather than antagonistic, cynical, and unfeeling. People high in **Openness** are imaginative, intellectually curious, behaviorally flexible, and open-minded (Costa & McCrae, 1992). Which of these dimensions do you think might relate to environmental attitudes and behaviors?

As it turns out, all five of these personality traits predict a proenvironmental orientation, with stronger evidence for some than others. Openness and Agreeableness appear to be the strongest predictors of a proenvironmental orientation (e.g., Hirsh, 2010; Hirsh & Dolderman, 2007; Markowitz, Goldberg, Ashton, & Lee, 2012; Milfont & Sibley, 2012; Nisbett, Zelenski, & Murphy, 2009). Individuals who score high on Openness tend to show flexibility in thinking, curiosity, and willingness to try new things, which may make them more likely to challenge conventional ways of doing things by adopting novel behaviors (e.g., composting) or products (e.g., an electric car) (Markowitz, et al., 2012). Being open to experience is an asset when it comes to making lifestyle changes.

Researchers speculate that one reason Agreeableness is positively correlated with proenvironmental concern is that Agreeableness is associated with higher levels of **empathy**, the cognitive and emotional understanding of another's perspective. Empathy with nonhuman nature predicts proenvironmental thinking and behavior, both when intentionally manipulated by researchers (e.g., Berenguer, 2007; 2010; Schultz, 2000) and when measured as a trait (Sevillano, Aragones, & Schultz, 2007; Tam, 2013). Another explanation for why Agreeableness predicts proenvironmental concern is that, like Openness, it is related to **self-transcendence**, prioritizing concern for others over oneself (Milfont & Sibley, 2012; see Figure 7.6). Self-transcendence encompasses benevolence, which is concern for the enhancement of other members of one's in-group, and universalism, which is appreciation and concern for the interests of *all* others, including nature (Schwarz, 2012).

Given the worrisome forecasts about our ecological future, it is perhaps no surprise that researchers have found a link, albeit small, between ecological concern and Neuroticism (Hirsch, 2010; Swami, Chamorro-Premuzic, Snelgar, & Furnham, 2010; Wiseman & Bogner, 2003). Neurotic individuals are prone to fret about negative outcomes in general, so ecological concern could represent just one form of overall apprehension. Conscientiousness, too, has emerged as a positive predictor of environmental concern (Hirsch, 2010; Milfont & Sibley, 2012; Swami, et al., 2010). One possible explanation stems from the fact that Conscientiousness is related to future time perspective; conscientious individuals tend to plan for the future, and environmental engagement often involves long-term planning (Milfont & Sibley, 2012).

Among the Big Five, the weakest predictor seems to be Extraversion; two studies that predate the development of Big Five assessment found evidence suggesting extraverted tendencies might be predictive of environmental concern (Borden & Francis, 1978; Pettus & Giles, 1987). Subsequent studies have not found a significant association between environmental concern and extraversion in individuals, but a cross-cultural analysis did find the link at a societal level. Using aggregate data on the Big Five personality profiles of 51 different countries, the researchers found a pattern suggesting that more extraverted cultures (such as those in Europe) exhibit higher environmental concern than more reserved cultures (such as those in Asia) (Milfont & Sibley, 2012). Perhaps extraverted cultures foster environmental concern, or perhaps there is an alternative explanation for this correlational finding.

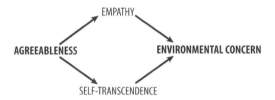

FIGURE 7.6 *Agreeableness predicts environmental concern.*
Agreeableness is related to both empathy and self-transcendence, each of which is related to environmental concern.

Some personality researchers support a *six*-dimensional model which includes Honesty-Humility in addition to the Big Five factors (Ashton, et al, 2004; Ashton & Lee, 2005; 2007). This dimension describes the extent to which a person is sincere, honest, trustworthy, humane, generous, and self-sacrificing. One study found that this factor explained variability in environmental concern above and beyond that accounted for by the Big Five; specifically, individuals higher in Honesty-Humility tended to have a more prosocial orientation, which, in turn, predicted proenvironmental attitudes and behaviors (Hilbig, Zettler, Moshagen, & Heydasch, 2013). Other researchers found a prosocial orientation highly characteristic of environmentalists, and more characteristic of them than of individuals who are less environmentally engaged (Kaiser & Byrka, 2011).

The low end of the Honesty-Humility dimension is associated with a "dark triad" of antisocial tendencies: psychopathy, Machiavellianism, and narcissism (Lee & Ashton, 2005). A conceptually related trait is the **social dominance orientation (SDO)** (Pratto, Sidanius, Stallworth, & Malle, 1994). Individuals high in SDO prefer social hierarchies rather than egalitarian arrangements, and are attracted to positions of authority and superiority. A recent cross-cultural study of participants in 27 countries found that SDO was negatively correlated with environmental concern, suggesting that individuals high in SDO perceive and endorse human dominance over nature (Milfont, Richter, Sibley, Wilson, & Fischer, 2013; see Figure 7.7).

As noted earlier, personality traits are relatively stable and unchangeable. You might, therefore, be wondering why it's useful to study the associations between traits and environmental attitudes and behaviors. Why bother if personality is a variable that can't be budged? It's useful to study because behavioral geneticists claim that only about 50% of the variability in personality that we see across the human population is directly due to genetic inheritance; the rest is shaped by experience, particularly during childhood development. In fact, major influences such as parenting interact with genetic predispositions in such a way that experience ends up contributing more than genetics to expressed personality (Kruger, South, Johnson, & Iacono, 2008). Perhaps you have heard the expression "it takes a village to raise a child." It seems that the village can potentially do a lot toward raising prosocial, environmentally concerned children.

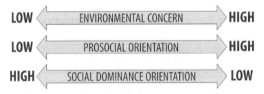

FIGURE 7.7 *Environmental concern is prosocial.*

Identity

As described earlier, *identity* refers to your sense of who you are. Identities have both personal and collective facets. For example, you may be a straight-A student (personal identity), as well as a member of a soccer team (social identity). Because personal and social identities are not mutually exclusive, one way to think of the distinction is that social identities are typically defined by social context (e.g., affiliation with particular groups) and are more subject to social influence (Clayton & Opotow, 2003).

Recall the student described at the beginning of this chapter who insisted that he was "not a campus liberal": Here is an example of a political social identity having a distinct influence over the student's environmentally relevant behavior. Importantly, his reaction revealed that he felt acting in a sustainable manner *threatened* his self-concept. When behavioral change threatens our identities, we are resistant to it (Murtagh, Gatersleben, & Uzzell, 2012). Can you think of other social identities besides political ones that might also contribute to unsustainable behavior? How

about a sense of oneself as a "fashionista" or a "high-tech gadget geek"? Both of these examples represent social identities grounded in material consumption.

Other social identities with implications for environmentally relevant behaviors are tied to occupation; consider the contrasting roles of "logger," "rancher," and "environmental lobbyist." Although all three are tied to nature, they perhaps seem to be incompatible, and even antagonistic. This is because social identities are sometimes associated with stereotypes that fuel prejudice between individuals who differ. For example, one study showed that ranchers and environmentalists in conflict over rangeland perceived each other quite negatively, each claiming the moral high ground, and dubbing the other group as hypocritical and undeserving (Opotow & Brook, 2003).

The identity "environmentalist" has received particular attention from researchers, as you may recall from Chapter 6. Not surprisingly, when people see themselves as environmentalists, they act in more environmentally friendly ways, participate in environmental groups, and believe the environmental movement is important (e.g., Dono, Webb, & Richardson, 2010; Dunlap & McCright, 2008b). Therefore, one potentially useful strategy for promoting sustainable behavior involves **identity campaigning**, focusing proenvironmental messages on "those aspects of a person's identity that either lead them to demand more ambitious change . . . or that underlie their motivation to engage in proenvironmental behavior" (Crompton & Kasser, 2009, p. 4). In other words, identity campaigning is about encouraging people to find their "inner environmentalist." This approach contrasts with environmental messages that support the status quo of identities based on money, image, and popularity, such as, "You'll be a hip trendsetter driving a hybrid car!" Some research suggests, however, that even environmentally concerned individuals may be reluctant to adopt the identity "environmentalist," presumably because of a disinclination to be affiliated with the negative stereotype that characterizes environmentalists as antisocial radical activists (Amel, Scott, Manning, & Stinson, 2007; Bashir, Lockwood, Chasteen, Nadolny, & Noyes, 2013; Scott, Amel, & Manning, 2011).

Whether individuals label themselves "environmentalists" or not, many people do have a sense of identity that includes a connection to the natural environment. In contrast to those who mistakenly perceive themselves as separate from, and independent of, their ecological context, some individuals experience a sense of common identification with the rest of nature that supports sustainable action. When people construe themselves as ecologically interdependent, rather than independent, they show more ecological concern, more cooperation over resources, and report more environmentally friendly behaviors (Arnocky, Stroink, & DeCicco, 2007).

In recent years, several researchers developed psychological instruments intended to assess the extent to which individuals experience a sense of self that is connected to, embedded within, or interdependent with, nature. Although these researchers' conceptual and operational definitions differ in some important ways, each of these measures is intended to tap into an individual difference that is theoretically related to environmental concern. Empirical support for this idea is growing. All of the measures listed in Table 7.1 are positively correlated with

TABLE 7.1 *Measures of nature-connected identity*

MEASURE	RESEARCHERS
Inclusion of Nature in Self (INS)	Schultz (2001)
Environmental Identity (EID)	Clayton (2003a)
Connectedness to Nature Scale (CNS)	Mayer & Frantz (2004)
Implicit Association with Nature (IAN)	Schultz, Shriver, Tabanico & Khazian (2004)
Connectivity to Nature (CN)	Dutcher, Finley, Luloff, & Johnson (2007)
Nature Inclusive Measure (NIM)	St. John & MacDonald (2007)
Nature Relatedness (NR)	Nisbet, Zelenski, & Murphy (2009)
Disposition to Connect w/ Nature (DCN)	Brügger, Kaiser, & Roczen (2011)

proenvironmental attitudes and behaviors. Recent debates center on the relative strengths, weaknesses, and potential contributions of each of these conceptualizations and tools (e.g., Brügger, Kaiser, & Roczen, 2011; Perkins, 2010; Perrin & Benassi, 2009; Scott, et al., 2014), but for now, there is no standard way to describe or measure a nature connected identity.

Some researchers focus on what could perhaps be called nature identity "subtypes," such as "gardener" (e.g., Kiesling & Manning, 2010) and "organic consumer" (Bartels & Onwezen, 2014). Specific environmental identities predict specific behaviors. For example, participants who identify as "energy savers" show greater intention to purchase carbon offsets (Whitmarsh & O'Neill, 2010) and willingness to pay more for sustainably produced energy and products (van der Werff, Steg, & Keizer, 2013).

So, how does one come to have a nature-connected identity? Research reveals that many adults who are actively engaged in proenvironmental efforts, personally and professionally, report having had **significant life experiences** in nature as children and adolescents, often accompanied by adult role models and mentors (e.g., Chawla, 2012; Tanner, 1980). Two important limitations of this body of research are worth mentioning. First, very few of the studies include comparison groups, so they do not rule out the possibility that one may have positive early experiences in nature and *not* develop a proenvironmental orientation. Second, nearly all of the studies rely on participants' *memories* of nature experiences, rather than following participants through the years to see what they experience and how they turn out. Of course, memory is not unbiased, so it may be the case that adult participants inadvertently distort their memories of childhood to be consistent with their current feelings about nature (Chawla, 2012). Still, it makes theoretical sense that repeated, positive, immersive experiences in nature during one's upbringing may foster "the development of an environmental identity, which includes a feeling of connection to nature and may include care for the environment as part of a person's goals and self-definition" (Chawla, 2012, p. 549).

Importantly, it may be the case that whether nature experiences lead to a feeling of affiliation and identification with nature depends upon the *context* of the experiences (Chawla, 2006; Scott, et al., 2014). Some of the ways people interact with nonhuman nature have a decidedly adversarial feel with the goal of *conquering* nature; it seems that these types of activities may be unlikely to promote empathic identification.

In one study, participants were surveyed about their past experiences with a variety of nature activities including basic survival skills (such as foraging, hunting, shelter building, and fire starting) and were also measured with the EID and CNS. As would be expected from the significant life experiences literature, most of the highly experienced participants also had high identity scores; however, there were some participants who displayed a mismatch. That is, they were highly experienced in nature but had low identity scores, or they had high identity scores, but little experience in nature. Interestingly, the majority of the highly experienced individuals who scored low on the identity measures were men, while the majority of the individuals who lacked experience but scored high on the identity measures were women (Scott, Manning, & Amel, 2008). One possible explanation is that in the culture where these individuals grew up (the United States), there is more opportunity, and more encouragement, for boys and men than for girls and women to *challenge* nature, as opposed to relate to it. For example, it is more likely to be a *Boy* Scout patch than a *Girl* Scout patch that is earned by building a friction fire, and the popular television show is not "*Woman* vs. Wild." A difference in how nature is likely to be experienced may be one explanation for gender differences in environmental concern.

Gender

Women tend to show greater environmental concern than men despite the fact that, as described earlier in the chapter, men tend to score better than women on surveys of environmental knowledge (Eisler, Eisler, & Yoshida, 2003; Zelezny, Chua & Aldrich, 2000). Women usually score higher on the New Environmental Paradigm (NEP) scale and are particularly concerned about hazards that pose risks to the local community or health of their families (e.g., Mohai, 1992; Stern, et al., 1993). Why is this the case? Is it because women have a special connection to nature, such as has been attributed to them in mythology, theology, and the writing of the transcendentalists? An alternative explanation is that women are more likely to see the connection between environmental conditions and harm to others because their traditional roles as family caretakers means they need to be more concerned about, and in touch with, the health of their families (e.g., Logsdon-Conradsen & Allred, 2010). They may be likely to develop environmental concerns because they are first to notice the damaging effects of polluted water, food, and air on their family's health.

Importantly, this gender difference in risk perception is not universal. Research in China and Egypt has found a *reverse* gender gap in environmental concern (Mostafa, 2007; Shield & Zeng, 2012; Xiao & Hong, 2010). And one study in the United States found that the gender gap was absent in communities sited within 50 miles of a nuclear power plant; in these potentially high-risk communities, both women and men showed similar risk tolerance, suggesting that both sexes had become desensitized to the chronic risk (Weiner, MacKinnon, & Greenberg, 2013).

More often, researchers in the United States have found what has been labeled **the white male effect**, a general tendency of white men to be less concerned about all types of risk than women and minority men (Flynn, Slovic, & Mertz, 1994). Numerous studies document this pattern with regard to perceived risk of environmental toxins such as industrial pollution in "Cancer Alley," a stretch of the Mississippi River in Louisiana (Marshall, 2004), and toxic debris left by Hurricane Katrina along the Gulf Coast of Louisiana and Mississippi (Campbell, Bevc, & Picou, 2013).

Researchers attribute the white male effect to the cultural cognition phenomenon described earlier in this chapter (Kahan, Braman, Gastil, Slovic, & Mertz, 2007; McCright & Dunlap, 2013). That is, when activities integral to the social identity of "hierarchical and individualistic white males" (Kahan, et al, 2007, p. 1) are said to pose environmental risks, those individuals discredit and dismiss the risks so as to reduce the threat to their identities. One study extended this further to investigate whether *conservative* white males (CWMs) would be even less worried than other adults about ecological problems. Based on their analysis of nine national surveys conducted between 2001 and 2010, the researchers concluded that "the white male effect is due largely to CWMs, and [their] low level of concern with environmental risks is likely driven by their social commitment to prevent new environmental regulations and repeal existing ones" (McCright & Dunlap, 2013, p. 1).

Gender differences appear in environmental behaviors as well as concern, and sometimes these differences reflect a private versus public realm distinction. For example, women are more likely than men to recycle at home and make ecologically responsible dietary choices (e.g., Tobler, Visschers, & Siegrist, 2011), but men are more likely to attend political meetings about environmental issues (e.g., Shields & Zeng, 2012). That is not to say, however, that women are not involved in environmental political action. In Chapter 2, you read about Lois Gibbs, the woman in New York who led a grassroots campaign to get the polluted Love Canal subdivision in which she lived declared a Superfund site. Women have led similar grassroots efforts throughout the world. For example, in India, women hugged trees to prevent logging, giving rise to the derogatory term "tree-hugger" (Mishra & Tripathi, 1978; Shiva, 1994, 1988); in Kenya, women planted trees throughout the Green Belt (Maathai, 2003); and in the Philippines, pregnant women, nursing mothers, old women, and children formed a human barricade to prevent tractors from clearing land that was used for community food (Ayupan & Oliveros, 1994).

CONCLUSION

Based on the material in this chapter, you might conclude that the following person would be most likely to behave in proenvironmental ways: A well-educated, ecologically literate, and conscientious woman, who had emotionally meaningful experiences leading to empathy and self-transcendent concern for the natural world; who believes that environmental crises like climate change are real, and that she has the ability to do something about them; who is open to new experience, agreeable, and holds biospheric values. While those might all be relevant variables, it is important to keep in mind that *correlations* between individual differences and environmental concern or behaviors do not necessarily mean that the individual differences are the *reasons for* the concern or behavior. In other words, the relationships may be incidental rather than causal. This means, of course, that many people *unlike* the conscientious woman described above are also prone to behave sustainably.

Situational contingencies you learned about in Chapter 5, and thinking biases described in Chapter 6, constitute significant influences on our behavior. As sustainability initiatives become more mainstream, and as we challenge and reframe conventional ideas and paradigms, the influence of personality and related individual differences may be weakened by ever stronger situations. It is the *combination* of who we are, the way we think, and the physical and social systems within which we operate, that mixes to create behavior. With that in mind, the next chapter will address the complex blend that creates motivation to join the swelling tides of sustainable behavior.

CHAPTER 8

To Be (Green) or Not to Be (Green) . . . It's a Question of Motivation

- Motivation Grows from Within
  - Seeded by Basic Needs
  - Rooted in Core Values
  - Planted in Perceptions of Control
- Situations Can Nurture Motivation
  - Fairness Helps It Sprout
  - Goals Direct Its Growth
  - Feedback Is the Fertilizer
- Cultivating Change at Different Stages of Growth
- Conclusion

People are inclined to create simple explanations for behavior: *She didn't know any better; he wasn't paying attention; they were pressured by their peers; we can't afford it.* . . . We also tend to assume that what motivates our own behavior must surely motivate others'. Yet after reading the last few chapters, you understand that behavior stems from an extensive number of external and internal causes. How do all of these variables fit together? Theories can help organize complexity into meaningful patterns, and **motivation theories** specifically combine variables that predict when people are driven (or not) to behave (or not). These theories articulate the relationships between behavior and how people think and feel (Chapter 6), as well as "boundary conditions," such as the situation (Chapter 5) and each person's individuality (Chapter 7).

Motivation encompasses what we choose to do (directing effort), how intensely we do it (mobilizing energy), and how long we keep it up (maintaining effort) (Mitchell & Daniels, 2003). Motivation theories provide meaningful structure for building research projects and measurement tools, as you will soon see. Plus, despite their abstractness, these theories can help solve real-world problems. Just knowing which personal and situational characteristics are likely to provoke behavior, and in what combinations, can help us see them operating in the world

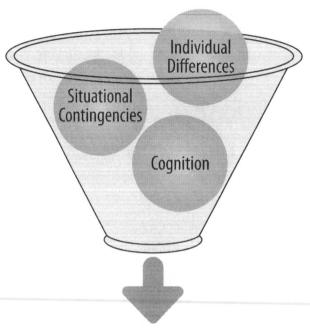

Environmental Behavior

around us. This kind of diagnosis is the first step for designing *informed* strategies for increasing sustainable behavior.

This chapter will focus on theories that explain how factors *combine* to inspire or affect motivation. The theories derive from a wide variety of psychological perspectives, including evolutionary, cognitive, social, and organizational psychology, so they each involve a slightly different set of concepts. Because these theories comprise a variety of internal and external variables, there is no perfect order in which to present them. To some degree, therefore, the organization of this chapter is arbitrary. You will first read about theories that feature internal characteristics as the core. Then, the chapter will describe theories that highlight motivational features of the situation. Finally, it will turn to theory describing how motivation varies depending upon where individuals are in the process of change. The review should help you see how motivation is multiply determined.

MOTIVATION GROWS FROM WITHIN

A good entry point into the realm of motivation is to consider one of the deepest internal sources of motivation: human needs. There is a reciprocal interplay between needs and the situation. Our evolutionary conditions molded these needs and our current circumstances determine whether these needs are fulfilled.

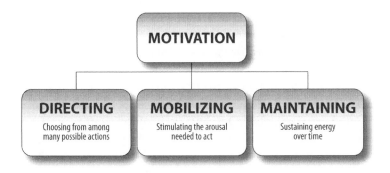

Seeded by Basic Needs

Several human **needs** are fundamental in driving our behavior, including basic physiological necessities (water, food) and psychological requirements such as social belongingness and **self-actualization**, which means to become one's best self (Maslow, 1943).

The scientific study of needs is challenging. What constitutes a "need" is subjective, and the literature reveals as many human needs as researchers studying them (hundreds, actually). Handling this many needs is unwieldy (a good theory should be as simple as possible) and makes it hard to predict with much accuracy which needs people will be experiencing at any given point in time. Despite these theoretical challenges, it is clear that *needs influence what people value and the behaviors they pursue*. Consequently, if people don't believe that their needs will be served by environmentally friendly behavior, attending to their own needs will likely trump any inclination to act sustainably.

In fact, it is these hard-wired needs (e.g., belonging, safety) that underlie many of the deeply embedded social and cognitive tendencies (e.g., seeking status, focusing on the present) that propel environmentally harmful behavior. Recently, psychologists have made substantial progress toward understanding the evolutionary origins of the needs that drive unsustainable activity (Griskevicius, Cantú, & van Vugt, 2012). As mentioned in previous chapters, certain behaviors actually reflect ancient survival strategies, such as hoarding and devouring food when it is available as a way to prepare for lean times. These strategies helped to keep our ancestors alive and enabled them to pass along these very tendencies to successive generations. Now that human living conditions have changed drastically and rapidly, the *behaviors that were once adaptive are now counterproductive*. Over time, our needs are likely to stay the same. It's how we choose to fulfill them that requires adjustment in order to live sustainably.

But what happens to motivation when people are asked (or forced) to change their behaviors? It may be a matter of whether people feel like these requests (or demands) are a crushing blow to their ego or, instead, people may feel that they clearly speak to core needs. **Self-determination theory** (SDT) posits three universal needs that must be fulfilled in order for people to function optimally (Deci & Ryan, 1985). People need **autonomy**, to feel they have choices and are free to decide what, when, and how to proceed, rather than being forced by rules, deadlines, or evaluation by others. We seek **relatedness**, a sense of being socially connected and accepted by others; this can take the form of wanting to fit in, following a desired role model, or participating in a broader, meaningful movement. **Competence** is the need to accomplish an action with grace and achieve the desired outcome, as opposed to feeling uninformed, or not having the requisite skills or abilities to achieve success.

SDT suggests that humans naturally strive to learn, grow, and be mentally healthy through fulfilling these needs, and that certain situations either support or impede this healthy development, resulting in different motivational states (Ryan & Deci, 2000; see Figure 8.2). SDT describes a spectrum of motivation ranging from its absence, **amotivation**, which occurs when the environment stifles need fulfillment, to **intrinsic motivation**, when the environment is supportive. Intrinsic motivation is our instinctive drive to engage *on our own terms* in situations that are novel, interesting, have value, and are challenging. Intrinsic motivation is the most natural and desirable means for achieving psychological health (e.g., joy, satisfaction), and it leads to sustained, highly engaged action. When people engage in proenvironmental behaviors of their own choosing, which take advantage of their specific skills and abilities, and provide a platform for engaging with others, they are

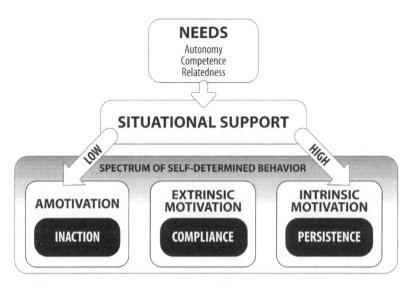

FIGURE 8.1 *Self-determination theory.*
If the situation supports fulfillment of a person's needs for autonomy, competence, and relatedness, then intrinsic motivation follows, which leads to high levels of action and personal satisfaction.

likely to flourish and the work is likely to get done. In contrast, amotivation generally results in no action whatsoever.

Between the extremes of amotivation and intrinsic motivation is **extrinsic motivation**. When people are extrinsically motivated, they conform to others' expectations so as to obtain incentives (e.g., money or praise), even if the behaviors are inconsistent with their own values. When the behaviors happen to be in harmony with the person's values, extrinsic motivation can look like intrinsic motivation. The difference is that, when extrinsically motivated, people act in line with their values *because of external rewards,* such as social approval. In contrast, when people are truly intrinsically motivated, they choose to act according to their values *no matter the situational consequences.*

Proponents of SDT believe that over time, people can transition, or self-regulate, away from the more extrinsic forms of motivation toward intrinsic motivation by taking "social values and extrinsic contingencies and progressively transform[ing] them into personal values and self-motivations" (Ryan & Deci, 2000, p. 69). Once that occurs, we act because we believe it's the *right thing* rather than because someone else tells us we *have to* or we *should*.

Empirical studies indicate that SDT is useful for understanding and predicting environmentally relevant behavior. In this cross-cultural body of research, levels of self-determination are typically measured using the Motivation Toward the Environment Scale (MTES; Pelletier, Tuson, Green-Demers, Noels, & Beaton, 1998; see Figure 8.2) and are correlated with a variety of proenvironmental behaviors. One Canadian research project assessed the links between perceptions of Canada's environmental policies, levels of self-determination, and general proenvironmental behavior (Lavergne, Sharp, Pelletier, & Holtby, 2010). Specifically, Canadian university students were surveyed about their government's style of regulating individual behavior (e.g., infrastructure for recycling, tax credits for transit passes, rebates on efficient appliances), their personal level of motivation (using the MTES), and self-reported proenvironmental behavior, measured by a set of 44 actions ranging from

---

The MTES provides 24 possible reasons people do things for the environment, including:

...I enjoy contributing to the environment (intrinsic motivation).

...For the recognition I get from others (extrinsic motivation).

...Don't know; I have the impression I'm wasting my time (amotivation)

People rate each of the 24 reasons for doing environmental things using a 7-point scale ranging from 1 (does not correspond at all) to 7 (corresponds exactly).

FIGURE 8.2 *Sample statements from the Motivation Toward the Environment Scale (MTES).*
Adapted from Pelletier, et al. (1998).

recycling paper to reading educational books about the environment. Students who perceived their government's policies to support individual autonomy (*the government gives me the freedom to make my own decisions in regards to the environment*) tended to report intrinsic motivation and more frequent proenvironmental behavior. In contrast, people feeling constrained by the government more often displayed signs of amotivation and were significantly less likely to report pursuing proenvironmental behaviors. A similar online survey in Australia correlated MTES scores with responses about specific household energy-saving behaviors such as *How often do you turn off the lights when not in the room?* (Webb, Soutar, Mazzarol, & Saldaris, 2013). As levels of self-determination increased, so did engagement in household energy-saving behaviors.

Self-determined, needs-based motivation seems to become especially important as behaviors get more difficult to execute (Green-Demers, Pelletier, & Ménard, 1997). When actions require very little time, energy, or resources (e.g., recycling), people perform them frequently regardless of whether motivation comes from within, or from external rewards. In contrast, people are unlikely to tackle difficult actions, such as educating oneself about environmental problems, unless they are intrinsically motivated (see Figure 8.3).

Because of the unique and high level of challenge associated with solving climate change and other contemporary problems, this finding means that facilitating intrinsic motivation, by supporting autonomy, relatedness, and competence needs, is likely to be crucial to the success of efforts to shift people's behavior. Of course, some sustainable actions are not likely to be inherently self-fulfilling, socially bonding, or confidence building for anyone. Hanging out laundry might be an example. How can motivation to do less-than-compelling but ecologically worthwhile tasks shift from extrinsic to intrinsic?

Some psychologists suggest that communicating choices in a way that respects people's autonomy, and encouraging people to identify those that are personally

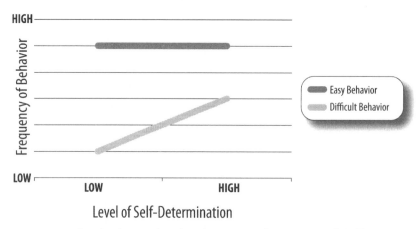

FIGURE 8.3 *Intrinsic motivation increases frequency of difficult behaviors.*

meaningful, helps motivation. For example, instead of telling people they *should* hang their clothes to dry, it is better to emphasize that they *could* do this; it is an option that they have. Providing a rationale for a boring activity (e.g., "you'll have the whitest whites if you hang them in the sun!") can create personal relevance, highlight its significance, and increase its intrinsic value as well. Also, posing questions instead of answers (e.g., "how much energy will you save if you ditch your dryer?") can pique curiosity and encourage further investigation (Moller, Ryan, & Deci, 2006).

Yet another idea is to use technology and social media to enhance intrinsic motivation. Competence is fostered by websites providing "how to" information for activities like composting, installing low-flow shower heads, and cooking local, seasonal produce. Relatedness is fulfilled by the sense of community that emerges in discussion forums and interest groups. Finally, people's sense of autonomy is enhanced when they are connected by way of the World Wide Web to an abundance of proenvironmental opportunities from which they can choose (Webb, et al., 2013).

In sum, needs drive many of our choices, energize action, and inspire sustained effort until they are satisfied. Needs are so deeply embedded in our nature that, most of the time, we don't even know they are operating, but they are influencing us nonetheless. Comparably, values also shape the relative importance of activities and outcomes, and, like needs, values are pretty resistant to change. Unlike evolutionarily based needs, however, values are guiding principles that each person develops over a lifetime of experience.

Rooted in Core Values

Many motivational theories emphasize values, and for good reason. You might have noticed that it is difficult to coax people to reduce their ecological footprints if they don't think the environment is relevant and important. But valuing the environment does not guarantee proenvironmental behavior. Why is that? The reality is that values actually seem to work *indirectly* to influence behavior. In other words, values and behavior are connected through a long "causal chain," a process that translates one's general living guidelines into very specific actions. The **Values Beliefs Norms (VBN) theory** lays out the chain of psychological processes thought to precipitate proenvironmental behavior (Stern, 2000; Stern, Dietz, & Kalof, 1993). As suggested in its name, VBN theory begins with values as the most fundamental feature. Values influence beliefs, then beliefs activate personal norms, and it's this activation of personal norms that is most directly related to behavior (see Figure 8.4).

The VBN theory underscores the pivotal role of values by positing them as basic in two ways (Dietz, Fitzgerald, & Shwom, 2005). First, values are basic because they are the most *stable* determinants of environmental behavior, and thus the hardest to change. Second, values have the most *widespread* effects. Values influence the other

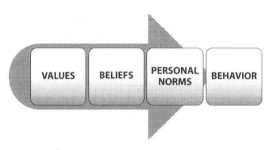

FIGURE 8.4 *VBN theory.*

Values influence beliefs, which in turn influence personal norms. Personal norms are most directly linked to behavior.

elements in the model, affecting what people believe, how they interpret the situation, and what they do about it.

Consider how the values described in Chapter 7 might affect responses to a planned urban development. If you hold biospheric values, you value the existence of other species for their own sake; as a result, you are more likely to believe you are responsible for their wellbeing and to feel morally compelled to act on their behalf. Thus, you will likely work toward preserving biodiversity by attending community planning meetings. Alternatively, people with egocentric values will likely get involved in the planning only if the development has the potential to affect their personal or business interests. With egocentric values, people are less likely to feel concern or personal responsibility for preserving biodiversity (i.e., *it's not a problem*, or *it's not* my *problem*).

Empirical research supports the notion that values underlie our beliefs about how the world operates and our role in it (Steg, Dreijerink, & Abrahamse, 2005). Altruistic and biospheric values both provide a focus beyond one's own needs and desires. These self-transcendent values compel people to think about how their own behavior impacts others, and to feel responsibility for those consequences (Schwartz, 1968). Beliefs about consequences and responsibility, in turn, influence the activation of personal norms and the feelings of having a moral obligation to do the right thing. In other words, beliefs operate as the *link* between values and personal norms (Steg, et al., 2005).

Recent research conducted at a U.S. national park illustrates how VBN captures the differences between people who act in proenvironmental ways and those who do not (van Riper & Kyle, 2014). At an island park in California, researchers distributed surveys that measured a range of characteristics, including:

- values based on the Value Inventory Scale (e.g., *Humans should live in harmony with nature*) (Schwartz, 1994);

- NEP worldview beliefs (e.g., *When humans interfere with nature it often produces disastrous consequences)* (Chapter 4) (Dunlap, Van Liere, Mertig, & Jones, 2000);

- norm-activating beliefs (e.g., *I am aware of the consequences of spreading non-native plants and animals,* and *I feel jointly responsible for the spread of non-native species*);

- personal norms (e.g., *I would feel guilty if I were responsible for the spread of non-native plants*); and

- self-reported proenvironmental behavior (e.g., minimizing the spread of invasive species by cleaning boating and camping equipment, and using boot-scraping stations).

As predicted by the VBN theory, people with self-transcendent values believed they were deeply connected to the environment, felt responsible for actions that impact it, would feel terrible about negative environmental impacts, and thus, were more likely to act proenvironmentally (van Riper & Kyle, 2014).

Because of the important, albeit indirect, impact of values on proenvironmental behavior, one would think that framing environmental messages in ways that resonate with people's value orientations might help accelerate environmentally responsible behavior change (Schultz & Zelezny, 2003). While this assumption remains largely untested, organizations such as the World Wildlife Fund have used VBN theory as grounds for designing values-based messages (Crompton, 2008; see Figure 8.5).[1]

FIGURE 8.5 *Appealing to altruistic values.*
In 2013, the Earth Day Network created a photo montage, called "The Face of Climate Change," depicting who and what will be affected by climate change using images similar to this one.

Although VBN identifies important processes that influence environmentally relevant behavior, it does not acknowledge the more conscious and rational dimensions of our behavior, nor does it recognize the situational constraints that often keep us from acting in concert with our values. People have the capacity to think in complex ways about their choices. When we do put in the effort to assess our odds of success, we sometimes come to the conclusion that a behavior is just not worth it (*Why waste my precious time and energy if it isn't likely to make a difference?*). When this happens, motivation falters because people generally will not risk failure simply to act in line with their values. A more complete understanding of environmental behavior requires venturing beyond the VBN's focus on values to address cognitive processes and contextual variables (Abrahamse & Steg, 2011; Aguilar-Luzón, García-Martínez, Calvo-Salguero, & Salinas, 2012; Chen & Tung, 2010). Two theories that do are the Theory of Planned Behavior and Expectancy Theory.

Planted in Perceptions of Control

Recall from the previous chapter that values inform our beliefs *and attitudes* (Schwartz, 1992). Many people do value the environment and, as a result, have positive attitudes toward environmental preservation and restoration. Behaving consistently with these attitudes is easy when the behaviors are **low cost** (i.e., quick, easy, and cheap). The attitude-behavior connection breaks down, though, when behaviors are deemed difficult, or **high cost**, by an individual (Diekmann & Preisendörfer, 1992). Energy conservation, for instance, entails many high cost actions, perceived to be expensive (purchasing solar panels or a more efficient vehicle), inconvenient (walking instead of driving), and potentially uncomfortable (e.g., adjusting room temperature to be cooler in the winter and warmer in the summer). When barriers like these interfere with our intentions, we drop our pie-in-the-sky

BEHAVIOR

FIGURE 8.6 *Theory of Planned Behavior.*
Attitudes, norms and perceptions of control combine to motivate behavior.

ideals and get practical. We start asking ourselves questions like: *Can I actually afford solar panels? Is it really that bad if I drive instead of walk?* Or *Just how cold can my room be in the winter before my friends no longer want to visit me?*

The **Theory of Planned Behavior** (TPB) positions the *intention to act* as the most direct cause of actual behavior, and posits that this intention depends on three psychological elements: attitudes, norms, and **perceived behavioral control** (Ajzen, 1991, 1998; see Figure 8.6). Having a sense of behavioral control means believing that a behavior will be easy, and that one has autonomy about whether to pursue it (see how these variables are typically measured in Figure 8.7). When a behavior is high-cost in time or effort, people perceive less behavioral control, and, consequently, are less likely to establish environmentally responsible intentions. For instance, perception about the availability of recycling facilities affects intentions to recycle (Chen & Tung, 2010; Knussen, Yule, MacKenzie, & Wells, 2004). While behavioral control is a perception, it is impacted by real **constraints**. For example, you can't easily modify room temperature if you don't have access to the thermostat.

For me, to drive to work would be	Good…	Bad
For me, to drive to work would overall be	Pleasant…	Unpleasant
Most people who are important to me would support my driving to work	Likely…	Unlikely
Most people who are important to me think I should drive to work	Likely…	Unlikely
For me, to drive to work would be	Easy…	Difficult
My freedom to drive to work is	High…	Low

FIGURE 8.7 *Theory of Planned Behavior measurement.*
Items are modified to assess perceptions about specific behaviors, in this case driving to work (adapted from Bamberg, Ajzen, & Schmidt, 2003).

Examining attitudes, norms and perceived control *in combination* illuminates why our behaviors often contradict our own attitudes and social norms: Even if one wants to be environmentally responsible, and it's socially acceptable to do so, it is not always *possible* due to low levels of real or perceived control. In other words, when the going gets tough, attitudes and values are not enough to ensure behavior (Stern, 2005).

Constraints include task difficulty and inconvenience. If a wetland conservation meeting is held 800 miles away or at the same time as your sister's wedding, you won't attend the meeting no matter what your attitude is about the issue. Conversely, if the meeting is being held next door just as you pass by and have time to drop in, you'll be more likely to attend. In either of these extreme cases, the strength of your attitude won't be the main factor driving your behavior. The structure of the situation is just too strong.

Research supports the impact of constraints. For instance, regardless of attitudes or awareness about the implications of driving, people who live in areas without mass transport are more likely to drive (Tanner, 1999). Household energy use depends more on income and household size than attitudes or values (Abrahamse & Steg, 2011; Gatersleben, Steg, & Vlek, 2002). In these and similar cases, reducing barriers to change will be more effective than trying to alter people's moral choices. Several methods for promoting proenvironmental behavior use this approach of identifying specific barriers that prevent action and designing interventions to minimize those specific impediments (e.g., in Chapter 11, you will read about Community-Based Social Marketing; McKenzie-Mohr & Smith, 1999).

TPB does a good job of predicting intentions and behaviors in a variety of environmental domains, such as use of water-saving technologies (Lynne, Casey, Hodges, & Rahmani, 1995); household recycling (Boldero, 1995; Taylor & Todd, 1995); home energy use (Abrahamse & Steg, 2011); participation in political action (Cordano, Welcomer, Scherer, Pradenas, & Parada, 2010); choice of travel (Bamberg, Ajzen, & Schmidt, 2003; Bamberg & Schmidt, 2001); and ecological behavior in general (Kaiser, Wölfing, & Fuhrer, 1999). It also predicts employee intentions, such as switching off computers every time they leave their desks for an hour or more, using video-conferencing for meetings that would otherwise require travel, and recycling as much waste as possible (Greaves, Zibarras, & Stride, 2013).

Theories are modified over time as they are tested by scientists, and TPB is no exception. Based on empirical evidence, scholars have added components that enhance its ability to predict intentions. Notably, a good predictor of future behavior is past behavior. For instance, knowing people's previous recycling behaviors and habits improves estimates of future recycling intentions (Knussen, Yule, MacKenzie, & Wells, 2004). The power of the TPB to predict recycling behavior is further strengthened by assessing attitudes about the typical nonrecycler (Mannetti, Pierro, & Livi, 2004). If people think negatively about a nonrecycler (e.g., lazy, immoral, clueless), they are personally more likely to recycle (Ohtomo & Hirose, 2007). Knowing whether people self-identify as environmental activists also enhances TPB predictions about proenvironmental behavior (Fielding, McDonald, & Louis, 2008).

Although people like to believe that they behave in line with their attitudes, the evidence indicates that proenvironmental behaviors are only moderately related to proenvironmental attitudes (Kaiser & Schultz, 2009). The Theory of Planned Behavior lends some insight into this gap by examining attitudes alongside situational variables that can inhibit action. TPB shows that we think our way into (or out of) some of our behavior for very practical reasons.

Expectancy Theory is another motivation theory that identifies perceived control as a critical feature of motivation, but defines it differently (Vroom, 1964). Control is a function of the belief that you have the *ability* to accomplish a particular action (recall self-efficacy from Chapter 7), and the belief that doing so will *lead to* specific outcomes deemed good, valuable, or otherwise important.

FIGURE 8.8 *Expectancy Theory.*
Perceptions of valence, instrumentality and expectancy all need to be positive in order to create motivational force.

In this theory, motivation is conceptualized as stemming from a combination of three perceptions: valence, instrumentality, and expectancy, which is why Expectancy Theory is often represented by the acronym **VIE** (see Figure 8.8). The extent to which an outcome is deemed "good, valuable, or otherwise important" represents **valence**. Valence is influenced by many variables including individual differences like values (i.e., is it important to me?) and attitudes (i.e., how do I feel about it?). It is also built on reinforcing consequences (e.g., do I want what I get from it?). For some people, reducing their carbon footprint is an important outcome (high valence), while for others, it is not nearly as important as having fun driving around in a cool, new car.

According to the VIE model, it is important for people to believe that their behavior is actually connected to getting a desired outcome. For instance, some people believe that driving less will reduce their carbon footprint. Others don't. The extent to which one perceives a link between action and outcome is known as **instrumentality**. Instrumentality is lower when the connections aren't obvious or when one has an external *locus of control* (Chapter 7). A person who lacks instrumentality may think, "My driving doesn't have any impact on climate change." Increasing a person's knowledge of ecological systems and how they operate can clarify connections and, therefore, increase instrumentality. Another factor that impacts instrumentality is a sense of *trust in the process*. A person who lacks this aspect of instrumentality may think, "If I take my car off the road, someone else will just start driving and diminish the impact of what I do. Nothing I do will actually help alleviate climate change."

The third component of VIE, **expectancy**, is the perceived relationship between effort and likelihood of successfully completing an action. A person considering bicycle commuting needs to believe that with enough effort, it *is possible* to get to work by bike. Expectancy is a function of personal abilities (e.g., fitness) and constraints (e.g., distance, condition of bike paths). High expectancy comes from confidence in one's abilities and a lack of external constraints.

Wanting to reduce your carbon footprint, believing that riding your bike to work will reduce your carbon footprint, *and* believing you can get where you need to go on your bike are all necessary for inspiring that bike ride. According to VIE, any of these three variables can be "deal breakers" for motivation. In other words, motivational force is diminished if the worth of any one perception is negligible. So, even if one sees the value in a particular result, if there doesn't seem to be any chance of completing the necessary actions, or there's no apparent connection between one's actions and the desired result, then motivation is zilch.

In sum, VIE suggests that we weigh the pros and cons of a behavior, estimate how likely it is our expectations will come true, and choose to behave if the scenario seems promising. The VIE theory not only helps us understand why a person demonstrates a proenvironmental behavior in a particular situation, it can help us design interventions to motivate behavior change by ensuring that valence, instrumentality, and expectancy are all positive (the more positive, the better). VIE Theory has yet to be applied broadly to environmental problems, but has the potential to make more precise predictions than TPB alone (Amel, Scott, Manning, & Forsman, 2009).

This alphabet soup of theories (SDT, VBN, TPB, and VIE, oh my!) highlights the role of internal factors in motivation. Yet, these theories acknowledge that the situation also plays an important part. The next section describes theories that focus primarily on how situations impact motivation. Although the situational variables are in the forefront, they are filtered by our brains to become the perceptions that drive our decisions to act.

SITUATIONS CAN NURTURE MOTIVATION

Many features of the situation can stimulate or undermine motivation. People are more likely to pursue sustainable behavior if they believe the situation is fair, when their actions are guided by smart goals, and when they have access to useful feedback.

Fairness Helps It Sprout

How often do you hear the words, "That's not fair!"? It doesn't feel good to encounter even a small injustice, and people go to great lengths to rectify perceived unfairness. Sustainability, by its very nature, provokes judgments about fairness because it involves decisions about how to allocate shared, limited natural resources. Perhaps you yourself have thought, "If other people aren't reducing their ecological footprint, why should I?" Whether something seems fair is a matter of perspective; two people can perceive the same situation very differently. Therefore, it is not enough for a situation to actually be fair; one must also consider how individuals are *interpreting* the situation.

For most of us, living sustainably will require changing our lifestyles and incurring costs in time and money. People can feel threatened by change because it entails taking risks, which makes them feel vulnerable. However, people are motivated to accept this discomfort and take the risks if they trust that the situation is fair (Saunders & Thornhill, 2003). But how do people determine what makes something fair? Here, the focus is on three types of fairness: how *outcomes* are distributed, the *process* by which decisions are made, and the *interpersonal treatment* one receives (Cohen-Charash & Specter, 2001).

Distributive justice refers to whether "outcomes" such as resources are allocated fairly among people. Whether the distribution of outcomes is considered fair depends on which distribution rule a person adheres to. And, when people interact using *different* assumptions about the appropriate distributive rule, you can bank on misunderstandings and arguments.

There are many widely accepted distribution rules, the most typical being equality, need, and equity. If following the rule of **equality**, everyone should receive the same benefit or pay the same cost. Everyone, for instance, should have the same access to, receive the same amount of, or pay the same sum for a resource such as water or energy. Others might suggest that people who **need** more of a resource should get more of that resource. This rule of need is commonly employed by opposing parties during debates about scarce resources, as seen in recent arguments in the U.S. Southwest, where farmers *need* water to grow high-demand crops, and people in the fast-growing cities nearby *need* water to run their households. The third common rule is **equity**. Using this rule, people who work harder, pay more, or do "better" should receive more benefit. For example, if a citizen uses her appliances during nonpeak hours (e.g., overnight while most people are sleeping), then she should be charged a cheaper rate than someone using the same amount of energy during peak hours (e.g., late afternoon when many people are returning home from work).

The idea of equity is so compelling that a large body of research exists examining its effects on behavior. **Equity Theory** posits that a sense of unfairness brings forth negative emotions and feelings of discomfort that lead us to seek change (Adams, 1965). Equity theory is based on two social psychology concepts introduced in Chapter 5:

- We regularly engage in *social comparison* to make sense of a situation (Festinger, 1954); and,

- When there is a discrepancy between what we expect and what we experience, we feel a type of *cognitive dissonance,* and this tension leads us to try to restore balance (Festinger, 1957).

People believe a situation is equitable if their contributions and benefits are *proportional* to someone else's (see Figure 8.9). People who put in the most time, effort, or money should get the most in return. If the comparison is out of balance,

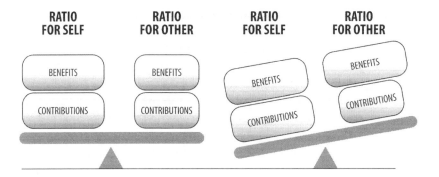

FIGURE 8.9 *Equity Theory.*

Scale on left represents an equitable situation, while scale on right represents inequity.

especially if someone else is getting "more than they deserve" or worse, a "free ride," people will purposefully act to alter the situation.

If people believe they are not getting as much benefit as they deserve, they can adjust the proportions by reducing their own contributions or by trying to change the other's behavior in some way. For instance, a web-based survey of Danish households found that men who feel they do more than their partners to save electricity will pressure their partners to increase energy-saving efforts (Thøgersen & Grønhøj, 2010). Sometimes people really cannot change their own or others' behavior, in which case they can resolve feelings of inequity by altering their *perception* of the situation. For instance, they can rationalize that they may be doing more to save energy, but that it's ok because they are probably in a better position financially or have more skills to contribute. When all else fails, it may feel better to compare oneself to others who are in a similar, inequitable position of putting in a lot of effort.

Reflect on how equity perceptions might motivate an individual's behavior. Say that a young man decides to lower his ecological footprint by staying home rather than flying to an exotic location for vacation. However, no one else seems to be holding back on long-distance travel plans; in fact, it seems his friends are all heading somewhere glamorous. Given that everyone shares the same planet, it seems rather unfair that others get to fly off to unusual places while he's stuck at home. In other words, he's putting in more effort than others to reduce greenhouse gasses, and yet the results of his effort will benefit everyone. It is likely that the young man will work to rectify his sense of injustice, either by persuading others to increase their contributions to sustainability (e.g. *C'mon guys, stay here and go camping with me!*), by decreasing his own contributions (e.g., *Forget the staycation, I'm booking myself an exotic cruise package!*), or by increasing his perception of his personal benefits (e.g., *By reducing my share of greenhouse gasses, I'm not only contributing to a healthier planet, I'm also saving money, and I have a clear conscience about my behavior. Besides, I'm still having fun!*). If none of those things work, he may just compare himself to some other people so it feels more balanced (e.g., *Hey, at least I don't have to be inside like my coworkers.*)

Sometimes *how* a decision is made can be more important to people than what they actually get from it. **Procedural Justice Theory** (Thibaut & Walker, 1975) suggests that people are more likely to accept an outcome, even one that is lopsided or disadvantageous, if they believe the process used to get to it was fair. Fair processes are those that are *unbiased, consistently applied* to everyone, and *correctable* if they prove unworkable. People are also more likely to think a process is fair if they are given an opportunity to voice their opinions or participate in the decision making. In contrast, when rules mysteriously appear after "closed door" meetings, it is easy to assume that they were built with dishonest intentions, and the affected people will likely reject the results.

Here's an example: During summer months in the arid U.S. West, many cities have instituted rules like "lawn watering is *not* allowed between 10 a.m. and 6 p.m." (Denver Water, 2014). If residents don't know how the rule was determined, who was involved in making the decision, and how rules will be implemented and enforced, they may not believe it is fair, and are likely to continue watering their lawns whenever they please. Whereas, even if residents don't like the restriction, if they feel like they were a part of the process that generated the rule, they're more likely to support it. For example, people might be willing to use less water or stop watering altogether if they were invited to speak up at community meetings or were able to vote on it.

A third, distinct form of justice focuses on how people think they've been treated. **Interactional justice**, being treated with dignity and respect, is especially important for building trust in ongoing relationships (Bies, 2001). An important feature of respectful communication is acknowledging the challenges and inconvenience that people will encounter. Providing transparent communication about *why* decisions are made shows respect and builds trust. Once it is broken, trust is hard to regain. Thus, decision makers (e.g., governmental representatives, employers) need to communicate openly and frequently with those who will be affected.

The U.S. Environmental Protection Agency (2009) defines *environmental justice* as the "fair treatment and meaningful involvement of all people regardless of race, color, national origin, or income with respect to the development, implementation, and enforcement of environmental laws, regulations, and policies." Do you see the presence of distributive, procedural, and interactional justice in this definition?

Extensive research about fairness perceptions in the workplace reveals that, when people believe they are treated fairly, they work harder, are less likely to defraud the organization, are more likely to help others, and feel more satisfied about their organization (Cohen-Charash & Specter, 2001). Do you think these same types of benefits are likely in the context of environmental behavior? Initial research attempts have revealed supportive results. For instance, justice perceptions predict peoples' willingness to pay for improvements in public goods such as bus service and landscaping common spaces (Ajzen, Rosenthal, & Brown, 2000). Also,

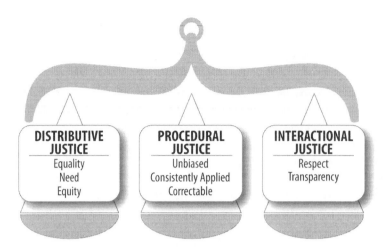

a survey of 616 Australian adults found more acceptance and support of carbon-related policies if participants perceived the policies to be fair (Dreyer & Walker, 2013). While evidence is growing, more empirical work is needed to understand the full impact of psychological justice on motivation to engage in environmental behavior (Clayton, 2000; Syme & Nancarrow, 2012).

Goals Direct Its Growth

Becoming sustainable often involves changing one's lifestyle or job, a large and potentially overwhelming prospect. Motivation can suffer as a result. Some of the most robust tools for helping people accomplish such challenging undertakings are derived from **Goal Setting Theory**. A goal is a desired end point or future state. Goals can be general, such as bringing down one's ecological footprint to be "one planet," or targeted, like becoming a vegetarian. Goal setting is the process of intentionally establishing objectives and then developing strategies that will lead to reaching those objectives. Goals are extremely effective since they support all three features of motivation: directing our attention, regulating how much energy to put forth, and sustaining momentum over long periods of time (Locke, 2000; Locke & Latham, 1990).

Goal setting works best as a motivator when people commit to specific, task-relevant goals that are relatively difficult but still realistically attainable. Having adequate information about progress and a clear time frame for finishing enhance people's ability to achieve their goals. These key features, that goals should be specific, measurable, attainable, relevant, and timebound, can be remembered using a handy acronym: S.M.A.R.T. (adapted from Meyer, 2003).

A key characteristic that makes goal setting effective is **specificity**, the ability to break projects down into small, concrete subgoals (Lock & Latham, 1990; 2002). For example, a specific goal when someone is trying to reduce carbon emissions

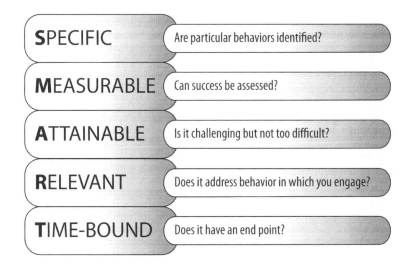

could be, "I will have one car-free day per week." Compare this to "I will do my best to have fewer driving days." Which do you think will lead to greater achievement? Putting a goal in specific behavioral terms provides clear direction, and allows people to create appropriate strategies for accomplishing it, such as biking to campus on Fridays.

Specific behavioral goals can easily be assessed, and **measurability** is critical for knowing when a goal has been reached. Measurement provides information about a project's starting and end points, as well as the progress made along the way. Information about incremental change provides a feedback loop that allows people to modify their behavior appropriately (e.g., knowing whether to speed up or slow down, drill deeper into one task or move on to a different task, fix mistakes, add new ideas, or change strategies) (Bandura & Cervone, 1983; Becker, 1978). So, someone who is aiming to reduce driving could record the amount of carbon emissions saved over time by bicycling one day per week. This would be a way of evaluating whether the goal is being met, or whether the behavior needs adjusting so as to better meet the goal.

It is important to consider what types of goals are motivational and which are paralyzing. To feel **attainable**, people need to have the skills and abilities to perform an action (Latham & Locke, 2006). For initiatives as novel as those needed for sustainability, people may believe they need training and support to gain appropriate levels of competence and feelings of self-efficacy (Amel, Manning, Forsman, & Scott, 2011). For example, a person unaccustomed to biking in urban areas may benefit from some coaching about the rules of the road, or some practice loading a bicycle onto the rack on the front of the city bus.

Although committing to an easy goal may increase performance slightly, *challenging* goals are more likely to substantially "move the needle" on behavior. But they can't be too challenging, because people avoid goals that they perceive as too difficult (Manning, Amel, Forsman, & Scott, 2009). When a goal seems attainable, and people see value in achieving it (think VIE), they will commit to pursuing it. Research shows that the highest levels of accomplishment occur when goals are challenging *and* people are committed to working toward them (Klein, Wesson, Hollenbeck, & Alge, 1999; see Figure 8.10). Bicycle commuting once a week is a good goal if, to the cyclist, it is in the sweet spot between too easy and too difficult, and feels like it serves the important purpose of reducing carbon emissions. The cyclist is even more likely to follow through when feeling determined to succeed, maybe even telling friends and family about it, "Honk when you see me biking on Fridays!"

There are exceptional circumstances when an extremely challenging "stretch goal" can facilitate achievement. For example, someone could aspire to bicycle commute *every day* for the whole school year. But unless a person has unlimited resources, a lot of control over circumstances, and the freedom to make radical decisions, stretch goals can end up being unreachable (Thompson, 1998). The ambitious cyclist's "every day" plan may be foiled by extreme weather, bike breakdowns, or obligations that don't allow the time required for a two-wheeled commute. Having a flexible schedule, owning extreme-weather biking gear, and possessing expertise about bike repair would allow a cyclist to achieve a "stretch goal."

In order for goals to enhance performance, they need to feel personally **relevant**. Relevance is high when goals address local issues rather than those that seem psychologically distant (Amel, Manning, & Scott, 2013; Chapter 6). For instance, when the bicyclist considers each ride to impact regional wildlife, the act seems more relevant than when thinking about impacts on polar bears and ice caps.

It may seem intuitive that a goal set by oneself will seem more relevant than a goal assigned by someone else. It is true that participating in setting one's goals

FIGURE 8.10 *High goal commitment and difficult goals together increase achievement.*
Adapted from Klein, et al., 1999.

can increase one's sense of procedural fairness, however, it does not generally affect the power of goals to motivate behavior. For instance, people with self-set goals demonstrate similar energy savings as those with assigned goals, with both groups using much less energy than those with no goals (McCalley & Midden, 2002). The bicyclist will likely be just as inspired by a campus-wide goal to decrease automobile commuting as by a self-set goal to reduce car use.

The most motivational goals are **time bound**. You probably know that deadlines can be helpful. They get people moving (especially the procrastinators among us), and keep people on task (Fulton, Ivanitskaya, Bastian, Erofeev, & Mendez, 2013; Herweg & Müller, 2011). However, people interpret deadlines differently. Some people find them to be onerous external controls, and others appreciate that deadlines assist them in focusing their attention and making plans. Self-determination theory would suggest that the latter outlook is psychologically healthier than the former because it preserves a sense of autonomy. The dark side of deadlines is that they can undermine people's motivation to take the time to try new skills, and may cause people to avoid challenging tasks that could potentially slow them down (Beck & Schmidt, 2013). For example, if the bicyclist feels pressured to get a regular commuting schedule in place *no later than the second week of classes*, testing a variety of routes may be out of the question.

The S.M.A.R.T. features of setting goals combine to direct, energize, and maintain action toward a desired end. One feature, measurability, is so critical for managing our actions that it deserves a more extensive look.

Feedback Is the Fertilizer

You might have noticed that it is easier to understand and reduce how much garbage you generate than the amount of electricity you use. This is because you can *see* garbage, but energy use is more of a mystery (recall humans' *visual dependency* described in Chapter 6). Its invisible nature makes energy use hard to manage, as is clear from the statistic that U.S. households use 20% to 30% more energy than they need (Ehrhardt-Martinez, Donnelly, & Laitner, 2010). If we could see how much energy is used when electronics are left plugged in or on "standby," we would be able to better understand the benefit of actions like unplugging them when not in use. Many environmental impacts are hard to see, and so we need additional information to guide our behavior.

Feedback is information about the effect of one's actions. Useful feedback informs people about their current status so that they can evaluate progress toward a goal. This helps people "self-regulate," by adjusting their effort, correcting errors, or even revising the goals themselves if necessary. Feedback is thought to create cause-and-effect connections in one's brain: If an action gets you closer to your goal, then you keep it up; or, in contrast, if an action does not work, then you will discontinue that approach and try something else.

Seeing positive results through feedback can be reinforcing. When people notice that feedback is valuable for making progress, they tend to seek it out more often (Tuckey, Brewer, & Williamson, 2002). Seeking (and finding) information that helps with task success stimulates people's feelings of control and autonomy, and levels of intrinsic motivation. There is evidence that its motivational effects can persist for weeks, even months, beyond when the feedback is provided (Houde, Todd, Sudarshan, Flora, & Armel, 2013; Sierio, Bakker, Dekker, & Van Den Burg, 1996).

Feedback can vary in terms of how frequently it is available and whether it is general or specific, immediate or delayed, positive or negative. These attributes all impact how well it works. For instance, in an effort to help their customers use less energy, some power companies provide information to consumers about their energy use via more frequent billing and enhanced energy usage information. Research supports that the type of detail provided in these energy bills impacts people's motivation to conserve (Ehrhardt-Martinez, et al., 2010). For instance, people respond well to appliance-specific information. They find it useful to know how much energy is expended for using their clothes dryer, and whether it is more than what's consumed by using their washing machine or running a refrigerator (Vine, Buys, & Morris, 2013). Knowing which appliances are energy hogs can help people direct their conservation efforts to make the most impact. Graphs that display historical comparisons of energy consumption are useful, too (Karjalainen, 2011). People like to see how much they've conserved compared to the previous year. People also prefer feedback that is presented in familiar units such as cost, or in ecological terms such as number of trees needed to offset carbon emissions, rather than in scientific units such as kilowatt hours (Jain, Taylor, & Culligan, 2013).

Technology can be used to deliver immediate feedback (Houde, et al., 2013). For instance, many newer model cars provide real-time information by displaying the current miles per gallon (mpg) as people drive. Research suggests that people reduce aggressive driving behavior such as speeding, sudden starts and stops, and idling, when feedback indicates these behaviors are linked directly to poor gas mileage. For instance, in a randomized field experiment, researchers installed fuel usage displays into the experimental group's personal vehicles and compared their average mpg to that of a control group (Brannan, 2011). Real-time information increased fuel economy overall by about 5%, and was especially effective for drivers with strong environmental leanings.

A critical feature of many new, energy-efficient buildings is a digital "dashboard" that displays real-time information about occupants' energy and water use (including number of toilet flushes!). The dashboards also display the impact of design features such solar panels. Websites and other gadgets facilitate energy conservation when they provide real-time information about energy use (Senbel, Ngo, & Blair, 2014). Access to real-time energy feedback reliably leads to substantially reduced household energy consumption (Hargreaves, Nye, & Burgess, 2010). In fact, a comprehensive review of the energy feedback literature demonstrates that people can max out on their energy saving at home when they can see how much energy each appliance is consuming in the moment (Ehrhardt-Martinez, et al., 2010).

Technological data that contains social feedback (Chapter 5) can further enhance motivation. Falling behind others who are in the same situation can motivate people to revise their own goals and change strategies. For instance, employees given information about their own usage in comparison to fellow employees were inspired to save more energy than those who merely saw their own results (Sierio, et al., 1996). Social feedback is not just effective at the individual level. For instance, U.S. college students cut their energy use in half when provided with real-time feedback at the dorm level (Petersen, Shunturov, Janda, Platt, & Weinberger, 2007). Research also shows that work group level feedback in organizations led to a 7% reduction in energy use compared to a control group receiving no feedback; without feedback, the control group actually *increased* energy use by 4% during the same time period (Carrico & Riemer, 2011).

Using negative feedback can be tricky because people react differently depending on various individual characteristics (Kluger & DeNisi, 1996). When researchers studied college students' reactions to negative information about their ecological footprints (e.g., "Your ecological footprint was larger than average"), it enhanced the motivation of some people but not others (Brook, 2011; Chapter 4). Negative feedback elevated the motivation of participants whose self-worth was linked to proenvironmental behaviors, but the same negative information was not effective for increasing motivation in participants whose self-worth was not linked to proenvironmental behaviors. This is likely because the latter group was not motivated to begin with, so the information was irrelevant to them.

Importantly, learning that one is not doing very well can be discouraging, especially to individuals who pride themselves on achievement. Take, for example, a comment from an employee reflecting on ecological footprint results: "As an academic, I want to get an A+ on everything. So how do I get an A+ on my footprint? I would score best if I were dead" (Hancock, 2002, p. 11). Disheartening feedback can cause people to give up.

In general, goal setting and feedback are a winning combination for motivating people to follow through on their behavioral intentions. One clever experiment from the Netherlands investigated the effect of combining specific goals and feedback by designing a computer-simulated control panel for a washing machine (McCalley, 2006). Local residents were recruited to test the convenience of its features, but were actually randomly assigned to conditions receiving energy feedback (or not) as well as setting specific conservation goals (or not). Responding to a variety of scenarios such as "wash a load of very soiled jeans," participants who both set conservation goals and received immediate energy feedback used the most conservative settings on the control panel (see also Abrahamse, Steg, Vlek, & Rothengatter, 2007).

The situation is a powerful force for or against motivation. However, even if the situation appears to promote sustainable behavior, not all individuals will be equally ready to jump on the bandwagon, as described in the next section.

CULTIVATING CHANGE AT DIFFERENT STAGES OF GROWTH

One reason why people behave differently when faced with the prospect of change, is that change is a *process* and each individual is starting from a different point. Consider roommates: One might not ever consider the environmental consequences of his or her behavior, leaving lights on everywhere even when away from home, while another might be obsessed with reducing his or her impact, using a smartphone to ensure that all appliances are turned off when not in use. Each roommate is likely to respond differently to information and opportunities to become more sustainable. **Stage theories** assume that there is a predictable, ordered series of steps that people go through in the process of changing their behavior. Motivation is enhanced by support tailored to the unique demands of each stage.

There are many different stage models with slightly different names, definitions and numbers of steps. Some models include a **precontemplation** phase in which someone doesn't even know or believe that change is necessary. However, recent

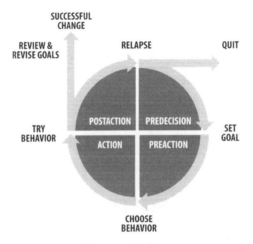

FIGURE 8.11   *Stages of change.*
Adapted from Bamberg (2013).

evidence supports four basic phases, beginning with a person considering the need for change (Bamberg, 2013). Each stage involves a critical task that facilitates transition to the next (see Figure 8.11):

- During the first, **predecision stage**, people consider the pros and cons of changing, and whether change is feasible. To move forward, a person has to decide specifically what change is needed, and set a goal to pursue. For example, one may decide to not use the car for short trips.

- The second, **preaction stage**, is when one is committed to change but has not yet acted. Here one must choose among possible behaviors to pursue, such as fixing up a bike or buying a bus pass.

- An **action stage** follows, in which one makes a plan for how to implement the change (e.g., I will watch a series of bicycle repair videos and purchase new bike parts online), and executes it.

- Once a person tries a new behavior, a **postaction stage** begins. The critical tasks here are monitoring and self-regulation. Assessing the effectiveness of the actions and modifying them accordingly will help keep the new behavior going (e.g., *Yeah! My bike works great but, if I'm going to continue, I have to find a faster, less dangerous route to the store*).

Achieving successive goals like these enhances self-efficacy, and stimulates further goal setting (Locke & Latham, 2002). For instance, once the cyclist is biking for short trips, longer commutes may feel achievable. However, along the way, relapse to former behavior is possible. Change is a rough ride because we often encounter barriers and temptations (*the weather has been really bad; my bike brakes aren't working correctly; some days I just don't feel like riding my bike*). Even when people are in the postaction stage, their motivation to stick with it can be threatened by habits rearing their ugly heads.

Based on this stage model, Bamberg (2013) designed a set of interventions to help residents successfully decrease driving in Berlin, and conducted a sophisticated field experiment to test whether it worked. First, based on their responses to a diagnostic question, participants were grouped according to stage. Within each stage, participants were randomly assigned to one of three conditions: a control group, which received no information, a group given typical informational pamphlets about automobile alternatives (e.g., bus schedule), or an experimental group that received stage-specific instructional support. The tailored support that the experimental groups received was as follows:

- At the *predecision* stage, a support team helped participants understand the impact of driving cars, as well as develop and commit to S.M.A.R.T goals for reducing their driving;

- For the *preaction* stage, researchers helped residents brainstorm and select from among alternative transportation plans;

- *Action* stage support included helping participants create a concrete implementation plan. This involved identifying potential barriers and strategies for overcoming them; and

- For those in the *postaction* stage, researchers provided feedback about participants' efforts, supportive praise for their accomplishments, and additional resources to help avoid relapse.

As predicted, people made more progress through the stages when they were coached using stage-specific strategies than when simply given informational pamphlets or no information at all. Additionally, car use dropped and public transportation increased significantly among the experimental groups. In contrast, the control groups failed to reduce their auto use appreciably. These results reinforce the notion that not everyone is ready for immediate action (Bamberg & Möser, 2007). There are some preliminary steps. Resources tailored to each stage help people translate intention into action (Abrahamse, et al., 2007).

CONCLUSION

Psychology can explain the process of motivation and predict when people will be motivated to act. Each person's motivation to behave stems from a complex mix of individual differences, cognitions, and situational constraints. In general, we all feel the strong pull to fulfill basic physical and psychological needs, and strive to live in sync with our values. People are motivated when we perceive that we have options, believe we are capable, and consider a challenge to be feasibly overcome. The external situation can entice or discourage our motivation to act sustainably. Motivation theories further account for individual variations in the specific values we each espouse, the attitudes and beliefs we develop through experience, and the different phases of understanding and readiness to change.

When we are motivated to redirect our choices, increase the energy we put forth, and maintain sustainable behavior for the long haul, the health of the planet benefits, and so does ours, as described in the next section of the book.

NOTE

1. Another clever example is from the "We can solve it" campaign http://www.youtube.com/watch?v=lTVxF8ILJaU

PART 3

What's Good for the Planet Is Good for Us

The goal for this section of the book is to make salient the reciprocal relationship between planetary and human wellbeing. Chapter 9 documents negative effects of industrialized lifestyles and polluted environments on mental and physical health, while Chapter 10 describes how connection with non-human nature plays an essential role in optimal human functioning.

Making Ourselves Sick
Health Costs of Unsustainable Living

- **Stressors in the Human Zoo**
  - **Sleep Deprivation**
  - **Overactivity and Inactivity**
  - **Malnutrition**
  - **Our Own Worst Enemy**
- **The Toxic Sea around Us**
  - **Detecting Effects**
  - **Disabilities and Disorders in Children**
  - **Reproductive Abnormalities in Adults**
- **Health Hazards of Climate Change**
- **Conclusion**

The modern human animal is no longer living in conditions natural for his species. Trapped not by a zoo collector, but by his own brainy brilliance, he has set himself up in a huge restless menagerie where he is in constant danger of cracking under the strain.

(Morris, 1969, vii)

Have you ever visited a zoo and witnessed a confined animal acting in a bizarre way that made you suspect it had gone a bit "stir crazy"? Maybe it was continuously pacing the same path in its enclosure, or engaging in a repetitive ritual such as rubbing its head against the wall. You do not have to be an expert to know that this type of behavior is not what you would witness if the animal was living in its natural, ecological context. Now, consider the assertion above, made by zoologist Desmond Morris. This quote concisely captures a theme of this chapter: *The industrialized environment is not humans' natural habit*. Indeed, as you read in Chapters 1 and 2, this unnatural habitat is a very recent creation in the timeline of human history, and scientists are only now beginning to understand its negative impact on ecological systems. So, too, are health experts now coming to recognize the repercussions it is having on the *psychological and physical functioning* of those of us who dwell within. Like the captive zoo animal with its self-soothing compulsions, we are mentally and physically stressed by our surroundings. Unlike the zoo animal, these surroundings are our own creation.

Certainly, humans have always experienced sources of stress, but numerous **stressors** encountered today were unknown to our ancestors. Ironically, many, if not most, of these modern stressors were created by people in pursuit of *improved* quality of life. Consider the example of the airplane, an innovation that allows people to travel long distances very quickly. How much stress does the average airline passenger experience while taking a flight from St. Louis to Portland, and how does this compare to the stress experienced by pioneers who endured weeks of hardship along the Oregon Trail in the 1840s? The comparison seems almost silly, yet the advent of air travel ushered in its own set of stressors, including air pollution and noise. These contemporary stressors differ from a broken wagon wheel or uncooperative oxen in that they are *chronic*, rather than acute (immediate and temporary). They are aspects of modern *lifestyle* rather than things that come and go, and they are experienced as stressors because they are incompatible with how our bodies and minds naturally function best.

Like other species, humans **evolved** (and continue to evolve) **via natural selection** in response to environmental pressures (Darwin, 1859). In a given

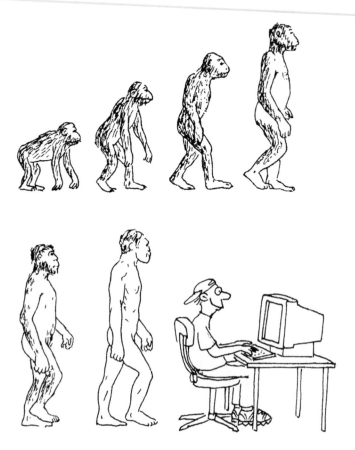

Something, somewhere, went terribly wrong.

environment, some characteristics help organisms survive and reproduce; thus, the genes for those traits are passed on to offspring in greater numbers than genes for less helpful characteristics. Over time, therefore, the predominant traits in a population change. Biologists concur that evolution by way of natural selection can take a *long* time, hundreds of thousands or even millions of years. There is evidence to suggest that humans may have experienced some more rapid spurts of change, such as during the 10,000 years since the development of agriculture (Zuk, 2013); however, the two centuries that separate the preindustrial era from today have not been a long enough period for any widespread evolutionary change in humans. The result is a mismatch between our preindustrial physiology and industrial lifestyle that is proving hazardous to our health (Gluckman & Hanson, 2008).

In this chapter, you will see that many ecologically harmful aspects of modern living (Chapter 1) are also detrimental to human development and functioning. What's bad for the planet is bad for us.

STRESSORS IN THE HUMAN ZOO

Stressors come in a variety of forms. They can be physical (heat, illness) or psychological (deadlines, insults). They can be internal (illness, anxiety) or external (noise, crowds). External stressors that are part of our surroundings are called **environmental stressors**. Stressors are not necessarily bad. Positive events can be stressors (graduation, a new job) and so can positive emotions (excitement, anticipation). Research with nonhuman animals suggests that a little bit of stress can even be good for the brain (e.g., Kirby, et al., 2013). The problems arise when stress is prolonged.

All stressors activate the **sympathetic nervous system**, a division of the autonomic nervous system, which regulates respiration, cardiovascular functions, and other life-support mechanisms. Sympathetic nervous system activity is often referred to as the **fight-or-flight response** because the body is mobilizing its energy stores to respond to the threat. Familiar signs of this response include a pounding heart, sweaty palms, and a flushed face. The autonomic nervous system works in conjunction with the **endocrine system**, a network of glands that release **hormones**, chemical messengers such as **adrenaline** that travel through the bloodstream and affect tissues and organs in various parts of the body. During the stress response, the release of **cortisol** increases glucose (sugar) levels in the bloodstream, providing energy to fight or flee.

In the moment, a burst of stress hormones is wonderfully adaptive. But, in the face of prolonged stressors, we end up with chronically elevated hormone levels that:

- raise blood sugar and risk for diabetes;
- increase blood pressure and the risk of associated illnesses such as heart disease and stroke;
- cause digestive problems (ulcers, irritable bowel syndrome, reflux);

- increase fat storage, especially in the belly;

- impair memory and concentration;

- contribute to anxiety and depression; and

- leave us vulnerable to infections and other diseases.

Clearly chronic stress is not good for our health (see Figure 9.1).

Take a moment to reflect on the chronic stressors endemic to the modern industrialized lifestyle. Think about the ways you feel stressed by your daily routines. Like many people today, you probably feel rushed, tired, hungry, and as if there aren't enough hours in the day. Chances are that the stressors making you feel this way are things that did not exist a century ago (and maybe not even a decade ago). Many aspects of modern living *neglect or disrupt basic biological needs, rhythms, or systems.* The following sections focus on how this lifestyle contributes to sleep deprivation, inactivity and overactivity, and malnutrition.

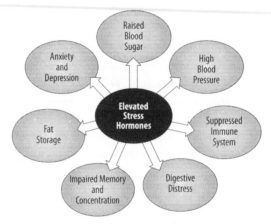

FIGURE 9.1 *How the body responds to chronic stress.*

Sleep Deprivation

Although many people push sleep way down the priority list, and even equate the need for sleep with weakness, the reality is that we all need sleep as vitally as we need water. Have you heard the expression, "I'll sleep when I'm dead"? As one sleep expert puts it, "If you fall asleep involuntarily (in class, in a car, at a movie), that is your brain taking over because you are not doing a good enough job keeping yourself alive" (Prichard, 2014, personal communication). In fact, people can survive weeks without food, but most of us can only make it a few days without any sleep. Many essential bodily functions occur while we are asleep, including cellular regeneration and repair, heightened immune system activity, and the consolidation of memories. Sleep may also help to protect against neurodegenerative disease (Xie, et al., 2013).

The consequences of sleep deprivation are serious and include impaired memory, learning, and attention; increased stress hormones; heightened aggression; lowered pain threshold; weight gain; increased risk for diseases such as cancer, diabetes, and heart disease; and shortened lifespan (Orzeł-Gryglewska, 2010; see Figure 9.2). These negative effects result from both occasional missed sleep (e.g., when you pull an all-nighter), and from the accumulation of a chronic sleep debt. A recent analysis of health data from more than 43,000 college students found that poor sleep was a strong predictor of significant academic problems, even after controlling for other factors such as depression and health issues; in fact, the negative effect of sleep problems on grade point average was about the same as the effect of binge drinking and marijuana use (Prichard & Hartmann, 2014).

Those of us living in the industrialized world are chronically sleep deprived (Lockley & Foster, 2012). You probably have heard that humans ideally need eight hours of rest per night, and chances are you fall short sometimes . . . or frequently. It may surprise you that eight hours may actually underestimate, by about 20%, what is optimal for humans; historical records suggest ten hours in bed was the norm prior to the late nineteenth century, *before the advent of electric lights* (Lockley & Foster, 2012). No doubt our ancestors sometimes "burned the midnight oil," but they were literally burning oil, or tallow, or wood; these energy sources cast a yellowish firelight that does not interfere with the production of **melatonin**, a hormone that promotes sleep onset. Electric light, especially that emitted from computer, tablet, television, and smartphone screens, includes blue wavelengths that suppress melatonin and disturb normal sleep (e.g., Chellappa, et al., 2013). Even a small dose, such as what you receive when you check your phone in the middle of the night, can be disruptive.

FIGURE 9.2 *Consequences of sleep deprivation.*

Light is a primary **zeitgeber**, an external cue that synchronizes **circadian rhythms**, physiological processes such as sleeping and eating that roughly follow a 24-hour cycle. Disruption of circadian rhythms due to artificial light exposure at night is associated with major depression and other mood disorders (Bedrosian & Nelson, 2013). If you have ever gone camping in a wilderness area with no artificial light in the evening, you have probably experienced a refreshing resetting of your sleep-wake cycle, finding yourself yawning shortly after sunset and raring to go at dawn.

Artificial light outside the home is also a problem. **Light pollution** is the ambient illumination coming from streetlamps, headlights, airports, office buildings, shopping malls, and numerous other sources that cast light upward or sideways, brightening the night sky. As you might expect, light pollution is especially prevalent in urban areas, but light from rural areas also contributes. This is a postindustrial phenomenon:

> For most of human history, the phrase "light pollution" would have made no sense. Imagine walking toward London on a moonlit night around 1800, when it was Earth's most populous city. Nearly a million people lived there, making do, as they always had, with candles and rushlights and torches and lanterns. Only a few houses were lit by gas, and there would be no public gaslights in the streets or squares for another seven years. From a few miles away, you would have been as likely to *smell* London as to see its dim collective glow. Now most of humanity lives under intersecting domes of reflected, refracted light, of scattering rays from overlit cities and suburbs, from light-flooded highways and factories. Nearly all of nighttime Europe is a nebula of light, as is most of the United States and all of Japan.
>
> (Klinkenborg, 2008)

More than a dozen years ago, in the first and only official report on global light pollution, two-thirds of the U.S. population and over half of the people living in Europe were no longer able to see the Milky Way with the naked eye (Cinzano, Falchi, & Elvidge, 2001).

Besides keeping us awake at night (and ruining our view of the constellations), light pollution poses other significant health risks. Living in more light polluted areas is associated with higher rates of breast and prostate cancer, even when controlling for other variables; this relationship is probably due to disrupted circadian rhythms, the inhibition of melatonin, and increased stress-related hormones (reviewed in Haim & Portnov, 2013). And it is not just human health that is being compromised by the bright lights of the big city; light pollution has been shown to negatively impact feeding and breeding behaviors of insects, birds, fish, reptiles, amphibians, and nonhuman mammals. It even interferes with the ability of trees to adjust to the changing seasons (Rich & Longcore, 2005). In 2013, France introduced light pollution legislation requiring shops and offices to be dark during the middle of the night. The government cited energy conservation, reduced carbon emissions, and improved health of humans and other species as the reasons for making the change (Bogard, 2013; see Figure 9.3).

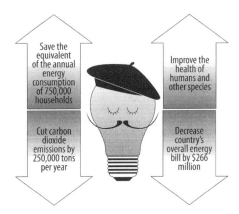

FIGURE 9.3   *Anticipated benefits of 2013 light pollution law in France.*

Artificial light is not the only thing disrupting our sleep; urban environments are also noisy. Jazz Age composer Cole Porter sang of "the roaring traffic's boom, the silence of my lonely room," yet the reality is that we can't shut the traffic noise out of our bedrooms. Nor can we completely stifle the reverberations of a passing jet, the rumble of nearby railroad tracks, the hum of the refrigerator, or the racket coming from the speakers of the musicophile living next door. Although to some extent, people habituate to constant or recurring sounds, both laboratory and field studies show that **industrial noise** causes sleep disturbance (Tassi, et al., 2013). In addition to interfering with sleep, noise itself is an environmental stressor; like other stressors, it produces sympathetic nervous system and hormonal activity associated with adverse effects on physical health and psychological functioning (Stansfield, Clark, & Crombie, 2012).

So here we are in our 24/7 culture: bathed in electric illumination and bombarded by background noise, partaking in round-the-clock screen entertainment, patronizing establishments that are open all night, traveling across time zones in a matter of hours on red-eye flights, and relying on coffee, sugar, and energy drinks to keep us on our feet. The infrastructure that supports this sleepless lifestyle is resource-intensive and polluting, and the lifestyle itself is wrecking our health.

Overactivity and Inactivity

One of the curious paradoxes in modern, industrialized society is that we are simultaneously the *most* active and the *least* active humans ever. We are the most active in that we are sleeping less and multitasking more; we are the least active in that so many of our hours are spent engaged in sedentary pursuits.

How often do you find yourself doing more than one thing at a time, such as texting while studying, or eating while driving, or checking Facebook while writing a paper, or . . .? Maybe the question should be how often you find yourself *not* doing more than one thing at a time! A recent observational study of college students studying alone found that participants distracted themselves an average of 35 times in a three-hour period, adding up to a total of 25 minutes off-task (Calderwood, Ackerman, & Conklin, 2014). This might sound pretty normal to you, but consider this: Except for listening to music, the electronic media with which these students were interrupting themselves did not even exist when we authors were in school, and some forms didn't exist as recently as a few years ago.

In Chapter 2, you read about surges in the tempo and complexity of daily life in the industrialized world during the 1920s and 1950s, attributable to wartime innovations in manufacturing and technology. Since the mid-twentieth century, however, the rate at which life has become faster and busier has increased exponentially, to the point where our technologies have outpaced our brains' ability to keep up (Klingberg, 2008). As you read in Chapter 6, our brains have limited attentional capacity. Though we may feel like we are able to do multiple things at the same time, the truth is that rapidly switching our attention back and forth from one thing to another usually results in doing all the tasks less well. For example, college students who engage in online chatting while taking lecture notes recall less from the lecture and take poorer notes (Wei, Wang, & Fass, 2014). And sometimes the consequences are more severe than suboptimal performance: Consider the well-documented, life-threatening hazard of texting while driving.

Perhaps you're thinking, "But I'm different; it's easy for me to do two (or more) things at once!" As with many cognitive limitations, people tend to overestimate their own abilities when it comes to dividing attention (Finley, Benjamin, & McCarley, 2014). There is some evidence that well-practiced multitaskers do not show deficits to the same degree as those who are less adept at shifting their attention (Alzahabi & Becker, 2013), but in general, the attentional demands of our fast-paced lifestyles are qualitatively (and perhaps quantitatively) different than what our brains have so far evolved to handle.

People readily become preoccupied with electronic technology because there are just so many things to watch, post, tweet, and blog about. The amount of information and entertainment available at one's fingertips today is staggering compared to what was accessible a mere decade ago. The wealth of available content paradoxically makes it more difficult to stay current because of information overload. Add to this the fact that electronic hardware and software are examples of planned obsolescence (Chapter 2), creating pressure to continually learn and use new technologies. It is no wonder that people end up overwhelmed trying to keep up with it all (Rosen, 2012). Back in the 1980s when home computers were first gaining popularity, the term **technostress** referred to an inability to effectively cope with the burgeoning computer culture (Brod, 1984); at that time, however, few households actually had computers and there were no mobile devices such as laptops, tablets, or smartphones (because, there was no Internet!)

Cornered
by Mike Baldwin

"That article you sent me about how technology causes a lot of stress crashed my computer."

Social networking, gaming, and other web-based delights can be so reinforcing that some individuals are displaying what could be considered "addictive" online behavior (e.g., Kardefelt-Winther, 2014). Although the latest version of the *Diagnostic and Statistical Manual for Psychiatric Disorders* (*DSM-5*; American Psychiatric Association, 2013) does not include Internet addiction among its official diagnostic categories, it does identify "Internet Gaming Disorder" as a condition warranting more clinical research to determine whether it should be a formal diagnosis (Ko, et al., 2014). Whether it is technically an addiction or not, many people suffer from a compulsion to constantly monitor their devices (e.g., Lee, Chang, Lin, & Cheng, 2014). An informal survey of our students revealed that nearly all of them *sleep with* their phones (and about half claimed they had texted in their sleep!).

Even very young children are engaging with digital technology to an unprecedented degree. While a daily dose of television remains a constant presence in most children's lives, what's rapidly increasing is their use of mobile devices. Two large, nationally-representative surveys in the United States found that between 2011 and 2013, the proportion of children with access to a mobile device at home rose from half to three-quarters, and the overall amount of time children spent using such devices tripled (Common Sense Media, 2013). In the 2013 sample, 72% of children under eight, and 38% of children *under two*, had used a mobile device for media. The American Academy of Pediatrics (2013) recommends no screen time of any kind for children under two.

Chapter 1 described the negative ecological impacts of electronic technology in the forms of energy consumption and e-waste. Clearly, the pervasive use of these devices is unhealthy for the planet, and evidence is mounting that it may also be detrimental to us. For example, a survey of more than 1,000 parents in California found that daily technology use by children, preteens, and teenagers predicted psychological distress, behavior problems, and attention difficulties, even when factoring out the effects of other variables, such as poor diet and inactivity (Rosen, et al., 2014).

Of course, while those of us in the industrialized world are spending so much time cyber-living, we are missing out on the real events unfolding around us, and the real people participating in them. It remains to be seen what all of this will mean for people's ability to form and maintain relationships, provide and seek social support, communicate effectively face-to-face, and do myriad other things that are fundamental to us as social creatures; but, right now, it doesn't look good

(Rosen, 2012). Preoccupied parents are mindlessly using iPhones to placate restless toddlers instead of playing with them. Friends hang out "together," each individually absorbed by a gadget. Couples sit side-by-side on the couch typing away on matching laptops. Students walk across campus texting, oblivious to the fact that they are about to collide with another pedestrian, a bicycle, or a car.

The irony in this multitasking, technology-driven society is that while our brains are working overtime, our bodies are on vacation. Google the phrase "sitting is the new smoking," and you will get scores of hits for headlines dating back to 2010 when health experts' warnings about the dire consequences of sedentary lifestyles started attracting international media attention. Obviously, sitting is not a new thing; even the ancient Egyptians had chairs (at least the royalty did), but these days, people in the industrialized world are spending the majority of their waking hours sitting—at desks, in cars, and on couches (Wilmot, et al., 2012). We sit more now than people did a century ago because of a decrease in walking as a primary mode of transportation, a shift in the workforce toward more inactive occupations, more labor-saving gadgets in the home, and an increase in stationary forms of recreation. Current trends reveal that sedentary pursuits among children and adolescents increase by about 30 minutes per day each year they grow older (Tanaka, Reilly, & Huang, 2014).

Sedentary lifestyles are associated with a variety of negative physical health outcomes including increased cancer risk (Schmid & Leitzmann, 2014), poorer cardiovascular health (Crichton & Alkerwi, 2014), sexual dysfunction (Cabral, et al., 2014), metabolic disorders and inflammation (León-Latre, et al., 2014), and even premature death (Patel, et al., 2010; Seguin, et al., 2014). One study followed over 17,000 Canadian men and women for more than a dozen years and found that those who sat for most of the day were 50% more likely to die from heart attacks than those who spent most of the day on their feet (Katzmarzyk, Church, Craig, & Bouchard, 2009). Another study of nearly 12,000 Australians estimated that for every hour adults sit watching television, their life is shortened by 22 minutes (Veerman, et al., 2011). Finally, a study of more than 120,000 adults in the United States found that longer time sitting predicted higher death rates from all causes, regardless of other factors such as weight, diet, and smoking (Patel, et al., 2010). Importantly, numerous studies suggest that sitting increases health risks *regardless of activity level when not sitting*. In other words, getting a daily workout at the gym does not buffer the negative effects of the seated lifestyle (e.g., Katzmarzyk, et al., 2009; León-Latre, et al., 2014; Patel, et al., 2010; Wilmot, et al., 2012).

It is becoming increasingly evident that to be healthy, our bodies should be moving throughout the day. This benefits not only physical health, but also mental health and cognitive functioning. Regular physical activity reduces stress and anxiety, improves mood, regulates hormones, and boosts intellectual performance (Ratey, 2013). An active lifestyle is a primary key to aging gracefully while preventing cognitive decline and dementia (e.g., Lövdén, Xu, & Wang, 2013; Zhao, Tranovich, & Wright, 2014). Some benefits accrue from intermittent spurts of activity, such

as the aforementioned gym workout, but to flourish, people need to incorporate more movement into everything we do; our biological heritage predisposes us to move frequently at a moderate intensity (Ratey, 2013). Therefore, to the extent that one is able, it is beneficial to take the stairs instead of the elevator, or a bike instead of the car. Ditch the leaf blower and pick up a rake! Using our bodies more helps us thrive and also *reduces the size of our ecological footprints*.

The modern lifestyle is proving hazardous for human health because it is both harried and sluggish. Health problems that were rare or nonexistent just a few generations ago (e.g., obesity and related ailments like diabetes) are now common and on the rise (Centers for Disease Control, 2014). Partly this is because people living in fast-paced, techno-centric societies are becoming increasingly overstimulated and unfit, but it is also due to another paradox of living in the developed world: We are simultaneously overfed and undernourished.

Malnutrition

The vast majority of the world's hungry live in developing countries in Asia and Africa where food insecurity and malnutrition are serious problems (FAO, 2013). Yet, it is not only the underfed that are undernourished. Even in the wealthy Western world, where there is an overabundance of available food, many people are not eating well. How often do you find yourself so pressed for time that your only options are a packaged snack from the convenience store, a burger and fries from a drive-thru, or nothing at all? Do you eat in your car, while walking across campus, or while working on your computer? Chapter 3 described the ecological benefits of the *slow food movement*, which promotes taking the time to mindfully and communally prepare and consume nutritious meals made from fresh, whole foods. In today's fast food culture, this approach may seem radical, but it actually hearkens

back to a time not so long ago when there was no other way to eat. In fact, eating-on-the-go first emerged with the rise of car culture in the mid-twentieth century (Schlosser, 2012). Around that same time, the first "TV dinners" appeared. Now, of course, there are all sorts of packaged goodies that make it easy to eat in front of all sorts of screens. It is probably no surprise to you that a recent study of Canadian children found that time spent watching television was negatively correlated with consumption of fruits and vegetables, and positively correlated with consumption of soft drinks, chips, fried foods, and pastries (Borghese, et al., 2014).

The **Standard American Diet (SAD)**, a term nutritionists use (along with "Western pattern diet") to describe what's currently typical in the industrialized world, is calorie-rich and nutrient-poor (Grotto & Zied, 2010). The ironic result is that some people in the developed world are suffering simultaneously from obesity and malnutrition, especially those living in poverty who lack access to high quality food (Centers for Disease Control and Prevention, 2012). Obesity among both adults and children has increased dramatically in the past 30 years and is becoming a significant problem; combined rates of overweight and obesity worldwide have risen 28% for adults and 47% for children since 1980 (Ng, et al., 2014).

Why are people getting fatter? One reason, described earlier, is that people are less active than they used to be. While activity has decreased, caloric intake has risen. Partly this is because commercial portion sizes have grown steadily since the 1960s (e.g., Young & Nestle, 2012), but partly it is because of how modern processed foods affect people's hunger motivation. In the 1970s, songwriter Larry Groce had a hit with *Junk Food Junkie*, a ditty that described his struggle trying to maintain a balance between his love for organic fare and his secret addiction to junk food: "In the daytime I'm Mr. Natural, just as healthy as I can be. But at night I'm a junk food junkie. Good Lord have pity on me!" (Groce, 1975). The song had audiences howling, but recent studies suggest that foods like potato chips actually affect the brain similarly to addictive drugs and alcohol. They activate reward centers that increase the secretion of **dopamine**, a neurochemical that gives us a "high" feeling. Just like with many drugs, people build up a tolerance so that the more chips we eat, the more we need to feel good, and the more likely we will experience withdrawal symptoms when we try to cut back (Andrews, 2013). On top of this, food that is nutritionally deficient makes people feel hungrier, compelling them to overeat (Fuhrman, Sarter, Glaser, & Acocella, 2010). Our biology makes it truly difficult to resist binging on junk food. This is obviously bad for our health, and it also stresses the environment by creating more demand for food produced and packaged via resource-intensive processes.

Even when people eat reasonable amounts, they still often end up getting fatter. This is because modern processed foods and beverages, such as those that fill the center aisles at the grocery store, are laden with *sugar* in various forms, including the pervasive **high fructose corn syrup**, one of the main products of industrial corn farming. Ingestion of sugar triggers the release of the hormone **insulin**. Overconsumption of sugar ultimately leads to **insulin resistance**, as the metabolic system is overtaxed. The result is **metabolic syndrome**, a condition that is a primary risk factor for diabetes and heart disease, and negatively affects cognitive functioning and brain structure (Yates, et al., 2012; see Figure 9.4) The United States Department of Agriculture (USDA) estimates that the typical U.S. resident consumes *90 pounds* of "added sugars" per year, and the U.S. Centers for Disease Control (CDC) estimates that about a quarter of U.S. residents suffer from metabolic syndrome (Taubes, 2011).

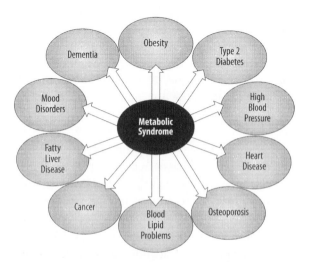

FIGURE 9.4 *Health problems associated with metabolic syndrome.*

"The 'Humungo Meal' comes with your choice
of 2 sides...heart disease, high blood
pressure, diabetes or obesity."

Most people are aware that junk food is a nutritionally deficient indulgence. What people may not realize, however, is that even many of the fresh, whole foods of today are nutritionally inferior to their counterparts from decades past. Fruits and vegetables grown in the 1950s contained significantly more **micronutrients** (vitamins and minerals) than they do today, because industrial agricultural practices strip nutrients from the soil and focus on maximizing yield rather than on nutritional quality (Sheer & Moss, 2011). A recent meta-analysis of 343 studies concluded that organic crops contain significantly higher levels of antioxidants than nonorganic crops (Barański, et al., 2014). A conventionally grown apple may look large and rosy in the supermarket, but chances are it is one of only a handful of varieties commonly produced, and is lacking micronutrients found in the 7,100 varieties grown in orchards across the United States in the nineteenth century, *6,800* of which are now extinct (Fowler, 2009). Sustainable farming involves routine replenishment of soil nutrients with compost, rotation of crops to allow for natural restoration, and propagation of heirloom seed varieties that help maintain genetic diversity in crops. These ecologically sound practices ultimately produce healthier food and healthier people.

Meat, too, is less nutritious when it is factory farmed than when it is pasture-raised or sourced in the wild. As described in Chapter 1, cows, pigs, and chickens in industrial operations are fed unnatural grain-based diets that not only contribute to their poor health but also to ours. For example, the fat found in meat and dairy from cows raised on feedlots lacks **healthy fatty acids** and **antioxidants** that are found in products from grass-fed cattle (e.g., Daley, Abbott, Doyle, Nader, & Larson, 2010). Meat produced sustainably (e.g., at Joel Salatin's Polyface Farms, described in Chapter 3), is better both environmentally and nutritionally.

The industrial food system is ecologically damaging and produces suboptimal nutrition. It is also wasteful, as described in Chapter 1. In more ways than one, nutrition and waste are inversely related. Convenience foods packaged in boxes, bags, plastic wrappers, cans, and jars are less nutritious than whole foods and contribute to the problem of solid waste. That much is probably obvious. What may be less apparent are the nutritional consequences of a wasteful approach to meat consumption. When you think of eating meat, what comes to mind? A steak, a burger, a drumstick, or a chicken breast? These are all examples of *muscle meats,* and they represent what typically appears on plates in much of the industrialized world. With some exceptions (e.g., France and Italy), modern Western cultures have moved away from the traditional practice of **eating nose to tail**; that is, making use of all parts of the animal including organs, bone marrow, blood, and, of course, heads and tails. This is detrimental to health as some of the most essential nutrients to be found in animals are located in these wasted parts (Price, 2009; see Figure 9.5). It is also ecologically irresponsible because, as you read in Chapter 1, meat is an especially resource-intensive food to produce. It is better for us and for the planet to make the very most of the meat we eat.

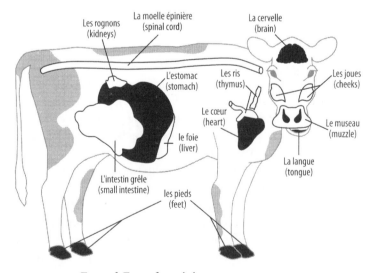

FIGURE 9.5   *Frugal French cuisine.*
French cooking traditionally features nutritious parts of animals that are wasted in many other Western cultures.

When you make dietary choices, do you think about **nutrient density**, how well the food will nourish you, or do you focus primarily on convenience, cost, calories, and flavor? Do you consider how various food additives, including sugars and preservatives, might affect your system? It is common for people to underestimate the importance of adequate nutrition for physical and mental functioning. Even doctors are guilty of overlooking diet when trying to diagnose why a patient feels badly, probably because conventional medical training involves little to no formal instruction in nutrition (Chen, 2010). Contrast this with veterinary medicine. What is the first thing the vet asks when you bring in your pet who seems unwell: "What have you been feeding him?" When it comes to animals other than ourselves, we seem to have no problem making the link between diet and wellbeing.

The implications of nutrition for human physical health have been studied for some time, but increasingly, researchers are making the connection to *mental and behavioral health*. For example, women who eat a standard Western diet of processed, fried, refined, and sugary foods score higher on a measure used to screen for psychiatric disorders as compared to women whose diet is based on whole foods such as vegetables, fruit, fish, and meat (Jacka, et al., 2010). Teenagers who eat more fast food exhibit more aggressive and delinquent behavior than their peers who eat leafy greens (Oddy, et al., 2009). Children with attention-deficit/hyperactivity disorder who are put on a sugar-restricted and additive/preservative-free diet experience a reduction in their symptoms (Millichap & Yee, 2012). Individuals who suffer from Type II diabetes, the kind that stems from metabolic syndrome, are more prone to depression, *and vice versa* (Fiore, et al., 2014). If current research trends continue, the vital importance of quality food for psychological functioning is only going to become more evident in years to come.

Our Own Worst Enemy

The infamous Ted Kaczynski (a.k.a. the Unabomber) declared, "[I] attribute the social and psychological problems of modern society to the fact that society requires people to live under conditions radically different from those under which the human race evolved" (quoted by Wright, 1995). Kaczynski's statement is consistent with the idea of evolutionary mismatch, but it also illustrates an unproductive way to think about modern living: It displaces personal responsibility for our lifestyles by invoking the abstract concept of "society." If we blame society or "the culture" for our unhealthy ways, we lose sight of the fact that these terms refer to the aggregate behavior of individuals.

Society isn't causing us to be stressed. We are stressed because we are making choices that support a stressful lifestyle, at home, in our neighborhoods, in our workplaces, when we are shopping, and in the voting booth. Perceiving modern stressors as outside of our control is essentially adopting an attitude of **learned helplessness**, which causes people to become depressed and unmotivated to try to change their circumstances (Evans & Stecker, 2004). People *do* have the power to influence or avoid the stressors described above. We can inspire systemic change (e.g., by supporting bike lanes and light restrictions in our communities) and can alter our own counterproductive patterns (e.g., by turning off the television and walking to the farmers' market).

It is important to acknowledge that people often engage in activities that stress bodies, brains, and ecosystems because there is some gain from doing so. Currently, there are personal, professional, and social reinforcers for many stress-inducing, ecologically unsustainable behaviors. Take, for example, the process of writing this book. We authors spent endless hours sitting in front of computers. Our sleep was disrupted by late night screen time and racing thoughts about content and dead-lines. Being aware of the environmental costs of the computers, electricity, paper, and other resources involved, we felt hypocritical. Yet, we were motivated to persist in spite of the stress on ourselves and the planet because of the reinforcing belief that the book may ultimately help change destructive patterns on a larger scale.

Those of us living in industrialized nations have become dependent on various technologies and routines that are difficult to give up despite their hazardous effects. Overhauling our stressful lifestyle may not be effortless, and it may not be quick, but it *is* possible. It may not be feasible to revert fully to preindustrial ways of doing things, nor would most people want to do so, but this does not mean we cannot turn back the clock in some ways that respect our biological needs. We can personally modify our sleep schedules, our activity levels, and our diets. We can work within our communities to transform infrastructure and systems so they will better support healthy living. Ultimately, changes in this direction will be more sustainable for people, physically and mentally, and for the planet as a whole.

However, some stressors in the modern world are less easily alleviated. Industrial living has significantly altered the natural environment in ways that make it less hospitable to humans and other species than it used to be. People have created cancer-causing chemicals that now permeate the planet. Emissions from burning fossil fuels have upset the carbon cycle, leading to shifts in weather and climate. No doubt some of the changes are permanent and will require us to cope and adapt as well as we can; still, the source of these changes is *pollution*, the generation of which is under people's control. We can choose to discontinue producing waste products that jeopardize our collective wellbeing. This chapter's main theme is *what's bad for the planet is bad for us*. Pollution is bad for the planet *because* it is bad for us, as well as all other species.

THE TOXIC SEA AROUND US

As you read in Chapter 2, Rachel Carson was the first to focus public attention on the link between pollutants and compromised health. In Carson's words,

> We have seen that [toxicants] contaminate soil, water, and food, that they have the power to make our streams fishless and our gardens and woodlands silent and birdless. Man, however much he may like to pretend the contrary, is part of nature.
> (Carson, 1962, p. 188)

A decade before she wrote *Silent Spring*, Carson published her award-winning *The Sea Around Us*, a poetic and highly readable treatment of oceanographic science. Were Carson still alive today (at well over 100 years old!), she likely would be dismayed to learn that environmental toxicants are more pervasive than ever, lurking in all living things including the oceans about which she wrote so lovingly.

To label a substance **toxic** means it is poisonous and poses a potential risk to health. Some substances toxic to humans and other animals include metals, such as lead and mercury; chemicals, both naturally occurring and synthetic; and radioactive waste. Historically, it was assumed that "the dose makes the poison"; in other words, the greater the exposure, the greater the effect. It is certainly true that severe poisoning can result in convulsions, unconsciousness, and death, while more moderate doses can produce symptoms such as hyperventilation, nausea, and vomiting as the body's defenses try to expel the poison from the body. Such dramatic responses are not typically triggered during lower level exposures, yet chronic exposure to small amounts may actually produce more insidious effects than larger exposures (Vandenberg, et al., 2012), especially when the developing brain is involved.

Detecting Effects

It is not easy to determine exactly how, and to what extent, environmental toxicants affect human health, alone or in combination with others (e.g., Bellinger, 2013). Most knowledge about specific impacts of toxic exposures is based on experiments using nonhuman animals. The research done with humans generally involves measuring concentrations of a substance in hair, blood, urine, or other tissues, and correlating those levels with various cognitive, behavioral, or physical health outcomes. As you know, *correlation does not necessarily imply causation*, and such studies can inspire even more questions. Several additional obstacles confront researchers trying to pin down the effects of potentially toxic substances on humans (reviewed in Koger, Schettler, & Weiss, 2005; Schettler, Stein, Reich, & Valenti, 2000):

1. There are no unexposed control groups with which to compare exposed populations, because environmental toxicants pervade the entire planet.

2. Humans are routinely exposed to a "toxic soup," rather than individual substances. Because experimental research is conducted on one or two compounds at a time, cumulative or interactive effects will typically go undetected by researchers.

3. The developing brain is particularly vulnerable during **critical periods**. Exposure to a substance at one point in time may appear to have no negative effects, even if the same exposure would prove toxic during a different, critical phase of development.

4. There can be a long **latency period** between exposure and when effects may be observable or measurable, making the connection difficult to detect.

5. The impact of toxicants is sometimes influenced by other factors. For example, toxic effects can be magnified when people have poor nutrition or fluoride in their drinking water. Chemical exposures also interact with genetic factors, as genes regulate the metabolism and excretion of substances. Thus, some individuals or populations may be more vulnerable than others.

> **WARNING:** This product contains chemicals known to the State of California to cause cancer and birth defects or other reproductive harm.

One thing's for certain: We can't know how substances affect us if we don't study them; yet, there are *tens of thousands* of unstudied substances being manufactured and dispersed in the United States alone. Regulatory efforts to reduce toxic exposures in the United States are reactionary rather than preventive; that is, chemicals are considered "innocent until proven guilty." There are more than 84,000 **synthetic chemicals** currently inventoried by the U.S. Environmental Protection Agency (EPA, 2014d). A quarter of these have been introduced since 1976, the year the government passed the Toxic Substances Control Act (TSCA), a law that requires manufacturers of chemicals to register them. Under the law, synthetic chemicals must be reported to the EPA, but they need not be tested for safety before they are put to use. In fact, the EPA has required testing of only about 200 of the chemicals on the list, and only nine of those have been subject to any regulation for posing an "unreasonable risk" to human health (Chameides, 2011; see Figure 9.6). Among the

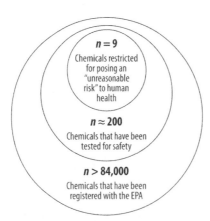

FIGURE 9.6   *Knowledge about chemical safety.*

Only a small fraction of more than 84,000 synthetic chemicals registered with the U.S. Environmental Protection Agency have been specifically tested for toxicity to humans Adapted from Grandjean & Landrigan, 2006.

regulated compounds are asbestos, chlorofluorocarbons (CFCs), polychlorinated biphenyls (PCBs), dioxins, and lead. Food, pharmaceuticals, pesticides, and cosmetics are excluded from the TSCA.

Do you feel alarmed by this information? How concerned should we be? The lack of testing on the majority of industrial and consumer chemicals leads to the unsatisfying conclusion that *we really don't know*. Although data may be sparse regarding the specific toxicity of many substances, there is abundant and growing evidence concerning how known toxicants affect human health (e.g., Grandjean & Landrigan, 2014), which is why legislators in the European Union adopted the Precautionary Principle (described in Chapter 4). Exposures cause several types of problems including (but not limited to) developmental disabilities, reproductive abnormalities, immune system suppression, neurodegenerative diseases, and cancer. The mechanisms by which they exert their effects vary:

- **Neurotoxins** destroy or damage neurons (nerve cells in the brain). Some substances disrupt chemical communication between neurons.

- **Mutagens** cause genetic mutations that can damage chromosomes, impair cell replication, and lead to cell death.

- **Carcinogens** are cancer-causing mutagens.

- **Teratogens** are toxicants that interfere with prenatal development, causing birth defects.

- **Endocrine disrupters** can mimic or exaggerate the effects of hormones, block receptors, preventing natural hormones from doing their job, or otherwise alter hormonal activity. Because hormones are involved in the function of virtually all bodily systems, endocrine disrupters have widespread effects.

Recently, investigators are describing **epigenetic** effects; that is, toxicants can alter gene expression and function (e.g., by switching genes on or off). Epigenetic effects can produce disease or disability much later in life, and even impact future generations (Perera & Herbstman, 2011; Weinhold, 2012).

Some of the most worrisome substances are those which act as **persistent bioaccumulative and toxic pollutants** (PBTs). They are *persistent*, meaning they can remain in the environment for long periods of time (years or decades) without breaking down or losing their potency, and they *bioaccumulate*, meaning that they become concentrated in the fatty tissues of living things. (Recall from Chapter 3 that the process of *biomagnification* occurs when bioaccumulated substances work their way up the food chain; i.e., the big fish eat the little fish and we eat the big fish.) Examples of common PBTs are mercury, certain pesticides, and the flame retardants used to treat mattresses, pajamas, upholstery, electronic appliances, and other consumer goods (Costa & Giordano, 2007).

Many PBTs act as endocrine disruptors (e.g., Gaspari, et al., 2012; Rana, 2014). Much of the recent focus has been on **phthalates** (pronounced THAL-ates) and **bisphenol A** (BPA), chemicals that are commonly found in plastic bottles, food wrappers and microwaveable trays, cosmetics, hairsprays, lotions, shampoos, and a variety of other everyday products (Table 9.1). In recent years, some states and countries have put limitations on the production and sale of products containing BPA; these days, it is common to see a label proudly declaring "BPA-Free!" Similar restrictions have been imposed on phthalates in some locations. However, neither of these is completely off the market yet, and it is unclear whether BPA-Free products are truly free of problematic ingredients.

Two other classes of compounds that act as persistent and bioaccumulative endocrine disruptors are among the few substances regulated by the U.S. EPA: **PCBs** (polychlorinated biphenyls) and **dioxins** (Bergman, Heindel, Jobling, Kidd, & Zoeller, 2013). PCBs are industrial chemicals that were used in lubricants, dyes, and electrical insulation from the 1920s to the 1970s. Although they were banned from production years ago, they continue to contaminate meats and fish, as well as other fatty foods (Colborn, 2004). Dioxins are an unintentional by-product of waste incineration, chlorine-based pulp and paper bleaching, and other industrial processes. They also derive from **triclosan**, a chemical that has been added to antibacterial soaps, deodorants, and other products since the late 1980s; it turns out that triclosan produces four novel dioxins when it is exposed to sunlight, such as when soap rinsed down the drain makes its way into a river (University of Minnesota, 2010).

TABLE 9.1 *Endocrine disruptors: BPA and phthalates.*

BISPHENOL A (ESTROGENIC)	PHTHALATES (ANTI-ANDROGENIC)
Baby bottles	Perfumes
Aluminum can linings	Hairsprays
Nail polish	Nail polish
Polycarbonate water bottles	Soaps, shampoos
Microwave ovenware	Skin moisturizers
Flame retardants	Plastic wrap
PVC stabilizers	Pesticides ("inert" ingredients)
Artificial teeth & tooth sealant	Detergents
Adhesives	Adhesives
Enamels & varnishes	Medical tubing
Returnable containers	Food packaging

From Weiss, B. (2007), p. 945. Permission to reprint from Bernard Weiss, personal communication.

Disabilities and Disorders in Children

Prenatal and childhood exposures to many environmental toxicants are associated with a variety of problems including impairments in learning, cognition, attention, emotion, and behavior. It is estimated that toxicants represent a larger risk to overall population intelligence (in terms of IQ) than other factors (Bellinger, 2012). Approximately 15% of U.S. children are affected by one or more developmental disabilities (Boyle, et al., 2011) and the rates are on the rise. For example, from the late 1990s to 2008, there was a 33% increase in attention-deficit/hyperactivity disorder and a *290%* increase in autism spectrum disorders (CDC, 2011). Although many variables are likely contributing to the rising prevalence, including lifestyle factors and changes in how the disorders are diagnosed, toxic exposures are an important and potentially preventable risk factor (Grandjean & Landrigan, 2014). Children are more susceptible than adults to toxic effects because their nervous systems and other organs are still developing (Giordano & Costa, 2012). Prenatal and postnatal neurodevelopment occurs in a precise sequence of anatomical and chemical events; thus, any chemical disruptions can cause dramatic impacts (e.g., Barouki, et al., 2012).

Exposures in children occur in ordinary ways (e.g., Weiss, 2000). Nursing infants consume toxicants that bioaccumulate in breast milk (still, experts agree that "the breast is best" nutritionally). Crawling infants and toddlers spend considerable time on floors, stirring up and breathing contaminated dust and residues. Many of the plastic bottles, sippy cups, and toys that children put in their mouths contain endocrine disruptors (Blake, 2014). Meat, eggs, and dairy products are often contaminated, as are popular (and nutritional) fruits and vegetables (see Figure 9.7). Accidental exposures to household cleaners and lawn chemicals occur because products are not stored properly outside the reach of curious children, or they are used in and around where kids live, play, and go to school.

Much of what is known about the effects of toxicants on children comes from decades of intensive study of a few substances including heavy metals, such as lead and mercury, and some pesticides. An overview of this information follows.

The harmful effects of exposure to **lead** are generally well-known due to intensive public awareness campaigns and U.S. governmental interventions in the 1970s to remove lead from gasoline and paint. Yet despite this legislation, exposures continue, particularly in urban communities where it persists in the dust and paint of older houses (Whitehead, et al., 2014), is found in contaminated soil (Morrison, et al., 2013), and shows up in school drinking water that has traveled through lead pipes (Lambrinidou, Triantafyllidou, & Edwards, 2010). Even where lead is strictly regulated, such as in North America and the European Union, it is found in imported toys and other consumer goods (Guney & Zagury, 2012; Shader, 2012), and in unregulated domestic items such as lipstick (Blum, 2013); lead in these products means that a loving kiss from Mommy or an action figure from Daddy could be dangerous.

FIGURE 9.7 *Shopper's guide to pesticides in produce, 2014.*

These lists of *least* and *most* contaminated fruits and vegetables were developed by analysts at the nonprofit Environmental Working Group (EWG). A detailed description of the criteria used in developing the rankings, as well as a full list of produce that has been tested, is available at www.foodnews.org. Reprinted with permission.

Detrimental effects of lead are wide ranging, and include:

- cognitive deficits and developmental delays, impacting intelligence, language, learning, memory, planning, and other complex skills (e.g., Needleman, 2004);

- attention-deficit and hyperactivity disorders (e.g., Yolton, et al., 2014);

- autism spectrum disorders (Gorini, Muratori, & Morales, 2014); and

- antisocial and aggressive behavior, including criminality and violence (Masters, 2003; Needleman, et al., 2002; Nevin, 2007).

Lead exposure may also be a risk factor for the development of schizophrenia, a severe mental illness marked by delusions and hallucinations (Guilarte, Opler, & Pletnikov, 2012). In 2012, the U.S. Centers for Disease Control declared that there is no known safe level of lead exposure for children (CDC, 2012).

Like lead, **mercury** is a naturally occurring substance, but industrial activities have tripled the levels found in air, water, and soil (Rice, Walker, Wu, Gillette, &

For more
information on
preventing lead
poisoning call
1-800-424-LEAD
or visit
www.epa.gov/lead.

LEAD
Awareness
Program

U.S. Environmental
Protection Agency

Blough, 2014). Mercury is released into the environment by coal-fired power plants, waste incineration, medical and dental practices, and mercury-based thermometers that end up in landfills and leach into water supplies. As little as 1/70th of a teaspoon of mercury can disperse to contaminate a 25-acre lake (McGinn, 2002).

Mercury contamination is widespread; a recent survey of nearly 300 streams in the United States found mercury in every fish sampled (USGA, 2014). In fact, the consumption of contaminated fish is the primary way children (and adults) are exposed to mercury (Rice, et al., 2014). For several years, the U.S. government issued fish consumption guidelines, advising limited portions for children and pregnant women. In 2014, the Food and Drug Administration (FDA) revised these recommendations because of concerns that children and pregnant women were missing out on the significant nutritional benefits of eating fish, especially for brain development. The new recommendations encourage the consumption of *low-mercury* fish, but at the time of this writing, the FDA has not implemented any labeling requirements that would help people to identify which fish fit that description.

Mercury's toxic effects in the body are extensive, causing problems in the functioning of the nervous, cardiovascular, pulmonary, digestive, renal, reproductive, endocrine, and immune systems (Rice, et al., 2014). Like lead, mercury has been linked to intellectual deficits and learning disabilities (e.g., Bellinger & Adams, 2001; Myers & Davidson, 2000), attention-deficit disorders (Boucher, et al., 2012), psychotic symptoms (Rice, et al., 2014), and autism spectrum disorders (Gorini, et al., 2014; Leslie & Koger, 2011). The popular press has paid considerable attention to a potential link between autism and vaccines containing *thimerosal*, a preservative that includes a form of mercury; however, empirical evidence for this relationship is inconclusive (Mitkus, King, Walderhaug, & Forshee, 2014; Rossignol, Genuis, & Frye, 2014).

Speed Bump

Pesticides are toxic by design. They are intentionally produced and released into the environment to protect crops and control diseases carried by mosquitoes, rodents, and other nonhuman creatures. Whether they are meant to kill weeds (*herbicides*), bugs (*insecticides*), rats and mice (*rodenticides*), microbes (*disinfectants*), or any other so-called pest, the properties that render them toxic can harm humans and other unintended targets (e.g., pollinators such as bees and bats, as well as our family pets). Despite their potential risks, pesticide use, primarily of insecticides and herbicides, has increased since 1995, and nearly three fourths of all U.S. homes use some type of pesticide (Ashley, et al., 2006).

Based on controlled testing of nonhuman species, it is likely that several different classes of pesticides are hazardous to humans. For example, many insecticides disable the nervous systems of insects through processes that are directly relevant to human physiology (Weiss, Amler, & Amler, 2004). In rodents, even small exposures to certain pesticides during critical periods of neonatal development permanently affect brain neurotransmitter receptors (e.g., Schettler, et al., 2000). Parallels in brain development between rodents and humans suggest the potential for similar effects in children.

Research on children confirms a variety of toxic effects of pesticides. Case studies suggest prenatal exposure interferes significantly with neurological development, resulting in brain atrophy and malformation (Bellinger & Adams, 2001). Children exposed after birth exhibit reduced memory, creativity, and attention (Grandjean & Landrigan, 2006; Polańska, Jurewicz, & Hanke, 2013). In one study, Mexican children living where pesticide use was common showed marked impairments in the Draw a Person task, a nonverbal measure of cognitive ability; some children's drawings were hardly recognizable as human figures (Guillette, Meza, Aquilar, Soto, & Garcia, 1998; see Figure 9.8). Research on other populations of rural children shows that frequent contact with pesticides is correlated with deficits in balance, hand-eye coordination, and reaction time (Bellinger & Adams, 2001; Grandjean & Landrigan, 2006). Exposure has also been linked to behavioral problems including impulsiveness and aggression (Ruckart, Kakolewski, Bove, & Kaye, 2004), as well as autism spectrum disorders (Roberts, et al., 2007; Rossignol, et al., 2014).

Drawings of a Person

5 year olds

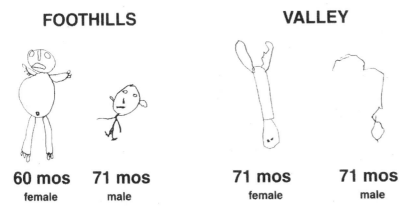

FOOTHILLS

60 mos
female

71 mos
male

VALLEY

71 mos
female

71 mos
male

FIGURE 9.8 *Drawings of a Person.*

A study of five-year old Mexican Yaqui children revealed effects of pesticide exposure on neurological development. Children in the valley region were regularly exposed to agricultural pesticides; children in the nearby foothills were not. Reprinted with permission from Guillette, et al. (1998), p. 351 (personal communication).

Adding to the wealth of data on lead, mercury, and pesticides, recent investigations focus on the endocrine disrupting and neurodevelopmental effects of phthalates, BPA, flame retardants, and the mix of chemicals that constitutes air pollution (e.g., Bellinger, 2013; Giordano & Costa, 2012). Although children are more vulnerable to environmental toxicants, adults are far from immune to their effects. Endocrine disruptors can adversely affect cognitive performance in adults and the elderly, and enhance the risk of Alzheimer's Disease (Stein, Schettler, Rohrer, & Valenti, 2008; Weiss, 2011). There is also abundant evidence that pesticide exposure contributes to the development of Parkinson's Disease (Fitzmaurice & Bronstein, 2011; Stein, et al., 2008). In addition to these problems in later life, people exhibit effects of toxic exposures during their prime reproductive years, as described in the next section.

Reproductive Abnormalities in Adults

Earlier, you read that nearly all bodily systems are under the influence of hormones. Whether you already knew that or not, you probably were aware that the *reproductive* system is under the control of hormones, including **estrogens** (e.g., estradiol) and **androgens** (e.g., testosterone). Although estrogens are considered "feminizing" and androgens are "masculinizing," they are not sex-specific hormones; that is, women and men differ only in the relative proportion of these hormones in their bodies. On average, men have much higher levels of androgens than estrogens, while, for women, the ratio is reversed. Factors including diet, exercise, and *exposure to endocrine disrupting chemicals* can alter these hormone levels.

Atypical levels of sex hormones can affect prenatal and pubertal genital development, secondary sex characteristics (e.g., beards and breasts), sex drive, and fertility. Some variation in humans is natural, but research suggests that exposure to endocrine disruptors is exaggerating this variation. Some of these changes are caused by **xenoestrogens**, endocrine-disrupting compounds that mimic the effects of estrogen in both men's and women's bodies, causing feminizing effects in males (Bergman, et al., 2013).

In females, documented effects include early onset of puberty, abnormal breast development, and ovarian disease (Costa, Spritzer, Hohl, & Bachega, 2014; Macon & Fenton, 2013). Effects on males include irregular prenatal development of the penis (e.g., "micropenis"; Gaspari, et al, 2012), testicular damage, and decreased sperm quality (Jeng, 2014). Many public health experts attribute rising rates of infertility in both men and women to endocrine disruptors (Marques-Pinto & Carvalho, 2013). Decades of research on nonhuman animals suggest that these chemicals may impact aspects of development relevant to gender identity and sexual orientation (Hines, 2011; Hood, 2005), and associated cognition and behavior (e.g., Colborn, Dumanoski, & Myers, 1997; Weiss, 2002, 2011).

In the preceding sections, you have seen how the introduction of synthetic chemicals can easily lead to unanticipated (and unwelcome) effects. Recall from Chapter 3 that within a system, even small actions can have big, unforeseen consequences. To date, researchers know a bit about the ecosystemic and health effects of some environmental

toxicants. They know less, however, about the stressful consequences to come as a result of climate change. For now, we can only speculate about what those will be.

HEALTH HAZARDS OF CLIMATE CHANGE

Global climate change is expected to create significant alterations in our environment and lives (IPCC, 2014). The frequency and intensity of extreme weather events such as floods, windstorms, and droughts will increase (as you read in Chapter 1, there is evidence that this is already underway). Coastal regions and islands will be threatened by rising sea levels and erosion, displacing people who currently inhabit them (also happening already). As urban environments warm, air pollution will worsen. The prevalence of infectious and insect-borne diseases will rise. Food and water shortages will strike many more parts of the world than they do now. Regions with already weakened economies will be bankrupted by property damage, insurance claims, loss of crops, and other costs. In a clear example of environmental injustice, the harshest and most chronic consequences will be experienced by the most disadvantaged members of society (Agyeman, Doppelt, Lynn, & Hatic, 2007). As temperatures increase and resources become scarcer, violent clashes between individuals, communities, and nations will escalate. Life expectancies will shrink. In fact, a panel of physicians concluded that climate change represents the biggest global health threat of the twenty-first century (Boseley, 2009; see Figure 9.9). If this all sounds catastrophic, that's because it is.

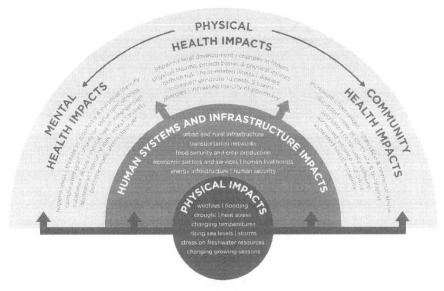

FIGURE 9.9 *Health impacts of climate change.*
Reprinted from Clayton, Manning, & Hodge, 2014, p. 13; with permission.

Feeling stressed? You're not alone. Because of the number and nature of climate change consequences, both traumatic and gradual, the prevalence and severity of stress disorders will increase, as will related problems such as anxiety, grief, depression, substance abuse, and suicide attempts (Clayton, Manning, & Hodge, 2014; Doherty & Clayton, 2011). Climate change offers a stark example of the interconnectedness between human and planetary health.

Fortunately, however, people have *some* control over the extent to which they are adversely affected by stressors. Philosopher Jean-Paul Sartre once said: "What is important is not what happens to us, but how we respond to what happens to us," and, to some extent, this is true. In general, coping with stress involves both behavioral and cognitive responses. People can take action, such as seeking social support and working to remove or reduce the stressor. They can also modify how they are thinking about the situation; one's interpretation of a threat can either increase or decrease the intensity of the stress response. An ability to maintain perspective and optimism, and not give in to helplessness or despair, contributes to **resilience**, the extent to which people are able to cope, physically and mentally. In the case of climate change, resilience will be determined by a complex set of factors including geographical vulnerability, economic resources, education, civic engagement, and individual emotional response (Reser & Swim, 2011; Swim, et al., 2009).

These days, the conversation among experts has shifted from causes of climate change (which are pretty clear) to strategies for dealing with the consequences. Efforts to *mitigate* climate change by reducing greenhouse gas emissions are more important than ever, but at the same time, we must prepare ourselves as well as we can for an altered world. The most recent report from the Intergovernmental Panel on Climate Change is titled *Impacts, Adaptation, and Vulnerability* (IPCC, 2014). The word "adaptation" in this context typically refers to technological developments, but psychologists recognize that human adaptation will be equally, if not more, crucial. No matter how much technological innovation occurs, our world is going to change. Anticipating and reacting to these changes will undoubtedly entail significant transformations in how we think and how we behave (Doppelt, 2012; Reser & Swim, 2011).

CONCLUSION

Because biological evolution proceeds at a snail's pace relative to the cultural evolution of the past 200 years, humans today find themselves physiologically at odds with a dramatically altered habitat and lifestyle. It is an understatement to say that this mismatch is adversely affecting people's health. Industrialized living, which taxes the planet's resources, jeopardizes other species, and disrupts natural systems, is rife with stressors that negatively affect our mental and physical wellbeing.

Still, it is not too late to move things in a different direction. We may not be able to undo the damage inflicted by chemicals and carbon emissions already released into the world, but we can choose whether to continue living in ways that unnaturally stress our bodies, brains, and biosphere. As you will read in the next chapter, returning to our roots by reconnecting with our natural habitat alleviates stress, promotes optimal child development, fosters a sense of connection to nature that promotes sustainable behavior, and represents a step towards healing the planet.

CHAPTER 10

Healing the Split between Planet and Self

We All Need to Walk on the Wild Side

- **The Ecological Unconscious and Biophilia**
  - **Our Preference for Natural Settings**
  - **Our Emotional Connection to Other Species**
- **Benefits of Contact with Nature**
  - **Improved Mental Health**
    - **Restorative Environments**
    - **Nature Therapies**
  - **Optimal Child Development**
    - **Playing in Nature**
    - **Learning to Love Nature**
- **Conclusion: Reawakening the Ecological Unconscious**

Living in the open air and on the ground, the lop-sided beam of the balance slowly rises to the level line; and the over-sensibilities and insensibilities even themselves out. The good of all the artificial schemes and fevers fades and pales; and that of seeing, smelling, tasting, sleeping, and daring and doing with one's body, grows and grows. The savages and children of nature, to whom we deem ourselves so much superior, certainly are alive where we are often dead.

(William James, 1899/2008, p. 135)

Harvard philosopher William James (1842–1910), whose words appear above, is frequently credited as the "father of American psychology" (Figure 10.1). James was not an empiricist, but he was a big thinker. His writings, including the pioneering and influential tome *The Principles of Psychology* (1890), foreshadow many of the topics psychologists have studied since.

The quote above comes from an essay titled "On a Certain Blindness in Human Beings." In this piece, James laments that people lack a true understanding of "the

FIGURE 10.1 *William James, The "father of American psychology."*
Licensed under public domain via Wikimedia Commons.

feelings of creatures and people different from ourselves," and yet, in our ignorance, we make judgments about the value of their lives and lifestyles. James's main point in this essay was akin to the adage about not judging a man until you've walked a mile in his shoes. But relevant to this chapter, James focused particularly on the problem of assuming that the sort of lifestyle he himself lived, as a middle-to-upper class city dweller, was superior to the lifestyles of rural folk and the "savages and children of nature" to whom he refers.

James worried that the "educated classes" (a term he used somewhat sarcastically) "have most of us got far, far away from Nature." He asserted that he and his peers were so caught up in the clamor of modern living that they had become desensitized to "life's more elementary and general goods and joys" (James, 1899/2008, p. 135). These are radical words coming at a time when the industrial revolution was in full swing. James went so far as to declare that the remedy to reawaken one's sensibilities amidst the increasing cacophony of material goods and economic goals was to "descend to a more profound and primitive level." Yet, he observed, success and worth as defined in his age of lumber barons, oil titans, and other gentleman capitalists were distinctly at odds with the sort of simple living that James believed offers insights into the significance of natural things. It seems that, even in these early days of the reign of the Dominant Social Paradigm described in Chapter 2, James was able to think critically about it and suggest that there might be something to be learned from people who had not jumped onto the bandwagon of industrial development. In fact, the quote about "savages and children of nature" continues as follows,

> . . . and, could they write as glibly as we do, they would read us impressive lectures on our impatience for improvement and on our blindness to the fundamental static goods of life.
>
> (William James, 1899/2008, p. 135)

In this part of the quote, James seems to suggest some value in sustainable living.

As influential as James's contributions were in the field of psychology, his concerns about the negative effects of living increasingly disconnected from nature did not catch on in the newly emerging discipline, a discipline that was born in the rapidly growing cities of late nineteenth-century Europe and the United States. As you read in Chapter 4, however, since the late twentieth century, environmental psychologists have studied human interaction with the natural world and ecopsychologists have expressed concern about the growing distance between humans and nonhuman nature. Their work suggests that James was characteristically prescient when he proposed that there may be something vital lost when we spend our lives in contrived settings, numbed by material comforts and consumed by cerebral pursuits.

Although the discipline of psychology has been generally preoccupied with the study of factors that cause psychological distress and impair functioning, it has taken nearly a century for psychologists to attend to the negative effects of the industrial lifestyle. In keeping with the traditional focus on the downside, the previous chapter described the *pathology of modern living*. In contrast, this chapter will follow the example of today's **positive psychology movement**, which emphasizes the study of factors that promote wellbeing (Aspinwall & Staudinger, 2003).

THE ECOLOGICAL UNCONSCIOUS AND BIOPHILIA

The idea that planetary and human wellbeing are mutually dependent is a fundamental premise of the *ecopsychological* perspective introduced in Chapter 4. The term "ecopsychology" was coined not by a psychologist, but by social historian Theodore Roszak (1992). Roszak invoked the idea of an **ecological unconscious**, a sense of interconnectedness between humans and other living things with roots in our ancestral past. Roszak suggested that modern living prevents most people from consciously recognizing this deeply embedded attachment. Nevertheless, he proposed that humans experience an "environmental reciprocity" so that "when the Earth hurts, we hurt with it" (Roszak, 1992, p. 308).

Although Roszak was not a psychologist, one of his inspirations was the work of psychoanalyst Carl Jung. In an essay entitled "Healing the Split," Jung wrote,

> Through scientific understanding, our world has become dehumanized. Man feels himself isolated in the cosmos. He is no longer involved in nature and has lost his emotional participation in natural events, which hitherto had symbolic meaning for him. . . . Neither do things speak to him nor can he speak to things, like stones, springs, plants, and animals. He no longer has a bush-soul identifying him with a wild animal. His immediate communication is gone forever, and the emotional energy it generated has sunk into the unconscious.
>
> (Jung, cited in Sabini, 2002).

A central goal of ecopsychology is to recover people's repressed connection to nonhuman nature.

Ecopsychologists theorize that this will not only benefit people's health, but it will restore their inherent sense of responsibility to the biosphere and all of its inhabitants. From the perspective of ecopsychology, ecological problems are not so much a crisis of technology as they are a crisis of insight (Fisher, 2002; See Box 10.1).

BOX 10.1 *Key Tenets of Ecopsychology*

1. **At the core of the human mind is the ecological unconscious**, repression of which causes madness in industrial society; to heal, people must become aware of their fundamental, primal connection to their ecological home (Glendinning, 1994; Shepard, 1998).

2. **Repression of the ecological unconscious means disconnection from the ecological self.** When people mistakenly perceive themselves as separate from, and independent of, their ecological context, they abuse the environment with which they feel no identification, connection, or empathy. They try to fulfill spiritual and intrinsic needs with extrinsic material goods (Kanner & Gomes, 1995; Kasser, 2009).

3. **Through ecologically based transcendent experience, people can reconnect with the ecological unconscious and reclaim their ecological selves.** Techniques to do so include mindful contact with nature, wilderness trips, reflective rituals, and ecotherapy (Roszak, Gomes, & Kanner, 1995).

4. **Recovery of the ecological self leads to sustainable behavior.** When people act from the ecological self, they do not have to try to make environmentally responsible choices. Instead, choices are naturally less intrusive and less toxic because people care about those whose wellbeing their behavior affects (Bragg, 1996; Naess, 1985; Thomashow, 1995).

The ecopsychological notion that the unconscious human mind harbors a profound bond with the biosphere is paralleled by Harvard biologist E.O. Wilson's **biophilia hypothesis**, which proposes that humans are born with an "innate tendency to focus on life and lifelike processes" (Wilson, 1984, p. 1; see also Kellert, 1997; Kellert & Wilson, 1993). According to the biophilia hypothesis, a feeling of kinship with natural environments and their inhabitants is universal and unlearned. Like the ecopsychologists, Wilson and his colleagues argue that human survival on the planet will require people to recognize this affinity for life that is a part of all of us.

Of course, people don't love nature unconditionally (Bixler & Floyd, 1997). Woods can be spooky. Sharks are scary. Mosquitos are annoying! Most people and other primates share an aversion or overt phobic response to potentially dangerous natural stimuli, such as spiders and snakes, a tendency some scholars call **biophobia** (Orr, 1993; Ulrich, 1993). Both attraction and aversion to nature can be explained evolutionarily. Furthermore, fear and fascination are not wholly incompatible. Frightening experiences in nature are not always experienced negatively, and some people seek them because they are thrilling (van den Berg & Heijne, 2005). For example, when researchers interviewed children leaving a bat exhibit at a zoo, the majority of the children said they liked and cared about bats, but about half also said they would feel afraid to sleep in a place where bats were present. When the frightened children were questioned further, nearly half of them clarified that the fear was different than what they would feel in a dark alley and they "kind of liked it" (Kahn, Saunders, Severson, Myers, & Gill, 2008). Biophilia can encompass both positive and negative affiliation with nature (Kahn, 1997; Wilson, 1984).

Ecopsychologists warn that contemporary urban–industrial living damages us at our cores. Biophilia advocates insist that we have an instinctive yearning to commune with nature. What do you think about these ideas? It is understandable if you feel skeptical. Not everyone is a "nature-lover," right? Well, as we say in science, that is an empirical question! As you read in Chapter 4, most work of ecopsychologists is theoretical rather than empirical, which is a primary reason it has been largely overlooked within the broader field of psychology. Still, there *is* research, conducted primarily by environmental psychologists, that is relevant for exploring the question of whether humans have an intrinsic attraction to nonhuman nature. Evidence comes in the form of our affective responses to natural settings and other species.

Our Preference for Natural Settings

Where do you desire to go on vacation? Although Paris or New York might be on your list, chances are your dream destinations also include a tropical beach, a remote cabin in the woods, a ski lodge nestled in the mountains, or the Grand Canyon. Where were your favorite places to play as a child? Perhaps you fondly recall a stream you explored, a lake you swam in, or a treehouse where you hid out. Across cultures and age groups, people display a strong preference for natural environments over built urban settings (Ulrich, 1983; 1993; Kaplan & Kaplan, 1989; Scopelliti, Carrus, & Bonnes, 2012).

Although there is some variability in which specific natural settings individuals prefer, numerous studies identify common features that tend to elicit positive responses. These include water, vegetation, an expansive view, and a place in which to take refuge. Picture yourself standing just inside the entrance to a cave looking down on the view of a river cutting through a wide, lush plain (see Figure 10.2). Does this image hold any appeal for you? For most people, it does. One explanation for why people respond positively to natural settings with these characteristics is that they are conducive to survival. With access to water and food, as well as a good vantage point from which to spot and hide from potential predators, ancestors who opted for such prime real estate likely had an evolutionary advantage; thus, their preferences were passed down to us (Appleton, 1975; Kaplan, 1992; Ulrich, 1993).

Interestingly, analogous preferences show up in urban settings. When you go to a restaurant, do you feel more comfortable in a booth or in the middle of the room, with your back exposed to the people sitting behind you? When you book a

FIGURE 10.2 *Preferred natural environments.*
People prefer landscapes that provide an expansive view and access to water, food, and a safe refuge. (A View of the Bodrak Valley from Inside One of the Bakla Caves. © Eugene Badusev.)

hotel, do you request a room with a view? Do you find indoor fountains and plants appealing? Research suggests that people respond more positively to built environments when they contain some of the same features as the natural environments we prefer (Ulrich, 1983). For example, people favor houses with integrated vegetation (e.g., ivy climbing the façade, grass on the roof) over houses without (White & Gatersleben, 2011). Participants judge photographs of built settings with water (e.g., ponds, rivers) more positively than settings without aquatic features (White, et al., 2010). A natural scene out the window is the primary draw for people choosing to purchase homes at the edges of urban developments (Kaplan & Austin, 2004).

Long before anyone was talking about biophilia or systematically studying people's landscape preferences, architects and designers were choosing aesthetically appealing natural sites and incorporating natural elements into their buildings. American architect Frank Lloyd Wright, who famously built a house over a waterfall, advised his students: "Study nature, love nature, stay close to nature. It will never fail you" (Lind, 1992, p. 23; see Figure 10.3). Today, some designers are taking it a step further by drawing inspiration directly from theory and research on biophilia; this approach is known as **biophilic architecture** (Beatley, 2011; Joye, 2007, 2012; Kellert, 2005; Kellert, Heerwagen, & Mador, 2008). Biophilic architects wouldn't build Wright's Fallingwater; although the house has undeniably marvelous

FIGURE 10.3 *Fallingwater.*
Architect Frank Lloyd Wright's design was built in the 1930s as a mountain retreat for a wealthy businessman. Photo by Daderot—Own work. Licensed under Creative Commons Zero, Public Domain Dedication via Wikimedia Commons.

views of its natural surroundings, its location on the waterfall is too ecologically disruptive. Biophilic design is informed by an ethic of environmental sustainability (Kellert, 2005).

Biophilic design takes two primary forms: organic (or naturalistic) and vernacular (or place-based). Organic design has long roots in architecture. The liberal use of lilies, shells, and sinuous vines in *art nouveau* structures from the early twentieth century is an example (see Figure 10.4). The iconic marble lions flanking the entrance to the New York Public Library are another. Elaborate landscaping around structures dates back at least as far as the ancient Romans' gardens at Pompeii. Of course, until recently in history, all structures employed natural materials and

FIGURE 10.4 *Organic design.*
This art nouveau subway entrance in Paris was constructed in the early 1900s. ("Paris Metro" by Barbara Mürdter. Licensed under Creative Commons Attribution-Share Alike 3.0 via Wikimedia Commons.)

lighting by default. What makes biophilic organic design new is that it is based on theory and research about how humans respond positively to natural materials, lighting, and shapes. The specific goal of biophilic architecture is to maximize human wellbeing while minimizing ecological impact.

Vernacular design, too, has been around for centuries. Indigenous structures throughout the world are examples of vernacular architecture in their use of local materials and cultural-specific decoration. What makes the biophilic approach to vernacular architecture unique is that reference to context is not simply a function of practicality, circumstance, or tradition, but is implemented intentionally to foster a sense of connection to place, a connection that, in theory, will inspire communities to care for the bioregion (Kellert, 2005). This assumption rests on a growing literature in environmental psychology on the human tendency to develop meaningful attachment to specific locations.

Place attachment is a feeling of connection to a specific locale, such as one's home, neighborhood, or city. Much of the work on place attachment focuses on built environments like these, but in recent years, researchers have devoted more attention to natural places such as lakes, rivers, beaches, mountains, and other recreational destinations (Lewicka, 2011). A general hypothesis is that attachment to place breeds caretaking for that place, but the results are mixed when it comes to caretaking in the form of environmental stewardship; some studies find an association while others do not (Korpela, 2012).

One explanation for the disparate findings is that researchers are not in agreement about how to define or measure the affinity people have for places. For example, many ecopsychologists prefer the concept **sense of place** over place attachment because to them, it communicates more than just a feeling of connection, and encompasses personal identity, meaning, and values associated with the location (Rogers & Bragg, 2012). A third concept, more directly tied to environmental stewardship, is **soliphilia**, a feeling of deep caring and responsibility for a particular bioregion and its inhabitants that may serve as a psychological foundation for sustainable behavior (Albrecht, 2012). Until the theoretical and operational definitions of place attachments become more standardized, it will remain uncertain whether and how attachment to place fosters sustainability.

Nevertheless, it is clear that people truly love some places. As in other close relationships, however, this affectionate bond can be threatened by changes in circumstance. Relocation presents an obvious disruption to physical attachment to place, yet research suggests that people sometimes retain cherished memories of significant places they have left; like longing for a lost love, we fondly and wistfully reminisce about places from our pasts (Korpela, 2012). Sometimes, however, yearning for place is due to the alteration of the place itself. **Solistalgia** refers to a melancholic feeling that arises when one's beloved landscape is threatened by environmental hazards (Albrecht, 2005; Albrecht, et al., 2007). For example, a case study of an Inuit community in Canada found that climate change is disrupting place attachment because it is interfering with the ways that these indigenous people interact with the land (Cunsolo Willox, et al., 2012). Changes in climate are negatively impacting traditional modes of travel and food gathering (i.e., fishing, hunting, foraging, and trapping). Subsequently, the residents are experiencing feelings of loss and alienation from their home.

Although it is possible for people to form attachments to buildings and other human constructed spaces (Korpela, 2012), our preference is for natural places filled with life and life-supporting elements. This is consistent with the biophilia hypothesis. Even more pertinent evidence for biophilia is our kinship with other animals.

Our Emotional Connection to Other Species

From a very young age, people display the desire to protect and help weaker beings, including nonhuman animals (Beck & Katcher, 2003; Myers & Saunders, 2002). Perhaps you can recall being concerned for animals when you were younger. Did you ever think you wanted to be a veterinarian or an animal rescuer? Evolutionary theorists suggest that because of biophilia, animals elicit love, affection, and caretaking similar to human babies. Young creatures of all species share many **neotenic** (baby-like) features (Gaulin & McBurney, 2003), including proportionately large eyes and head, high forehead, and small nose and mouth; these features are commonly perceived as cute (see Figure 10.5). Cute features naturally provoke tender reactions.

FIGURE 10.5 *Neotenic features.*
Baby-like characteristics in diverse species elicit caretaking responses from people.
With permission from Shutterstock

Although some children and adults are intentionally cruel to other species, cruelty is not typical and is generally a symptom of psychological disturbance (Shapiro, Randour, Krinsk, & Wolf, 2014).

People are intrinsically attracted to and fascinated by other animals. Tourists flock to zoos and aquariums. Birdwatching and wildlife photography are popular hobbies. Animals are an ever-popular subject in classical art and pop culture (think of those ubiquitous kitty pictures and videos on the Internet). Close encounters with wild animals are the primary enticement driving ecotourism (Bulbeck, 2005). Taxidermy, pelts, antlers, bones, and butterfly collections have graced homes since at least the sixteenth century when cabinets of curiosity routinely included assorted animal remains; though it may seem counterintuitive, these, too, may reflect people's affinity for other species (Poliquin, 2012).

Partly, humans are fascinated by other animals because they are different than we are, and partly, we are fascinated because they are so much the same. Many other species display behaviors that seem to indicate that their inner worlds are not so different from ours (Bekoff, 2007; Braitman, 2014). When people observe animals, they commonly attribute human qualities and behaviors to them. For example, when researchers eavesdropped on zoo visitors, they heard numerous comparisons to people, such as how the tiger was sitting like a person or the gorilla looked just like Daddy (Clayton, Fraser, & Burgess, 2011). When it comes to companion animals, this tendency is especially pronounced. If you are a pet owner, you probably narrate your pet's feelings and behaviors.

We see animals as similar to us, and we find it easy to become emotionally attached to them. In relationships with companion animals, especially dogs, people tend to feel unconditionally loved and accepted. One survey of pet owners found that compared to relationships with romantic partners, relationships with pets were perceived as more secure (Beck & Madresh, 2008). In another study, pet owners reported that their animal companions were as good, or better, than other humans at providing soothing support (Brown, 2007). People's attachment to their pets is perhaps most evident when they lose them; a common response is profound bereavement equal to what's experienced when losing human loved ones (Walsh, 2009).

Our emotional connection to other life forms is so strong that it can be provoked by animals that are not even real. When she was about five years old, a daughter of one of the authors had a toy Furby, a fuzzy electronic robot, vaguely resembling a cross between an owl and a chinchilla, that was equipped with a set of preprogrammed responses to voice commands and questions (see Figure 10.6). Upon returning from an extended trip, she switched on Furby and eagerly asked, "Furby, did you miss me?" When Furby answered (randomly) with an emphatic, "No!" she burst into tears. Under more controlled conditions, researchers have studied how children and adults respond to robotic life forms. For example, the majority of children who interacted with AIBO, a highly responsive robot dog, said they could be

FIGURE 10.6 *Artificial pets.*
Even simulated life forms can evoke emotional reactions in people.

friends with it, while 59% of adults in an online discussion described AIBO in ways similar to living pets (e.g., "I consider him to be part of my family" and "I care about him as a pal") (Melson, Kahn, Beck, & Friedman, 2009). As the biophilia hypothesis suggests, we have an affinity for life *and life-like processes.*

Still, artificial life forms are not the same as real ones, nor do they necessarily inspire all the same emotions and behaviors in people as living creatures do. In the AIBO studies, there was a difference between the younger and older children's inter-actions with the robot dog. Preschoolers treated the robot the same way they would treat a real dog (e.g., petting, talking, and gesturing to), and they showed similar behaviors toward a stuffed dog (Kahn, Friedman, Pérez-Granados, & Freier, 2006). In contrast, older children (ages seven to fifteen) were less likely to socially engage AIBO than the preschoolers, and they were five times more affectionate toward a real dog than they were toward the robot (Melson, et al., Children's behavior, 2009).

The pattern from these studies suggests that simulated life forms, although evocative of real animals, are not a complete substitute. Similarly, built spaces, no matter how biophilic the design, do not function identically to natural landscapes. This is true in terms of how we feel toward them, as described in this section, and in terms of how we benefit from them, which is described in the next.

BENEFITS OF CONTACT WITH NATURE

"Nature" can refer to a lot of different things. Wilderness is one obvious example. Managed parks, campgrounds, and groomed gardens might also qualify. To some people, any outdoor location counts as nature. Conservatories are buildings, but they are full of nature. A pet goldfish is a nature representative. What about the jungle cruise ride at Disneyland or the Rainforest Café in the Mall of America? These may be a form of nature, too.

Research documents a multitude of benefits humans derive from contact with "nature," but what is meant by the term varies across studies. In some cases, it means untamed terrain; while in others, it refers to manicured walking paths. The stimulus might be a tree, a potted plant, or a photograph of a flower. Interactions might be with wildlife or domestic pets. Research studies also differ with regard to what constitutes "contact" with nature. The setting might be a wilderness campsite or a psychology laboratory with a view out the window. Experiences may involve immersion or no actual physical interaction at all. The focus may be educational, therapeutic, or simply scientific. Overall, a pattern of results is emerging that suggests humans benefit from various forms of contact with all varieties of nature. In some studies, even simulated nature has proven more beneficial than no nature; still, the big picture suggests that intimate contact with real nature is the best (Kahn, 2011).

Improved Mental Health

The previous chapter discussed the importance of good nutrition for mental functioning, but it did not mention that one of the most prevalent nutritional deficiencies found in people living in the industrialized world is a lack of adequate **Vitamin D**. Three-quarters of teens and adults in the United States are deficient in this nutrient, which is needed for healthy development of bones, muscles, and neurons (Ginde, Liu, & Camargo, 2009). Vitamin D strengthens the body's defenses against illness and disease; insufficiency is associated with increased risk of Parkinson's disease (Lv, et al., 2014), Alzheimer's disease (Afzal, Bojesen, & Nordestgaard, 2014), and schizophrenia (Clelland, et al., 2014) among other maladies. The reason this nutrient wasn't mentioned in the last chapter is because only a minority of the Vitamin D in people's bodies comes from food. Instead, most of it is produced by the body in response to *sunlight exposure*. Many health experts are now warning that indoor living and the overzealous use of sunscreen jeopardize people's ability to produce this necessary and beneficial nutrient (e.g., Wacker & Holick, 2013).

Catching some rays is also associated with mental health benefits independent of Vitamin D synthesis. These include lessened depressive symptoms and increased feelings of vitality (Knippenberg, et al., 2014). It's no wonder that people report a preference for bright, sunny environments (Beute & de Kort, 2013).

Profiting from sunlight exposure is a passive process; all we have to do is bare our skin and let nature do the rest. In contrast, for some of nature's benefits, people have to get their hands dirty . . . literally. Healthy soil is brimming with **bacteria**, microorganisms that play a crucial role in our wellbeing (e.g., Flint, Scott, Louis, & Duncan, 2012; Raison, Lowry, & Rook, 2010). In a world of antibiotics and antibacterial cleaners, most people think of bacteria as bad, but the fact is there are *trillions* of friendly bacteria that comprise the body's **gut microbiome**, a colony residing in the intestines that affects metabolism and brain function. In the past decade, animal research on the **gut-brain axis**, a bidirectional communication pathway, has revealed that gut bacteria influence learning, memory, and decision making (Montiel-Castro, González-Cervantes, Bravo-Ruiseco, & Pacheco-López, 2013). In both animal and human studies, friendly bacteria, also known as **probiotics**, have been shown to beneficially affect mood (Foster, 2014). Increasing the number of probiotic organisms in the gut helps to alleviate symptoms of anxiety, depression, and chronic fatigue (Dinan, Stanton, & Cryan, 2013; Messaoudi, et al., 2011). These tiny, teeming therapists make their way to people's guts by way of their mouths, traveling on unwashed organic produce or hands that have been digging in the dirt.[1]

Research on the mental health benefits of simple exposure to sunlight and soil is in its infancy and is primarily the domain of neuroscientists. In contrast,

environmental psychology research on the health benefits of experiences in nature began more than three decades ago. Much of this work focuses on how natural settings reduce stress and help us feel refreshed.

Restorative Environments. The first empirical investigations of psychological restoration in nature were inspired by an unexpected finding that participants in a wilderness survival program felt emotionally better after spending more than a week in the woods (Kaplan & Talbot, 1983; Staats, 2012). Though this study was the catalyst, very little research since has looked at restorative effects of wilderness, leaving open the possibility that for some people, wild settings may not be soothing, but rather might increase stress due to anxiety about getting lost or encountering predators (Gatersleben & Andrews, 2013; Staats, 2012). Also, few subsequent studies involve nature experiences lasting as long as the survival program did. Instead, the bulk of the research on restoration in nature involves experiences lasting a few hours or less in managed settings such as gardens, parks, and other outdoor recreational areas.

Despite differences in the type and duration of contact, numerous studies show decreases in stress and increases in emotional wellbeing following time spent in natural spaces. For example, 95% of patients surveyed at a children's hospital reported a positive change in mood after spending time in the outdoor garden (Cooper Marcus, 2006). Young adults who spent time walking in a nature preserve displayed less anger than peers who spent the same amount of time walking in an urban setting (Hartig, Evans, Jamner, Davis, & Gärling, 2003). Adult gardeners who were stressed by a demanding mental task rebounded better, both in terms of subjective feelings and hormone levels, after spending half an hour gardening rather than reading (van den Berg & Custers, 2011). Many similar studies corroborate the general pattern of stress reduction and improved mood.

Specific characteristics of natural environments that make them restorative can be distilled from researchers' observations and participants' self-reported perceptions (see Figure 10.7). In general, people perceive wilder spaces, such as woodlands, as more restorative than tended spaces, such as urban parks and playing fields (Tyrväinen, et al., 2014; White, Pahl, Ashbullby, Herbert, & Depledge, 2013). Trees and other plants are often a dominant feature, but it is not only *green* space that is experienced as restorative; blue spaces (i.e., ones that include water) are high on people's lists (Ulrich, 1993; White, et al, 2010), and even white (i.e., frozen and snow-covered) landscapes may be restorative for some people (Korpela, Borodulin, Neuvonen, Paronen, & Tyrväinen, 2014). Visually complex and pleasant scenes are generally perceived as most restorative (Korpela, 2013). It is likely that visual features are not the only relevant characteristics, but researchers are only just beginning to study other aspects such as natural sounds (e.g., Payne, 2011; Ratcliffe, Gatersleben, & Sowden, 2013) and ambient temperature (e.g., Hipp & Ogunseitan, 2011).

FIGURE 10.7 *A restorative natural environment.*
Visual complexity and the presence of features such as trees and water increase the actual and perceived restorativeness of natural landscapes.
With permission from Shutterstock

Several preferred landscape characteristics described earlier in the chapter, including water and good views, are also associated with restorative effects, suggesting that people may be instinctively attracted to settings that are beneficial (Han, 2010; Hartig & Staats, 2006; van den Berg, Koole, & van der Wulp, 2003). People also seem to be able to evaluate which environments are likely to be most restorative (Korpela, 2012). One study on beaches in California found that visitors were three times more likely to perceive the setting as potentially restorative on days when the air quality was "good" by government pollution standards (Hipp & Ogunseitan, 2011); although air quality is not among the characteristics researchers identify as "restorative," it is certainly linked to wellbeing. Interestingly, although we may have an intuitive sense about which particular environments are likely to restore us and promote health, we have a tendency to *underestimate* just how good they will make us feel, and therefore may not take as many opportunities to visit them as we can, or should (Nisbet & Zelenski, 2011).

Even when people don't make the effort to venture out into a restorative environment, they may obtain somewhat similar benefits just by viewing one through a window. One of the first studies to demonstrate this took place in a hospital; surgery patients whose rooms had a view of trees recovered more quickly and required less pain medication than those whose rooms looked out on a brick wall (Ulrich, 1984). College students with a more natural view from their dorm rooms were better able to focus their attention than those with views of built structures (Tennessen & Cimprich, 1995). Office workers with views of nature from their workspace showed quicker heart-rate recovery after exposure to a low-level stressor than did workers with a real-time high-definition plasma screen view of nature or workers with no view at all (Kahn, et al., 2008). It seems that "a room with a view" is more than just an aesthetic preference.

When people lack views of nature, they sometimes try to compensate by bringing nature inside. For example, a survey of workers in Norway found that those in offices without windows were five times more likely than workers with windows to have brought plants into the workplace (Bringslimark, Hartig, & Patil, 2011). Some studies suggest that the presence of indoor plants can be restorative, but the experimental findings are mixed (Bringslimark, Hartig, & Patil, 2009). The office workers without a view were also three times as likely to bring in a nature photograph. Although images of nature are not as restorative as views of real nature, they can reduce stress, improve mood, and relieve mental fatigue (Berto, 2005; Hartig, Böök, Garvill, Olsson, & Gärling, 1996; Laumann, Gärling, & Stormark, 2003). If you are stuck in a windowless room working on a computer, a coral reef screen saver may be better than nothing.

Although it is well documented that contact with real and simulated nature is restorative, researchers are not completely in agreement about *why* this is the case. The two predominant theories are both consistent with the biophilia hypothesis, highlighting humans' evolved affinity for restorative features (Staats,

2012). Where they differ is in terms of how they conceptualize *restoration*. The **psycho-evolutionary theory** emphasizes recovery from stress and negative affect, and attributes these to a calming effect of visually appealing scenes (Ulrich, 1983). **Attention Restoration Theory (ART)** describes restoration as the revived ability to concentrate after mental fatigue. When we engage in prolonged **directed** (deliberate and sustained) **attention**, such as that required by schoolwork, we eventually lose the ability to concentrate and start feeling distracted, impulsive, and irritable. Natural settings give directed attention a rest; because they are innately fascinating, they capture our **involuntary** (or automatic) **attention**, which is essentially unlimited, and give directed attention a chance to become replenished (Kaplan & Kaplan, 1989; Kaplan, 1995). These theories are not mutually exclusive and empirical work supports both of them (see Figure 10.8).

Attention Restoration Theory is supported by experiments in which participants' attentional capacity is replenished by real or simulated nature after having been depleted by cognitively demanding tasks (e.g., Berman, Jonides, & Kaplan, 2008). Similarly, the psycho-evolutionary perspective is supported by experiments in which researchers have induced stress in participants before exposing them to restorative stimuli (e.g., Ulrich, et al., 1991). Participants in these studies are from the general population and are assumed not to have any particular cognitive or emotional needs for restoration. In other studies, however, researchers have specifically targeted **clinical populations** of people who display deficits or distress that might benefit from restoration in nature. Such therapeutic applications are addressed next.

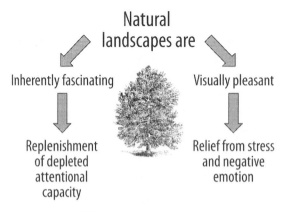

FIGURE 10.8 *Nature restoration theories.*
Attention Restoration Theory (Kaplan, 1995) asserts that restorative environments give depleted attention a chance to be replenished; the psycho-evolutionary perspective (Ulrich, 1983) posits that pleasant landscapes relieve stress and improve mood.

Nature Therapies. Both restoration theories are supported by quasi-experimental studies in which people with preexisting conditions, such as depression, experience improved mental wellbeing after spending time in natural environments (e.g., Berman, et al., 2012; Roe & Aspinall, 2011). The most studied clinical population is children with **attention-deficit disorders** (ADD/ADHD). To explore whether nature experiences might ameliorate attention problems, researchers conducted regional and national surveys of parents in the United States (Faber Taylor, Kuo, & Sullivan, 2001; Kuo & Faber Taylor, 2004). Parents reported that their children's symptoms were less severe following time spent outdoors engaged in either active recreation (e.g., soccer) or passive leisure activities (e.g., reading). The researchers followed up with an experiment in which children diagnosed with ADHD took 20-minute walks in three urban settings: a densely built downtown, a residential neighborhood, and a city park. The three different types of walks were spaced one week apart and the order was randomized for each child. Children displayed significantly better ability to concentrate following the walk in the park than the walks in the other two settings (Faber Taylor & Kuo, 2009). In another experiment, children with attentional disorders performed better on a concentration task when they were in the woods than when they were in a town (van den Berg & van den Berg, 2011).

Together, these studies suggest that experiences in nature may offer an appealing therapeutic alternative to stimulant drugs such as Ritalin or Adderall, which are currently the primary method of treating attentional disorders. Use of these medications in on the rise (National Institute of Mental Health, 2007), in spite of the fact that their benefits are often temporary and are accompanied by negative side effects such as appetite suppression and sleep disruption. Although there are no clinical trials of nature as an alternative treatment, additional survey data suggest that children with attention-deficit disorders who *regularly* play in natural settings have generally milder symptoms compared to their counterparts who play mostly in built outdoor settings or indoors (Faber Taylor & Kuo, 2011).

Contact with nature in a variety of forms is proving therapeutic for other mental health issues besides attentional disorders (Annerstedt & Währborg, 2011; see Figure 10.9). For example, **horticulture therapy** utilizes gardening to aid people with depression and other illnesses (Gonzalez, Hartig, Patil, Martinsen, & Kirkevold, 2010; Hartig & Cooper Marcus, 2006; Messer Diehl, 2009) Although there is no systematic research on exactly why it is helpful, observed effects may be due to the restorative properties of plants, sunshine, soil bacteria, and fresh air. **Animal-assisted therapy** (e.g., with dogs or horses) is being used to help problems ranging from behavioral disorders in children (Katcher & Wilkins, 1998) to dementia in the elderly (Nordgren & Engström, 2014). Anecdotally, animals make good therapists, presumably because of humans' innate emotional connection to them, as described earlier in this chapter, but at this point empirical support for their effectiveness is limited (Julius, Beetz, Kotrschal, Turner, & Uvnäs-Moberg, 2013; Marino, 2012).

The oldest form of nature therapy is wilderness immersion. Although **wilderness therapy** as a formal practice dates back only as far as the mid-twentieth century, the healing properties of wilderness figure prominently in the writings of

FIGURE 10.9 *Nature-assisted therapies.*

Interaction with animals and plants can help in the treatment of mental illness, behavioral disorders, and neurodegenerative diseases.

With permission from Shutterstock

John Muir and the transcendentalists who inspired him, as described in Chapter 2. Like the restorative effects of nature, the therapeutic potential of nature immersion came to the attention of mental health professionals by accident. During a tuberculosis outbreak in the early 1900s, mental health patients at a hospital in New York were quarantined; due to space limitations, they were relocated to tents by a local river. Soon, the hospital staff noticed marked improvement in their symptoms. When the patients were moved back inside as colder weather set in, their condition deteriorated, prompting some other hospitals to implement "tent treatments" for their patients (described in Russell, 2012). During the 1920s, as mental health workers became more aware of the therapeutic value of outdoor experiences, camps for emotionally troubled youth opened up in several locations in the United States. At about the same time, educators in Germany began using an outdoor challenge model that was the basis for the leadership program Outward Bound (founded in the 1940s) and for many wilderness therapy programs today (Russell, 2012).

There is no standard model for wilderness therapy (Friese, Hendee, & Kinziger, 1998; Russell, 2012). In general, the approach integrates individual and group counseling techniques with recreation in wilderness areas. Participants are guided in developing both personal and interpersonal skills while learning relevant technical skills (rock climbing, rafting, or simply preparing food and shelter), most of which require effective teamwork (Wilson & Lipsey, 2000).

Teenagers with emotional and behavioral problems are the most frequent participants in wilderness therapy programs. Typical teenage clients are in therapy due to substance abuse and/or antisocial behaviors such as defiance, impulsivity, and inappropriate anger (Clark, Marmol, Cooley, & Gathercoal, 2004; Russell, 2012; Wilson & Lipsey, 2000). Most are under 18 and over 60% are male (Russell, 2012). Research suggests that other potential target populations are military veterans struggling with post-traumatic stress (Dustin, Bricker, Arave, Wall, & Wendt, 2011; GSU Newsroom, 2014) and women recovering from abuse and eating disorders (Cole, Erdman, & Rothblum, 1994; McBride & Korell, 2005).

Historically, wilderness therapy programs were neither licensed nor accredited, and this is still the case in some places. Especially in the past, staff members were neither trained in mental healthcare, nor as wilderness guides. In fact, some wilderness therapy programs in the United States were condemned for alleged physical and emotional abuse and neglect of their clients, resulting in the deaths of at least ten teenagers between 1990 and 2004 (Krakauer, 1995; Kutz & O'Connell, 2007). Concern about the lack of consistency, professionalism, and oversight in wilderness therapy prompted the formation of the **Outdoor Behavioral Healthcare Council** in 1997; its purpose is to share best practices, conduct outcomes research, and develop accreditation standards.

Wilderness therapy is purported to provide a variety of benefits including improved physical health, social skills, self-control, and self-esteem, as well as decreased antisocial behaviors (Werhan & Groff, 2005); however, empirical evidence of its effectiveness is limited. A primary reason is because it is difficult and ethically questionable to conduct randomized experiments in which some troubled teens are assigned to a wilderness therapy condition while a control group receives an alternate treatment or none at all (Russell, 2012). Another challenge for researchers is assessment of outcomes. It can be difficult to operationally define "successful therapy" and to achieve consistency between observations made by different therapists (Lariviere, Couture, Ritchie, Cote, & Oddson, 2012). Still, there are a handful of empirically sound quantitative and qualitative studies that show positive benefits of some individual programs (e.g., Harper, Russell, Cooley, & Cupples, 2007; Russell, 2005).

Although evidence is accumulating that wilderness therapy is effective in treating some emotional and behavioral problems, it is not yet clear what specific aspects of the therapy are causing the changes. Some programs are based on the outdoor adventure challenge model (the programs in which the teens died were an extreme "boot camp" version of this). Others employ standard psychotherapeutic techniques, such as cognitive-behavioral therapy, in wilderness settings; this is the predominant model in most reputable programs today (DeAngelis, 2013). No matter which approach is used, it is not clear how much *wilderness immersion* actually contributes to the outcomes. Theoretically and anecdotally, wilderness experience is therapeutic, but no researchers have yet systematically isolated the wilderness effect (Greenway, 1995) from the benefits of personal challenges or traditional psychotherapy (Taylor, Segal, & Harper, 2010).

An umbrella term that is sometimes used to describe all nature-assisted therapies is **ecotherapy** (Buzzell & Chalquist, 2009; Chalquist, 2009; Clinebell, 1996). Importantly, this label also encompasses a newly emerging practice of helping people deal with psychological distress directly related to the ecological crisis, such as worries about climate change. This type of ecotherapy may or may or may not be nature assisted (Smith, 2010). Finally, the term can refer to ecologically embedded approaches to traditional psychotherapy, where typical problems (e.g., depression, anxiety) are treated by helping the client reconnect with the ecological unconscious (Macy, 2012).

Of course, most psychotherapists and other mental health workers do not practice ecotherapy of any variety. Current models of wellness in the mental health profession do not address the therapeutic benefits of nature or ways that nature can be integrated into traditional treatments (Reese & Myers, 2012), let alone the notion that much of the psychological distress people experience in contemporary industrialized cultures may be the result of their disconnection from nature and their ecological selves (Roszak, 1992). Clinicians who do practice ecotherapy tend to be individuals who have had significant personal experiences that raised their consciousness about the essential bond between humans and nonhuman nature (Rust, 2009). One survey of more than 200 mental health practitioners found that those who were most likely to adopt an ecotherapeutic approach had a history of positive experiences in nature, had received graduate training in environmental psychology, lived in locations where outdoor recreation was popular, and held a worldview consistent with the New Ecological Paradigm (Wolsko & Hoyt, 2012).

Research on restorative environments and nature therapies reinforces ecopsychologists' claim that the key to personal wellbeing is connection with nonhuman nature. These studies also buttress the biophilia hypothesis and the idea that people function best in an environment to which we are biologically suited. The findings clearly support a prescription for people to increase their contact with nature on a day-to-day basis, but what about across the lifespan? If contact with nature benefits us in small doses, what cumulative effect might it have on shaping the people we become? This is the question addressed in the next section.

Optimal Child Development

Childhood and adolescent recreational activities have changed dramatically over the last few decades. For many children, particularly those living in urban settings, riding a bike, climbing trees, building forts, and exploring the woods have largely given way to electronic pastimes.

In his bestselling book *Last Child in the Woods: Saving our Children from Nature Deficit Disorder*, Richard Louv (2005) sounded the alarm about the growing phenomenon of "indoor children" by describing negative health, social, and environmental impacts. His term **nature deficit disorder** is not an official diagnostic label, but perhaps it should be. Increasingly, researchers are growing concerned that children are missing out on experiences in nature that provide critical developmental support in the form of mental and sensory stimulation, physical challenges, and opportunities for creative play and exploration (Faber Taylor & Kuo, 2006). One could argue that video games provide stimulation, indoor obstacle courses are physically challenging, and children can play creatively in their bedrooms, but there is a theoretical basis for the idea that *natural* environments contain features that uniquely support development.

You probably are familiar with the **nature–nurture** debate in psychology: Historically, researchers have attempted to separate the relative influences of genes and learning on child development. The prevailing view today is that maturation is shaped by dynamic *interactions* between inherited and experienced factors (Keating, 2011). One evolutionary perspective suggests that genetic predispositions prepare humans for rapid learning and retention of particular associations and responses, especially those that promoted survival in early ancestors (e.g., Seligman, 1971; Ulrich, 1993). Evolutionary psychologists refer to the selection pressures that shaped our genetic predispositions as the **Environment of Evolutionary Adaptedness** (EEA; Gangestad & Simpson, 2007). This "environment" is not a specific time or place, but is a set of conditions to which humans are best adapted. As you learned in the previous chapter, modern society differs in significant ways from the EEA. Though it may be difficult to pinpoint the precise parameters of the EEA, it is certain that the backdrop was wild nature. Therefore, in theory, the natural landscape is the optimal venue for human development.

Empirical data on human development in relation to nature is relatively sparse and focuses primarily on how children come to understand and evaluate nature, rather than on how nature impacts their growth. Data derive mainly from observations of children or interviews with them (Myers, 2012). Most studies are qualitative rather than quantitative, and the quantitative ones are largely correlational or quasi-experimental (see Chapter 4 for a review of these research designs). This makes it challenging to draw firm conclusions about cause and effect. Of course, the reason that there are very few well-controlled, experimental studies is because researchers cannot either ethically or feasibly randomly assign children to different developmental contexts. Still, there is a growing body of descriptive and suggestive evidence consistent with the hypothesis that interaction with nonhuman nature is fundamental to healthy child development (Charles & Senauer, 2010).

Playing in Nature. Children's primary occupation is play. It is the mechanism through which much maturation happens. Where and how children play influences their physical, cognitive, social, and emotional development (Rubin, Fein, & Vandenberg, 1983). As described by Louv (2005/2008), the where and how have shifted dramatically in the past couple of decades. For example, an online survey of mothers of in the United States found that while more than two-thirds of the mothers played outside every day when they were young, only one-third of their three- to twelve-year-old children did (Clements, 2004). An overwhelming majority (85%) of the mothers in this study said television and computers kept their kids inside. Keep in mind, this study is more than a decade old and predates many of the gadgets that preoccupy children today, such as iPads, iPods, iPhones, and all of their entertaining apps. More recent large-scale, national surveys in the United States have found that children's participation in outdoor activities has fallen each year since 2006 (The Outdoor Foundation, 2010).

Back before there were so many varieties of electronic recreation, researchers used to find that children strongly preferred outdoor play to indoor pastimes (e.g., Raymund, 1995). This may be changing. In the more recent surveys mentioned above, lack of interest was one of the top two reasons children offered for not engaging in outdoor activity; the other was lack of time (The Outdoor Foundation, 2010). However, when children *do* play outside, they display a preference for climbing trees and splashing in mud puddles over playing on pavement (e.g., Lucas & Dyment, 2010). It seems they intuitively know what is best for them.

Natural landscapes differ from most built spaces in a variety of ways: The terrain is varied and uneven, irregular obstacles present unique physical challenges, and the supply of objects to study and manipulate is unlimited. Changing seasons

and weather fluctuations mean familiar places are never the same from day to day or moment to moment. Close encounters with other creatures are unpredictable. There are endless places to explore. All of these differences make natural settings beneficial for children's physical and cognitive development.

As you read in the previous chapter, inactivity is associated with poor health. Studies find that children of all ages are more active in natural landscapes than on asphalt or turf playing fields. Structured areas support structured play in the form of rule-bound sports and games; in contrast, natural areas promote diverse, unstructured activities such as climbing, digging, and building (Dyment & Bell, 2008). Even children who shy away from competitive sports are attracted to this type of play. Natural settings encourage children to get moving. For instance, a study of 11 preschools in Sweden found that children took more steps in settings that contained trees, shrubs, and uneven ground (Boldemann, et al., 2006). When staff at a daycare center in Australia redesigned play areas by adding vegetation and loose organic materials (e.g., rocks, stumps, and mulch), they reported an overall increase in activity among the children (Nedovic & Morrissey, 2013).

Free-form physical play in natural settings helps children to develop better motor skills. For example, a quasi-experimental study of kindergarten children in Norway found that those who played one to two hours per day in a forest near their school (experimental group) had better balance and coordination than those who played on school grounds (control group). The experimental group also showed more overall improvement in motor skills across the nine-month duration of the study (Fjørtoft, 2004).

Complex environments afford children more opportunities for diverse activity (Heft, 1988). Because of their inherent complexity, natural environments are particularly engaging and challenging, not just physically, but also mentally. It seems that nature sparks children's creativity and imaginative play (e.g., Faber Taylor, et al., 1998; Kirkby, 1989). For example, after the redesign at the aforementioned daycare center, children exhibited a significant increase in dramatic play; as one staff member described: "The children have become dinosaurs and the pebbles are their food. The children have become babies and the pine cones are their bottles. The children have become lizards and they must find water to survive" (Nedovic & Morrissey, 2013, p. 288–289). Though it may seem cute, imaginative play is serious business. It is through make-believe that children develop important life skills such as cooperation, negotiation, problem-solving, and emotion regulation (Singer, & Singer, 1990).

Learning to Love Nature. A significant part of a child's intellectual development involves learning to name, classify, and discriminate between different objects. As an ever-changing milieu featuring a vast array of stimuli, nature seems an ideal classroom. **Folkbiology theory** posits that humans possess an innate tendency to perceive, categorize, and think about living things. Even without formal schooling, children across cultures develop an everyday understanding of the natural world (Medin & Atran, 1999). For example, children recognize that there is a vital life force in food and water that gives people energy, helps them heal, and makes them grow (Inagaki & Hatano, 2004). Children as young as four understand that plants and animals can grow (Hickling & Gelman, 1995) and heal when injured (Backscheider, Shatz, & Gelman, 1993), but inanimate objects cannot.

The details and extent of children's naïve understanding of living things and natural systems are shaped by both the cultures in which they live and their experiences in nature (Medin & Atran, 2004; Ross, Medin, Coley, & Atran, 2003). For example, in one study, researchers told children that there was a substance called *hema* inside living things. The fact that this is an imaginary substance isn't important; what matters is that the researchers varied what they said about it. Some of the children were told that the substance is inside *humans*, and were then asked if other animals also possess it; the rest of the children were told that the substance is inside *other animals* and were then asked if humans also possess it. Some children made anthropocentric projections of similarity, extrapolating from humans to other animals ("If humans have it, wolves would have it too"), but not the reverse ("Wolves have it, but people don't"). Others made biocentric, bidirectional projections (from humans to other animals, and vice versa). Native American children raised in rural settings made more biocentric projections than did their urban, nonnative peers, presumably due to more exposure to a wider variety of wild animals and cultural traditions about affinity between species (Ross, et al., 2003).

Without sufficient experience, children do not develop an accurate or nuanced understanding of the natural world. For example, children living in cities are able to assign living things into simple taxonomic categories by the age of six (e.g., bird, insect, tree, and plant), but their grasp of animals' natural habitats is less developed than that of rural children (Coley, Freeman, & Blaszczyk, 2003). Older urban children do not necessarily fare any better. The 2010 documentary film *Play Again* includes a poignant

segment in which elementary and middle-school city kids are shown a variety of pictures and are asked to identify them. Some of the pictures are of common plants and some are of corporate logos. The children do not hesitate for a moment to identify the specific brand emblems, but when it comes to the plants, they feebly respond with "flowers," "plants," "some kind of weed." Several of the children refer to the dandelion as a "wish flower." For children living in industrialized culture, perhaps it is presently more adaptive to be able to distinguish between fast food outlets than edible and inedible wild plants, but this skill does not contribute to ecological understanding.

Children's understanding of nature may influence their moral reasoning about it. In general, research suggests that children tend to be more morally concerned about people than other species (e.g., Hussar & Horvath, 2011), but some studies have found evidence of biocentric moral reasoning in some children, especially those with a more sophisticated grasp on ecological systems. For example, interviews with hundreds of children from three diverse cultures (inner-city Houston, Texas; the Brazilian Amazon; and Lisbon, Portugal) revealed that most children viewed pollution as a moral violation, and the majority offered anthropocentric explanations, citing harm to humans as the problem (Kahn, 1999; Kahn & Lourenço, 2002). However, these interviews also revealed a developmental progression; older children were more likely than their younger peers to emphasize how pollution negatively affects wild animals and their habitat. These findings are consistent with another interview study that found that the older the children were, the more likely they were to recognize that animals need more than just food and water; they also need healthy habitat and protection from harmful human activities (Myers, Saunders, & Garrett, 2004). It seems that as children develop a fuller understanding of ecology, they are more likely to adopt a biocentric perspective and see nonhuman nature as having inherent value, with moral standing comparable to humans.

When children exhibit biocentric moral reasoning, they focus on *animal* welfare. They don't tend to express concerns about plant life, possibly because their awareness of plants as "alive" develops later than their awareness of animals (Melson, 2013b). As described earlier, children care about animals. Animals figure prominently in their worlds as live, stuffed, or imaginary companions; as captive or wild specimens; as characters in books and cartoons; and as roles the children themselves assume. Yet, developmental psychologists have largely neglected the study of children's relationship with animals (Melson, 2001). The sparse research that exists is primarily about children and their pets; research on children and wild animals is practically nonexistent (Melson, 2013a). Interaction with pets is a primary way that children learn about caring, and about biological processes, so it follows that children's concern for animals may serve as the foundation for broader ecological concern (Myers, 2007; Myers & Saunders, 2002).

In general, positive childhood experiences in nature and with nonhumans appear to be important for fostering proenvironmental values and behaviors during childhood and into adulthood (e.g., Chawla & Derr, 2012; Wells & Lekies, 2006). These experiences serve as the foundation for the development of an ecologically connected identity, as described in Chapter 7 (Kals & Ittner, 2003). However, in the industrialized world, children's direct interaction with wild places is decreasing, both because children are increasingly opting out, and because the places are becoming fewer and farther between. Consequently, children may be experiencing an **extinction of experience**: a sense of alienation from nature, and a loss of the intimacy that motivates concern and conservation (Pyle, 1993).

Even when children do have experiences in nature, the breadth and depth of their environmental concern may be stunted by the ecologically degraded condition of the settings they visit. When one's experiences are limited to environmentally damaged areas, such landscapes can come to seem ordinary. For instance, two thirds of the children interviewed in Houston (one of the most polluted cities in the United States) understood the problem of pollution in general, yet they did not believe their air or water was polluted. They exhibited **environmental generational amnesia**, the phenomenon of each subsequent generation perceiving current environmental conditions as the new normal (Kahn, 1999, 2002).

Environmental generational amnesia has the potential to eclipse the true extent of the ecological crisis in the minds of future generations. Children growing up in an increasingly degraded world may not realize what they are missing, both in terms of healthy habitat and in terms of a life-enhancing connection to nonhuman nature (Kahn, 2002). If humans lose sight of what it means for ecosystems to thrive, we may fail to adequately safeguard them; subsequently, we run the risk of undermining our own wellbeing, and may ultimately sabotage the ability of future generations to develop to their full potential.

Indeed, those of us who are already adults lack the perspective of our ancestors who lived before the days of polluted water and air, denuded landscapes, and climate change. Still, many of us have an intuitive (or learned) sense, increasingly supported by empirical evidence, that our personal welfare is intimately and inextricably linked to a healthy planet.

CONCLUSION: REAWAKENING THE ECOLOGICAL UNCONSCIOUS

> As a species, we came of age on the savannahs of East Africa and lived a life more wild than we do today. Much of that wildness still exists in the architecture of our bodies and minds, and needs to be rediscovered, re-engaged, developed, and lived—we need to be re-wilded—for us as a species to flourish.
>
> (Kahn & Hasbach, 2013, p. 207)

At the beginning of this chapter, you read about William James's recommendation that people "descend to a more profound and primitive level" so as to reawaken an appreciation for the simple joys of living close to nature. To many people, the word "primitive" implies crudeness, immaturity, and a lack of sophistication; from this perspective, primitive living is underdeveloped living. When James used this word, however, he was suggesting that a more primitive lifestyle may be superior to the *over*developed routines of industrialized culture. James advised that people stop and smell the roses. We would do well to take James's advice to heart. Because of our evolutionary history, we are predisposed to thrive in close union with the rest of the natural world.

So what does this mean, practically speaking, for those of us living a modern lifestyle in a decidedly altered environment? Here are a few suggestions, inspired by the work of ecopsychologists, environmental psychologists, outdoor behavioral healthcare professionals, wilderness educators, and others who focus on the reciprocal relationship between human and planetary wellbeing:

- **Spend more time outside**. This does not have to involve wilderness excursions (though those are highly recommended). Even twenty minutes strolling through a nearby green space can leave you feeling restored.

- **Tune in to your surroundings**. It is difficult to feel connected to nature when wearing earbuds or texting on your phone. Fully use your senses to experience the natural setting. Make it a *barefoot* stroll through the nearby green space. It is through our senses that we come to know the world (Abram, 1996; Sewall, 1999).

- **Learn about your bioregion**. Explore your area, tap into childish curiosity, educate yourself about endemic species, and sample the wild edibles that are all around you (even in the city). Familiarity with context fosters a sense of place.

- **Mindfully interact with nature**. Regularly invest time in an outdoor activity that absorbs your attention and focuses it on the nature around you. Mindful practices in natural settings enhance people's experience of the present moment and, thereby, their felt connection with surrounding ecosystems (Macy & Brown, 1998). Examples include gardening, wildlife tracking, meditating in a favorite "sit spot," or hunting for trash along a hiking trail.

- **Practice earth-living skills**. Make a fire without matches, sleep in a shelter you've built from forest debris, forage for your supper. Unmediated contact with wild nature helps people experience their ecological interdependence firsthand (Scott, Amel, & Manning, 2014).

As the research reviewed in this chapter suggests, modifying one's lifestyle to involve more frequent contact with nature is beneficial to health. It is also likely to contribute to a deeper sense of connection to nonhuman nature. Studies have found that being more connected to nature is robustly correlated with subjective wellbeing and happiness (Cervinka, Röderer, & Hefler, 2012; Howell, Dopko, Passmore, & Buro, 2011; Wolsko & Lindberg, 2013; Zelenski & Nisbet, 2014), and with environmental concern and proenvironmental behavior (e.g., Clayton, 2003a; Mayer & Frantz, 2004). If we heed William James's fatherly counsel and reconnect with our primitive selves, we will be better for it—and so will the earth.

NOTE

1. Another source of beneficial bacteria are **lacto-fermented foods** like sauerkraut, kimchee, and kvass that have been traditionally prepared by covering vegetables in a saltwater brine and letting them sit (unrefrigerated) for a week or more; the salt kills unfriendly bacteria that would cause the food to spoil, but the friendly *lactobacilli* thrive and transform the vegetables.

PART 4

Getting Psyched for Sustainability

The goal of this final chapter is to encourage you to take what you've learned and apply it to move behavior in a sustainable direction. The chapter presents a variety of theoretically and empirically grounded ideas for how to face this challenging task with positivity, wisdom, and enthusiasm.

CHAPTER 11

Getting Psyched for Sustainability
Being the Change We Want to See

- Pursue a Positive Path to Sustainability
  - Visualize an Ecologically Healthy World
  - Use Meaningful Metrics
  - *Work* for Sustainability
  - Foster Resilience
- Harness Human Nature
  - Act on Big Ideas via Small Steps
  - Go with the Evolutionary Flow
  - Develop New Heuristics
  - Leverage Moments of Flux
- Create a Social Avalanche for Sustainability
  - Methodically Plan Change
  - Seek Opportunities to Lead
  - Seek Opportunities to Follow
- Conclusion

I am only one. But still I am one.
I cannot do everything, but still I can do something;
And because I cannot do everything,
I will not refuse to do the something that I can do.
Edward Everett Hale (1822–1909), quoted in Gover (1909), p. 28.

Throughout this book, you have learned about the planet's serious ecological problems from the perspective that they are really *behavioral problems*, caused by the collective actions of human beings and their underlying thoughts, feelings, and values. By applying the psychological principles reviewed in this text, people *can* develop and implement solutions to shift behavior in a sustainable direction.

PURSUE A POSITIVE PATH TO SUSTAINABILITY

True sustainability will involve supporting the needs of current and future generations, while respecting the interconnectedness of environmental, economic, and social factors (Brundtland & World Commission on Environment and Development, 1987; Schmuck & Schultz, 2002). This means that no matter how it is accomplished, sustaining life on the planet will require substantial changes at both personal and societal levels. People's reticence to confront this reality is likely due to an understandable tendency to become emotionally overwhelmed upon learning about the extent and urgency of ecological problems (Macy, 1995; Norgaard, 2011).

Given that people generally strive toward experiencing positive emotions (Frijda, 2007), calls to proenvironmental action that provoke fear and anxiety are not likely to be motivating (Chapter 6). Instead, finding ways to tap into optimism and foster empowerment may be the key to getting people on a sustainable path. As you read in the last chapter, the lens of *positive psychology* shifts the focus from how things go wrong for people to how they go right (Csikszentmihalyi, 2004). Positive psychologists suggest that we are most able to thrive in the face of challenges when we experience agency, compassion, purpose, and resilience (Lopez, 2009; see Figure 11.1). Therefore, efforts to promote sustainable behavior will be most successful if they draw on and nurture these qualities.

Visualize an Ecologically Healthy World

The Civil Rights Movement provides an illustration of why it is critically important to build motivation for social change from a positive frame: "Martin Luther King Jr.'s 'I Have a Dream' speech is famous because it put forward an inspiring, positive vision that carried a critique of the current moment within it. Imagine how history would have turned out had King given an 'I have a nightmare' speech instead!"

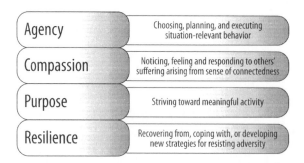

FIGURE 11.1 *Some positive psychology concepts important to human thriving in challenging situations.*
Adapted from Lopez, 2009.

(Shellenberger & Nordhaus, 2004, p. 31). Positive images of a healthy future are desperately needed to spur commitment to solving the ecological crisis. Here's one such vision:[1]

> Imagine for a moment a world where cities have become peaceful and serene because cars and buses are whisper quiet, vehicles exhaust only water vapor, and parks and greenways have replaced unneeded urban freeways. OPEC has ceased to function because the price of oil has fallen to five dollars a barrel, but there are few buyers for it because cheaper and better ways now exist to get the services people once turned to oil to provide. Living standards for all people have dramatically improved, particularly for the poor and those in developing countries. Involuntary unemployment no longer exists, and income taxes have largely been eliminated. Houses, even low-income housing units, can pay part of their mortgage costs by the energy they *produce*; there are few if any active landfills; worldwide forest cover is increasing; dams are being dismantled; atmospheric CO_2 levels are decreasing for the first time in two hundred years; and effluent water leaving factories is cleaner than the water coming into them. Industrialized countries have reduced resource use by 80% while improving the quality of life. Among these technological changes, there are important social changes. The frayed social nets of Western countries have been repaired. With the explosion of family-wage jobs, welfare demand has fallen. The progressive and active union movement has taken the lead to work with business, environmentalists, and government to create "just transitions" for workers as society phases out coal, nuclear energy, and oil. In communities and towns, churches, corporations, and labor groups promote a new living-wage social contract as the least expensive way to ensure the growth and preservation of valuable social capital.
>
> (Hawken, Lovins, & Lovins, 1999, pp. 1–2)

In their book, *Natural Capitalism,* Hawken and colleagues (1999) detailed numerous examples of how corporations, communities, and individuals were already (more than 15 years ago) accomplishing significant changes. Their conclusion was that an ecologically healthy human society is not a futile, utopian dream, but a real possibility.

There are already examples around the globe of people collaborating to realize the dream of a sustainable society. **Transition towns** are communities (neighborhoods, city blocks, and just about any other organized group) where people are acting on their imaginings. For example, residents of Lewes, England, created the *Ecotopian Grapevine Gazette*, which contained "imaginary news stories about events or innovations that had not happened yet . . . written as if they had happened. At the end of each article, [was] the name of someone readers could call and participate in making that story a reality" (Hopkins, 2008, p. 98). In the years since, Lewes residents have started community gardens and composting, held regular open house events during which homeowners can share their household innovations, and introduced a local currency. Transition towns use positive visualization as the basis for making communities more ecologically sustainable. As of this writing, 150 initiatives have emerged in 37 U.S. states, and over 475 communities in 16 countries are officially designated as transition towns.[2]

Having positive visions of the future while working diligently on making them a reality supports a sense of hope. Hope is a crucial psychological commodity that can help people stay motivated and keep negative emotions at bay (Alarcon, Bowling, & Khazon, 2013; Seligman, 1975). But perhaps more important than hope is the realization that *everyone is capable of making a positive difference in the community,* and with *ability* comes *responsibility.*

> When we realize the degree of agency we actually do have, we no longer have to "hope" at all. We simply do the work. We make sure salmon survive. We make sure prairie dogs survive. We make sure grizzlies survive. We do whatever it takes.
>
> (Jensen, 2006).

Joining with others to work on big problems reduces despair while enhancing interconnection and potency. Together with family members, neighbors, and other community members, we can each contribute to the transition to a sustainable world.

Use Meaningful Metrics

Goals are motivating as long as it's possible to assess whether they are being met. How will we know if we're making progress toward sustainability? What are the relevant benchmarks? In the United States, societal wellbeing is currently measured in

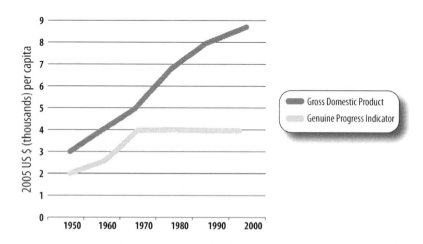

FIGURE 11.2   *Progress indicators.*

Gross Domestic Product indicates economic conditions while the Genuine Progress Indicator includes economic, human, and environmental conditions.

purely economic terms: Gross Domestic Product (GDP). However, if environmental and human health is to be seriously considered in the transition to a sustainable economy, alternative indices will need to be used. The current calculation of GDP not only hides the environmental costs of consumption, it obscures many other costs of inefficiency and environmental damage such as hazardous site clean-up, environmental restoration, medical care for environmentally based diseases, law suits, and so forth. Because these costs are not subtracted, they *contribute* to the GDP instead of lowering it. For example, the labor involved with oil spill cleanup is considered *a benefit* to the economy, rather than a cost incurred for doing unsustainable business (Kasser, 2009a).

A variety of new alternative indices include the long-term impacts of human actions (Costanza, et al., 2014). One such measure, the Genuine Progress Indicator (GPI), works by subtracting the human and environmental costs from the economic gains of producing goods (Clarke & Lawn, 2008; see Figure 11.2). For instance, permanent loss of wetlands due to construction, while raising GDP, would decrease GPI (Kubiszewski, et al., 2013). The states of Vermont and Maryland recently adopted the GPI and are basing governmental decisions on the predicted GPI impact. While imperfect, measures like this one reveal the long-term implications of economic activity.

Other indices, such as the Happy Planet Index, measure subjective wellbeing (Diener & Seligman, 2004; Seligman, 2012) combined with objective data such as life expectancy and ecological impact (Ng, 2008; see Figure 11.3). In 2012, the United States was ranked 116th on one such measure, the **Sustainable Society**

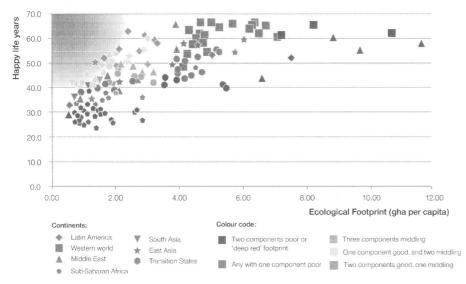

FIGURE 11.3 *Plot of happy life years by ecological footprint.*
The most sustainable societies (e.g., Costa Rica) fall in upper left. Many Western (e.g. United States) and Middle Eastern countries (e.g., Qatar) have high levels of happiness but at a large environmental cost (top right). (Abdallah, Michaelson, Shah, Stoll, & Marks, 2012).
Courtesy of the New Economics Foundation, Happy Planet Index 2012 report

Index, developed to evaluate 151 countries' performance in terms of *personal development* (health conditions, education, gender equality), *environmental health* (air, water, land quality), *societal balance* (governance, employment, income distribution), *resource use* (waste recycling, renewable energy, and water sources), and overall *sustainable world* measures (forest area, biodiversity, greenhouse gas emissions, ecological footprint, and international cooperation) (van de Kerk & Manuel, 2009).

Work for Sustainability

Movement toward sustainability can take many important forms: reducing one's own ecological footprint, organizing and volunteering in one's community, and engaging in political action, to name a few. The workplace is another significant place to engage. Every organization's policies, operations, products, and services likely need modification in order to function within ecological boundaries. If you are an employee, perhaps you can identify some ways that your workplace could become greener.

While traditional businesses are cleaning up their act, innovative industries are focusing specifically on promoting sustainability. Hence, the recent growth in the **green economy**, which is based on ecologically responsible services, manufacturing, and commerce (Dierdorff, et al., 2009). The U.S. green economy includes over 3.4 million public and private sector employees (Bureau of Labor Statistics, 2013) working in 12 areas of activity; some occupations are new, like the *capture and storage of carbon*, and some are classics with a new twist, such as *green construction* (Dierdorff, et al., 2009; see Figure 11.4).

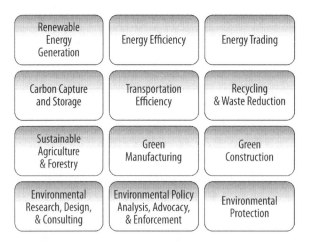

FIGURE 11.4 *Economic sectors that comprise the green economy.*
Adapted from Dierdorff, et al., 2009.

The green economy includes many environmentally related jobs that have been around for a long time but are now in higher demand. Some are highly specialized professions, such as environmental science and zoology; others are entry-level trades like insulation installation, which is predicted to be among the fastest growing occupations for the next decade (Bureau of Labor Statistics, 2014). Many occupations that are not traditionally associated with the environment are being redesigned to encompass "green" skills. For instance, architects, builders, engineers, and city planners are learning to plan buildings, neighborhoods, and cities in ways that alleviate the need for fossil fuel-based temperature control and transportation.

Finally, the green economy has generated over 90 completely new careers due to changing needs, technologies, and systems.[3] A great variety of specialists are needed to research, design, build, sell, install, operate, maintain, recycle, and regulate new energy technologies (e.g., wind, solar, geothermal). Experts are also needed to coordinate sustainability initiatives within organizations. Comprehending ecological systems can enhance performance of all of these jobs. "Thinking like an ecologist" (Chapter 3) is an important competency in the green economy.

A growing number of businesses and governmental bodies are recognizing that long-term economic vitality is completely dependent on healthy humans and healthy ecosystems (Giddings, Hopwood, & O'Brien, 2002; Willard, Upward, Leung, & Park, 2013; see Figure 11.5). In turn, a green economy promotes both environmental health and social welfare. For example, when businesses and communities collaborate to detoxify brownfields, plant gardens in vacant lots, install energy-efficient home retrofits, and improve public transit, everyone benefits. Efforts like these create vibrant and prosperous neighborhoods and local, livable-wage jobs (Baiocchi & Lerner, 2007; Dunn, 2010; EPA, 2013). In this way, urban transformations help to rectify economic and environmental injustices.

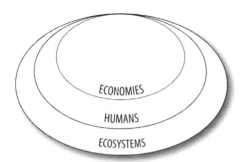

FIGURE 11.5 *The nested dependencies depiction of sustainability.*
A healthy economy fully depends on healthy humans, and both economic and human health depends fully on healthy ecosystems.

Foster Resilience

Whenever possible, it is best to tackle tasks feeling energetic, involved, and efficacious. It is reinforcing to feel that one is fully **engaged** in important work (Maslach & Leiter, 2008). However, thinking about changing behaviors, organizations, and infrastructure can be daunting. And the day-to-day effort of swimming against the tides can be stressful. As you read in Chapter 9, stress can negatively affect one's health and quality of life. Chronic stress can lead to **burnout**, the psychological state in which people become physically and emotionally exhausted, feel detached from others, and experience an eroded sense of accomplishment (Maslach & Jackson, 1981; Sohr, 2001). Burnout, thought to be at the opposite end of the spectrum from engagement, can lead people to quit trying altogether (Swider & Zimmerman, 2010). It is important to build one's resilience to stress.

There are a number of established methods for reducing stress. In addition to problem-based coping (Chapter 6), we can engage in **proactive coping**, which consists of anticipating and preparing for a stressful event before it occurs (e.g., Bloodhart, Swim, & Zawadzki, 2013). Try to imagine ahead of time what will happen when you pull out your own to-go container at a restaurant or take the stairs while your friends take the elevator. Even mundane scenarios like these are potentially stressful; mentally prepping ahead of time can make the real deal feel less nerve-wracking.

When the going gets tough, the tough rely on their support systems. **Social support** comes from other people in a variety of ways: It can be emotional (empathy), instrumental (practical help), informational (advice), or esteem-building (praise); ultimately, it can provide a reassuring sense of belonging (Gottlieb & Bergen, 2010). As you know from experience, social support helps us get through tough times. Connecting with people who are striving to become sustainable can be just the ticket when it feels like the effort is overwhelming. Share success stories, difficulties, and strategies. Commiserate about that compost heap. If you develop a network of people you can rely on, things won't seem so bad.

One's social network can be especially supportive when it is fun and friendly. **Humor** is related to lower stress and less burnout because laughter eases tension, shared jokes create a sense of camaraderie, and having an appreciation for life's little absurdities can keep us looking on the bright side (Mesmer-Magnus, Glew & Viswesvaran, 2012). When groups of people work together, they can maintain calm and increase resistance to burnout by interacting with **civility**, treating each other with courtesy and respect (Leiter, Laschinger, Day, & Gilin-Oore, 2011). When it comes to stress reduction, it really does pay to be Mr. Nice Guy.

Finally, in the last chapter, you learned about a fundamental way to relax and recharge: restoration in nature. When you are feeling stressed about sustainability, take a hike, swim in a lake, or meditate under a tree. If your opportunities are limited, find a window and look at the landscape. Let nature mother you.

HARNESS HUMAN NATURE

Envisioning a positive future, staying emotionally healthy, and finding meaningful opportunities to engage are all important for the big picture of change. We can also use psychology to help us manage day-to-day individual behavior. The more we can do to work with, rather than against, people's natural tendencies, the more successful our efforts are likely to be.

Act on Big Ideas via Small Steps

When faced with seemingly insurmountable challenges, most people are liable to turn their attention to addressing little day-to-day things instead of tackling the big stuff because doing so reduces anxiety. However, instead of distracting ourselves with small problems, we can take small *steps* to work on big problems.

This is a smart strategy because when we are emotionally overwhelmed, our high arousal compromises our problem-solving abilities; conversely, when we are tuned out, we become apathetic and inactive. Like Goldilocks, we need to seek that sweet spot in the middle. People perform optimally when experiencing moderate levels of arousal (Broadhurst, 1959; Yerkes & Dodson, 1908). Defining problems incrementally keeps arousal in the optimal range. Incremental challenges have immediacy, tangibility, and controllability that reverse powerlessness and apathy, but don't overwhelm. Small wins also offer intermittent bursts of success, fostering momentum and motivation for further change (Chapter 8).

If small steps seem trivial, recall from Chapter 1 that researchers have estimated the amount of CO_2 emitted in the industrial production of just one kilogram (2.2 pounds) of beef to be equivalent to the amount emitted by an average European car driven 155 miles (Bittman, 2008). That means a person can significantly help curb climate change by taking the relatively small step of implementing "Meatless Mondays." Such small changes, one day at a time, reflect small wins, but eventually (and collectively) make a huge contribution.

Go with the Evolutionary Flow

Efforts to change are most successful when they harmonize with, rather than thwart, our deeply embedded, evolutionarily based human tendencies (Griskevicius, Cantú, & van Vugt, 2012; van Vugt, Griskevicius, & Schultz, 2014; see Figure 11.6). For instance, environmental messages crafted to feel psychologically close are consistent with our inclination to focus on the here and now (Chapter 6).

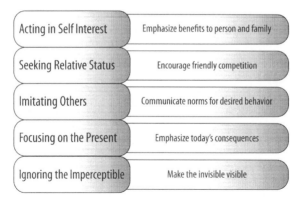

FIGURE 11.6 *Five evolutionary tendencies and strategies that tap into them.*
Adapted from Griskevicius, et al., 2012.

Providing informational feedback about home energy use and garbage creation makes the invisible visible and, therefore, easier for people to manage (Chapter 8).

Another example can be found in the phenomenon of *self-interest*. People are evolutionarily predisposed to do things that benefit themselves and their kin. Although this biological selfishness can contribute to unsustainable behavior, it can also be harnessed for good. To tap into this, rather than asking people to conserve energy on behalf of strangers or the planet, campaigns can focus on how conservation benefits oneself and close family.

The *green to be seen* phenomenon is driven by the basic human need to demonstrate relative status (Griskevicius, et al., 2010; see also Chapter 5). Effective change programs can forgo a focus on having more stuff, and instead confer other types of relative status such as being the winner of a competition. Publicly acknowledging energy savers, bicycle commuters, and zero-waste heroes not only generates feelings of high status in the highlighted individuals, it also showcases role models of appropriate behavior. And, according to the same evolutionary principles, other people will want to emulate the winners.

Develop New Heuristics

Eventually, sustainable behaviors will become the default and will be supported by policies, social norms, and infrastructure; hopefully, sooner rather than later. Until then, each of us can seek opportunities to do things differently. Given how overwhelming our daily lives can be, heuristics (Chapter 6) can serve as handy tools to keep us going in a green direction. One simple heuristic is to look at each decision we make as *an opportunity to reduce our ecological footprint*.

For instance, in a typical fitness gym, one is surrounded by sweaty people huffing and puffing on all manner of machines. Many of these (stair, running, climbing, elliptical) are powered by electricity. A treadmill workout consumes electricity

FIGURE 11.7 *Certifications can serve as heuristics.*
Nonprofit and governmental organizations certify many products to help shoppers identify options that are sustainably produced and conserve resources.
Right, Water Sense logo courtesy of Environmental Protection Agency; left, with permission from Shutterstock

and burns calories (food); it's a footprint double-whammy. A better alternative would be to get that run in on the way to campus, and leave the car at home. You can probably think of numerous ways we can utilize *our* energy to minimize the draw on energy resources.

Shopping is another domain where we can apply the footprint heuristic. Every shopping choice supports something, whether it's sustainable or not. When we choose environmentally responsible products and services, we not only reduce our ecological footprints, but we model behavior to others, and financially reinforce proenvironmental business practices. While shopping, we can employ other heuristics as well. All kinds of symbols, from brand names to certification tags, can make great mental shortcuts (see Figure 11.7). It is important to keep in mind, however, that not all labeling is as green as it seems.

The ecological principles discussed in Chapter 3 can be transformed into usable heuristics. Many of the principles themselves are already concise ideas that can be used to judge a situation (e.g., is this choice an upstream solution? Is this item part of a circular system?). However, people can generate lots of finer-grained questions, based on the principles, which can become heuristics too. Wording and principles can be adjusted according to your priorities and personality. And, as long as you bring your head along, they are portable. It does take practice and repetition to make heuristics stick. To aid ourselves while developing these new habits of mind, one can literally carry these ideas along by creating a handy checklist such as the Unshopping Card (see Figure 11.8).

Leverage Moments of Flux

Breaking habits can be frustratingly difficult, as you have probably discovered if you've ever gone on a diet or tried to reduce screen time. Most of the time, we are fairly oblivious to our habits because they are automatic behaviors. This is both the beauty and the challenge of habits. They allow us to do all kinds of things without having to think about it, but this also means we do not notice what we're doing when we're

The UnShopping Card

Do I _really_ need this?

Is it made of recycled or renewable materials?

Is it recyclable or biodegradable?

Could I borrow, rent or buy it used?

Is it worth the time I worked to pay for it?

Oregon State OSU **Extension Service**
UNIVERSITY

viviane.simon-brown@oregonstate.edu
http://www.cof.orst.edu/extended/sustain/

FIGURE 11.8   *The Unshopping Card.*
Reprinted from the Oregon State University Extension Service, Sustainable Living Project.

doing it. As you learned in Chapter 5, changing habits requires changing contingencies. We can alter the antecedents (e.g. ensuring visible and easy access to a bike, using a smartphone app that issues push notifications when it is good biking weather, and stashing the car keys out of sight) and eventually new habits may develop.

Sometimes change is enabled by naturally occurring windows of opportunity during which antecedent cues change by default (Moschis, 2007). New experiences in life, like moving, getting married or divorced, starting a new job, or even beginning a new school term, are all situations in which everything is being readjusted. Research shows that even a relatively simple change in life circumstances can be enough to disrupt a habit (Wood, Tam, & Witt, 2005; Verplanken & Wood, 2006). When our lives are in flux, adopting a new (sustainable) habit can be a lot easier.

CREATE A SOCIAL AVALANCHE FOR SUSTAINABILITY

Methodically Plan Change

Each proenvironmental behavior comes with a unique set of barriers and benefits, and each person approaches them with varying levels of motivation. Furthermore, while many effective strategies for behavior change were identified and reviewed in prior chapters (e.g., modeling, prompts, creating ease, rewards, feedback, and instructions), the effectiveness of each technique varies across situations depending on what and whose behavior is at stake (Osbaldiston & Schott, 2012). It can seem daunting, then, to plan and organize change interventions. However, taking it one methodical step at a time can help optimize the match between interventions and particular situations. One especially effective framework for navigating the complexities of human behavior change is **Community-Based Social Marketing** (CBSM). While not the only effective strategy grounded in psychological principles and methods, CBSM is flexible, comprehensive and effective. CBSM programs have been applied around the world to change behaviors that are most environmentally impactful: agricultural practices, transportation, home energy use, water use, overall resource consumption and waste, and toxic chemical production and use (McKenzie-Mohr, Lee, Schultz, & Kotler, 2012).

Rather than attempting to change underlying attitudes or values, CBSM involves learning about specific barriers that prevent people from changing behavior and designing psychology-based interventions that minimize or reduce these impediments (McKenzie-Mohr 2000a, 2000b). CBSM leverages the power of social networks and can be used by communities of all sizes and types, ranging from apartment complexes and work organizations to neighborhoods and watershed regions. The five basic steps of CBSM reflect the systematic process of psychological research you read about in Chapter 4: determining which behavior to change (DV), identifying behavioral contingencies (IVs), and designing a program to improve those contingencies. Programs are tested on a small group, fine-tuned, and then systematically applied. The final step is assessing the outcomes (see Figure 11.9) (McKenzie-Mohr, 2011).

The first step of the CBSM process is identifying and selecting environmentally appropriate behaviors to promote. Perhaps your campus is attempting to reduce its carbon footprint. This can be accomplished by changing food options, commuting patterns, and electricity demand. Many specific behaviors could be encouraged, such as commuting by

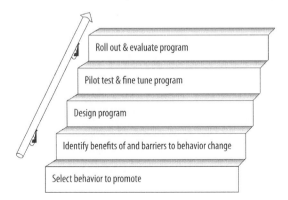

FIGURE 11.9 *Steps in Community-Based Social Marketing.*
McKenzie-Mohr, 2011.

bike, bus or carpool; turning off computers at night; and opting for meat-free meals in the cafeteria. Which behaviors will make the most significant and direct impact? Which are most likely to be adopted? Collecting some initial data through observations, interviews, and surveys can help determine the optimal focal behavior for the project.

The benefits and barriers associated with bicycle commuting are most certainly going to be different than those affiliated with eating a vegetarian meal. Thus, the next CBSM step, identifying the perceived benefits of, and barriers to, action, is critical. Barriers might include structural restrictions (e.g., no bike racks on campus), social norms (e.g., people are in the habit of eating meat with every meal), or personal costs of time, money, or status (e.g., I can get to school faster in my cool car; real men eat meat). Benefits are things that *will be perceived as benefits by the people whose behavior is changing.* Obviously bicycle commuting and reducing meat consumption benefit the environment, but not all individuals on campus will personally consider *that* a benefit; instead, benefits might be things like not having to purchase a parking permit and getting exercise on the way to school. Again, data collection can help pinpoint the barriers and benefits to be addressed.

Designing interventions that effectively decrease barriers to the target behavior and/or increase the benefits of it is perhaps the most challenging task within the process (you may recognize this as adjusting contingencies from Chapter 5; e.g., adding cues or removing potentially punishing outcomes of performing the behavior). This is challenging because situations are complex and varied. There are quite a variety of tools that can be employed, such as education, prompts, feedback, incentives, social norms and public commitments. Which ones will prove most useful depends on how high or low the relative barriers and benefits are. This balance is affected by factors such as how challenging, novel, and (un)popular the new behavior is, how deeply ingrained the old habits are, and the relative likelihood of changing the context; for instance, infrastructure often takes a long time to change. Luckily, Schultz (2014) has boiled down the current state of the science into a handy grid that can help with the intervention design process (see Figure 11.10).

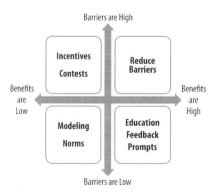

FIGURE 11.10 *Interventions that Match Needs of People and Situations.*
Adapted from Schultz (2014).

Let's look at it quadrant by quadrant:

- *Low benefit-low barrier.* If behavior isn't displayed when there are few benefits but also few barriers, it is likely due to low motivation. For example, reducing gas consumption by combining trips and avoiding rapid acceleration is free and completely under one's control but might not seem like it's important. Research has shown that highlighting social information (in the form of modeling and norms) works well for this situation.

- *Low benefit-high barrier.* When benefits are low and barriers are high, changing situational constraints (i.e., contingencies), especially by increasing benefits, can be helpful, if only temporarily. For instance, more expensive, infrequent behaviors such as purchasing a bus pass are often perceived as disproportionately high cost to minimal benefit. Incentives and competitions with rewards for the winners are especially effective to increase people's perceptions of value.

- *High benefit-high barrier.* In high benefit, high barrier situations, people are likely to be motivated, but the behavior may be perceived as too challenging. For instance, although bike commuting can significantly reduce carbon emissions and improve physical health, it takes time, can feel unsafe, and is not always comfortable. In this case, actually reducing barriers that affect difficulty (not just changing perceptions) through practice or facilities that support behavior (e.g., dedicated bike lanes) can help.

- *High benefit-low barrier.* High benefit, low barrier situations are when people are likely to be motivated but may need more information, may be forgetting or struggling with interfering habits, or don't know their current impact. An example might be carpooling. People might be agreeable to the idea and there may be few barriers, but they may have years of habits to overcome and might lack confidence reaching out to others. In these cases information, prompts, and feedback can be effective. For instance, a smartphone app might help initiate behavior through prompts, provide easy contact with a network of potential carpoolers, and deliver specific feedback about reduced carbon emissions.

Before applying a new program to a large audience, it makes sense to try it out on a small group first. Psychologists refer to this as a **pilot test**, which provides an opportunity to identify and correct problems without incurring significant expense. Often, instructions that seem completely clear and obvious to the writer are confusing to a person new to the situation, or web-links can malfunction. Once rough spots are ironed out, and instruction materials and measurement tools are honed and pretested, then the program is ready to be rolled out to the broader target community with confidence. Another key feature of this phase is estimating each program's cost-effectiveness. Some programs are effective, but cost prohibitive; sponsors and granting agencies will want to fund a more economically prudent option when available.

Launching a program within the broader target community takes substantial planning, time, and effort, and, even then, the process is not finished. While it is easy to assume that the new program has caused behavior change, it is important to formally evaluate its effectiveness through systematic observations or surveys. Using quasi-experimental methods (Chapter 4), such as including data from control groups, can strengthen the quality of findings. Evaluation is important for judging whether the program actually fulfilled its promise and whether further improvements should be considered.

Seek Opportunities to Lead

To create the necessary momentum, sustainable behavior must move beyond each individual's private sphere into the realm of organizations: work, faith, education, sports, neighborhood, and other communities. Within groups, whether formal or informal, someone has to step up and advocate for sustainable ideas, and the ideas need to be perceived as consistent with the group's values (Bansal, 2003). There are standard tools that organizations use to publicly communicate their values. A **mission statement** concisely expresses a group's purpose and aspirations. **Vision statements** tend to be future-oriented descriptions of the essential goals or injunctive norms of a group (Chapter 5); they provide a bridge between reality and aspirations. When clearly articulated, mission and vision statements provide guidance for daily decisions, especially among organizational leaders (Darbi, 2012). Many organizations already have mission and vision statements that encompass sustainability. When the connection is not so obvious, providing explicit links may be necessary. For instance, when mission statements include values such as community, citizenship, leadership, stewardship, and the common good, sustainability proponents can highlight how these relate to environmental sustainability (see Figure 11.11). If a group does not yet have a mission or vision

> ...the University of St. Thomas educates students to be morally responsible leaders who think critically, act wisely and work skillfully to advance the common good.

FIGURE 11.11 *Mission statement referring to the common good.*
Sustainability advocates at the University of St. Thomas in St. Paul, Minnesota (including two of the authors) have garnered institutional support for numerous sustainability initiatives by linking them to the school's mission to "advance the common good."

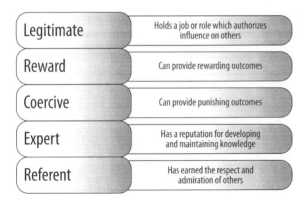

Legitimate	Holds a job or role which authorizes influence on others
Reward	Can provide rewarding outcomes
Coercive	Can provide punishing outcomes
Expert	Has a reputation for developing and maintaining knowledge
Referent	Has earned the respect and admiration of others

FIGURE 11.12 *Adapted from French and Raven's (1959) Bases of power.*

statement, writing them provides a significant opportunity to feature the value of contributing to a sustainable future. Although sustainability-consistent mission and vision statements are no guarantee of sustainable behavior, they are a critical first step toward walking-the-talk; a way to develop a shared understanding of "this is how we do it here."

It only takes *one individual* to suggest a sustainable idea within a group. Whether the idea gains traction or not is another story and often depends on the person's relationship with others. Most individuals have a network of people with whom they regularly interact; i.e., a sphere of influence. Anyone with some source of power, some method of motivating the others, can influence those within their network to believe in and contribute to an idea (French & Raven, 1959; see Figure 11.12).

Some people hold an official position that grants them **legitimate power**; it's their *job* to create opportunities for others, provide support, and generally make things happen. Examples include team captain, club president, CEO, manager, group leader, coach, teacher, director, parent, or minister. There are ways to link to sustainability within all of these diverse domains. For example, spiritual leaders have encouraged followers to become more involved in environmental issues and adopt more proenvironmental behaviors by invoking the sanctity of nature (Beyer, 2004; Tarakeshwar, Swank, Pargament, & Mahoney, 2001). In fact, surveys from over 150 Georgian Presbyterian churches revealed that the vast majority of ministers supported environmental stewardship (over "domination of nature"), and their personal proenvironmental behaviors had a significant influence on the behaviors and beliefs of the members of their congregation (Holland & Carter, 2005).

Even without an official leadership role, individuals can influence others by drawing on various other sources of power. For instance, being a reliable source of information generates **expert power**. The most potent way to influence others is to garner their respect and admiration through good work, honesty, kindness, humility, and generosity. People who do so hold **referent power**. As long as referent leaders do not exploit their power, others willingly follow their lead. CBSM makes the most of expert and referent power by identifying **community block leaders**, well-respected community members who already engage in a target behavior, and asking them to promote and mentor people in their network (McKenzie-Mohr, 2011).

The extensive societal changes required for sustainability necessitate **transformational leaders** who can get people to refocus, reprioritize, work at tasks above and beyond the call of duty, *and* feel good about it at the same time. Such leaders:

- energize and empower followers to take on risks and to overcome challenges;
- embody expert and referent power, creating trust by making sound decisions (Pierro, Raven, Amato, & Bélanger, 2013);
- consistently model proenvironmental behavior (Robertson & Barling, 2013);
- encourage others to think critically and creatively, enabling autonomy and developing new generations of leaders; and
- display empathy and encouragement, enhancing others' self-efficacy (Barling, Christie, & Hoption, 2010).

It may seem like a rare individual who can pull off this kind of leadership. However, *anyone* can learn and mindfully apply these leadership behaviors (Brown & May, 2012; see Figure 11.13).

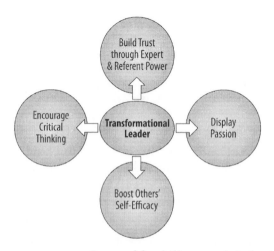

FIGURE 11.13 *Learnable skills associated with transformational leadership.*

While most people tend to focus their effort on private, individual behavior, **political advocacy** for the larger public good is equally essential because even when people know about and would like to choose responsible behaviors, it is not always possible to do so. Personal behaviors are significantly constrained by societal infrastructure. For instance, sometimes the post office, grocery store, and office are ten miles from where one lives and there is no mass transit system or the road is narrow and curvy, making bike riding dangerous for bikers and drivers alike. Driving a car becomes the only practical option.

Changing the infrastructure of our daily lives by designing contexts, processes, and products differently will allow people to behave sustainably without having to think about and assess how their choices impact ecological systems (Dolan, et al., 2012). For instance, legislation in 28 countries (and 25 U.S. states) requires manufacturers or retailers to **take back** products, components, or packaging for remanufacture or reuse (Electronics TakeBack Coalition, n.d.; EPA, 2012c). When consumers have a place to take their obsolete phones and computers, conveniently the very place where they are buying a replacement, recycling behavior increases significantly. But, just like in organizations, *somebody* has to take the leap and put forth an idea. Unlike changing private behaviors, changing infrastructure requires individuals to participate in public dialog and political action.

Individuals make a difference when they speak out about local, national, and international policy. Citizens can use their right to vote and participate in public forums. New governance models, such as participatory budgeting, give community members a more direct voice in forming priorities and funding projects that affect their neighborhoods (Baiocchi & Lerner, 2007). Traditional channels, too, can be employed in novel ways. For example, several young people recently filed lawsuits to require the U.S. federal government to initiate "national climate recovery" actions to avoid the most catastrophic consequences of global warming. The plaintiffs' claims are based on the "public trust doctrine," which dates back to Roman law and requires the "government to protect and maintain survival resources for future generations" (Our Children's Trust, 2014). In fact, the constitutions of 98 nations recognize a healthy environment as a basic right.

One of the best things about civic participation is that it may lead to more civic participation. When citizens are involved in community decisions (e.g., about regional environmental resources and land uses), they find themselves empowered by their responsibility. They also feel deepened ties to other community members and experience enhanced feelings of attachment to place (Chapter 10). Therefore, they feel motivated to remain involved.

Seek Opportunities to Follow

Some have argued that a large-scale **social movement** is required to bring about our sustainable future (Hawken, 2007; Meyer, 2007). Importantly, social movements are made up of leaders *and followers*, so even if people don't feel like they

FIGURE 11.14 *Diffusion of innovation throughout a social system.*

are the leader type, they can still inspire social change by being on the lookout for opportunities to jump on the bandwagon. The classic model, **diffusion of innovation**, describes the flow of ideas from their inception to adoption throughout a social system (Rogers, 1962/2010; See Figure 11.14).

First, an **innovator** must get the ball rolling with a useful idea, process, solution or object. These typically bold and imaginative people have relevant technical expertise, and are comfortable with novelty, risk, and uncertainty. They tend to be able to absorb financial consequences of failure. Innovators are often socially connected to other innovators, but are considered a bit deviant by the average Joe. Thus, while innovators initiate movement within the system and influence the next phase of the process, they are not likely to generate extensive diffusion. In rare cases, however, innovators can also be skilled and prolific communicators; in this case, social diffusion tends to progress more quickly (Tolba & Mourad, 2011).

For a social movement to gain momentum, the creative pioneer needs some intrepid followers to become **early adopters**. These first followers model enthusiastic support for the idea and facilitate "contagion" by inviting their networks to join in (Young, 2009). Early adopters like to try new things and have a knack for recognizing which inventions or strategies are likely to be successful, helping them develop a reputation as reliable sources of information (i.e., as experts). They tend to be more integral to the system than innovators so they can easily spread the word. In fact, it is in this category that you find **opinion leaders** who are highly connected and often operate at the center of communication networks. Because of their informal use of expert *and* referent power, opinion leaders have the most influence on diffusion, through word-of-mouth and modeling (Dearing, 2009). Opinion leaders do not necessarily have the *most* connections. In fact, having too many connections tends to bog down the diffusion. A recent analysis of Twitter behavior suggests that diffusion decreases as opinion leaders get too popular. Under these conditions, opinion leaders become overwhelmed and unable to keep up with incoming information (Harrigan, Achananuparp, & Lim, 2012). In order to bring along the next phase of diffusion, early adopters must behave in ways that are prudent, public, and easy to follow. When people in the **mainstream** see how "joining in" works, they feel competent and confident to do so.

Although people in the mainstream typically take longer to jump in, once they do, the movement's momentum is hard to stop. You might have observed this when a fad or video "goes viral." Contagion moves fastest in communities of people who share interests and personal characteristics (e.g., age, cultural background, political leanings). Within the mainstream, the **early majority** actively seeks new information and has rich social networks, while the **late majority** tends to enter the fray only when there is little risk and when they perceive peer pressure to do so. Often, the later adopters are cautious because they cannot afford to take financial and social risks. However, as the level of adoption grows, the risks of participation drop significantly so people can join in more and more confidently. Eventually, it will be riskier to be left out than to join in. This is when the latest adopters, people who are suspicious of changes, or those who are least able to "afford" risk, join in.

The diffusion of innovation model assumes that innovativeness is an individual difference variable. In fact, researchers have identified personality attributes that contribute to innovativeness, as well as distinct roles that group members play in the process of diffusion. However, while some of us may have general innovative tendencies or creative ideas, whether we act on them varies from situation to situation depending on our interest, available resources, and perhaps courage. But if you don't see yourself as a sustainability innovator, remember that the early adopters, opinion leaders, and early majorities can be more important than being the leader; it's actually the followers who transform the effort from an oddity into a real movement. So, if you see someone doing something sustainable, consider joining in, and using your networks to facilitate diffusion. Make your effort visible so that others can follow, provide praise for participating, and convey the excitement of making change.

CONCLUSION

Human behavior is not easy to change, but it does change. Consider how much human behavior has changed in the last 100 years. Because the shift to sustainability needs to happen sooner rather than later, it is necessary to intentionally accelerate the pace of our own and others' transition. If current and future generations are to have a chance to fulfill their potential, experience happiness, and delight in the diverse wonders of the natural world, people need to make significant changes in their behaviors, thoughts, feelings, and values. Change begins by learning about why it's needed and how it works, and hopefully this book has assisted you with that step.

No one approach to psychological inquiry can capture the complexity of human behavior. Fortunately, a diverse group of psychologists is working together to facilitate the "social avalanche" toward sustainability (Fischer et al., 2012, p. 159). Whether they identify as conservation psychologists, ecopsychologists, or environmental psychologists—or work within traditional areas like cognitive, developmental, or social psychology—researchers are conducting important scientific investigations into behaviors that promote or detract from ecosystem health. As you read about their work in this text, you gained an understanding of the basis of environmentally destructive behavior and some insight into what strategies will likely work best for the project of building a sustainable world.

Meeting current challenges will require a personal commitment to changing one's own behaviors as well as the larger structures that propel them. Inaction will ensure that business proceeds as usual, yet business as usual is already leading societies to environmental, social, and economic collapse. Alternatively, undertaking action will change us all, as well as the world around us. Cumulative individual efforts will create new and synergistic coalitions, leading to ecologically-harmonious organizations and policy. No matter what actions you decide to take, becoming aware of your behavior, thoughts, and feelings will significantly facilitate your effectiveness. As Deborah Winter concluded in the first edition of this text:

> it is essential to proceed gently, with conviction, patience, perseverance, and most of all, with trust; trust in yourself, as well as in the interconnected whole that embraces you. "If you take one step with all the knowledge you have, there is usually just enough light shining to show you the next step" (Williams, 1994). May your steps be steady, graceful, revealing, and rewarding, and may you know them as part of the greater ecological dance.
>
> (1996, p. 302)

NOTES

1. Another moving example is available in Alex Steffen's TED talk, http://www.ted.com/talks/alex_steffen_sees_a_sustainable_future
2. You can find tools and strategies at http://www.transitionnetwork.org
3. See http://www.onetonline.org/find/green

Appendix: Self-Change Project

You can follow the steps below to modify your own unsustainable behavior (see Figure A.1; adapted from Martin & Pear, 2009). Changing your own behavior can be a step toward changing the behavior of others; modeling is a powerful social influence. It can also serve as a springboard to systemic change by inspiring you to actively shape the societal contingencies that constrain your behaviors.

Step 1. Define the problem. As an illustration, a self-control project could focus on reducing paper waste. In this case, *the problem* is that approximately 35% of the world's timber harvest is used to produce paper. Paper is the largest component (constituting 28% or 70 million tons) of the municipal solid waste generated in the United States (U.S. Environmental Protection Agency, 2013). Although about 66% of used paper is recovered for recycling, every stage of paper production and consumption constitutes a significant environmental threat (e.g., Vanasselt, 2001): Most wood extraction procedures lead to deforestation and the loss of forest diversity; toxic liquid pollutants are released during pulp- and papermaking; paper products in landfills generate methane, a gas that contributes to climate change; and the incineration of paper bleached with chlorine releases dioxin, a highly toxic chemical that is a known carcinogen and endocrine (hormone) disruptor (see further discussion in Chapter 9). Thus, even small actions undertaken by many people could significantly reduce the harmful impacts of paper manufacturing, use, and disposal.

Step 2. Observe baseline behavior. Start by identifying the variety of different behaviors involved with your paper use. Think about their relative importance. For instance, maybe you throw out a lot of paper packaging, accumulate a lot of paper grocery bags, print a ton of assignments single-sided, and throw away scads of paper napkins. Which of these dominates your paper use? Which has the most downstream impact? Keeping it simple but significant will give you the most bang for the buck, so after you have learned where you can make improvements, pick one or two changes as your focus. After you have picked a behavior, count and graph it over a period of time, say, two weeks *prior to initiating any attempt at change*. This provides a **baseline** measure of behavior as it is occurring with existing contingencies.

Given this picture of unaltered behaviors, you can also monitor situational factors that influence them. For example, you may be good about photocopying on two sides of the paper *unless you are rushed*. Because the photocopy machine is more likely to jam when making double-sided copies, you might avoid doing

Steps of a Self-Change Project

1. Define the problem.
2. Observe baseline behavior.
3. Set a specific goal.
4. Make a public commitment.
5. Design a plan to alter the contingencies.
6. Create a contract.
7. Track behavior and adjust the plan as needed.
8. Generalize the change to related behaviors.

FIGURE A.1 *Steps of a Self-Change Project.*

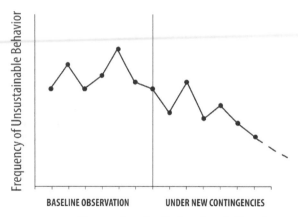

FIGURE A.2 *Example of a Behavioral Graph.*

it when you have limited time. Similarly, it might be easy to use cloth napkins instead of paper when at home, but not when you go out to eat. Analysis like this might reveal that your paper overuse occurs because of time pressure or convenience. Figure A.2 shows what a behavioral record would look like in graph form, both for the baseline period and the period where the contingencies are changed.

Step 3. Set a specific goal. An intention to reduce consumption or minimize paper waste is too vague. Defining the goal *specifically,* in behavioral terms, enhances your ability to track progress, which increases probability of success (for more on this, see Chapter 8). With respect to paper use, your goal might be to reduce it by 50%. The goal should represent a significant but realistic change. In behavioral terms, this could involve specific actions such as the following:

- Read articles online, only printing when absolutely necessary.

- Photocopy on both sides of each page of paper.

- Print drafts of manuscripts and informal notes on the backs of junk paper.

- Reuse envelopes for informal mail.

- Reuse gift wrapping received from others.

- Use scrap paper for lists, notes, and reminders.

- Use cloth towels and napkins instead of paper.

- Purchase secondhand books and pass your books onto others.

- Go to the library for magazines and books instead of purchasing your own.

- Bring a cloth bag to the grocery store rather than taking paper (or plastic) ones.

Step 4. Make a public commitment. Tell someone about the project. As you read in Chapter 5, people who have made a public commitment are considerably more likely to follow through with changing the behavior.

Step 5. Design a plan to alter the contingencies. Figure out ways you can change the antecedents and consequences of your undesired and desired behaviors. For example, to reduce napkin use, you could keep a cloth napkin in your backpack so you'll have it with you when you go to meals. If it is easier to reuse envelopes opened with a letter opener than ones that get torn when you open them by hand, place a letter opener where you open the mail. Designing such antecedent conditions enables more effective cuing of one's behavior.

Manipulating consequent stimuli can also be very helpful. When you successfully change your behavior, you can reward yourself with a special treat. Perhaps inviting a friend over for lunch, swimming for an hour, or watching a movie would be reinforcing events. (Make sure the reinforcer doesn't contradict your overall goal. For instance, a new magazine subscription would be contraindicated in this example!)

Similarly, you could institute punishers. You could ask a friend to scold you (or you could scold yourself) when you forget to bring your own bags to the store (positive punishment). By keeping a running account of the costs associated with your paper use (e.g., how much you spend on paper towels and napkins, printer paper, and giftwrap), you can experience more directly the adverse impacts of waste (i.e., negative punishers). If this loss of money is not significant enough to create discomfort, you could require yourself to do something more costly when you waste paper, such as donate money to an organization that plants trees, or plant some trees yourself.

Step 6. Create a contract. Drawing up a formal contract makes your plans concrete. The contract should specify the goal behaviors and consequent contingencies.

For example, it might say, "I will print all of my assignments double-sided. If I successfully do this for a whole week, I will reward myself with a movie. If I fail to do this for a whole week, I will make a $5 donation to an environmental group." Keep your contract visible (make it your computer wallpaper!).

Step 7. Track behavior and adjust the plan as needed. Continue to record and graph your behaviors. This way you can judge the extent to which the altered antecedent and consequent strategies are successfully changing your paper use. If you are experiencing less success than you had anticipated, revisit your contingency plan and make adjustments as necessary.

Step 8. Generalize the change to related behaviors. At some point, your new behaviors should become habitual, or at least be performed more often than not. Then you might start noticing ways in which you can include additional behaviors within the same domain. For example, the goal of conserving paper might broaden to include reducing food packaging. You might be inspired to make changes in other categories, too, like conserving energy by lowering the thermostat, turning off appliances when not in use, or carpooling and bicycling instead of driving alone.

Good luck (and good work)!

References

Abdallah, S., Michaelson, J., Shah, S., Stoll, L., & Marks, N. (2012). *The Happy Planet Index: 2012 report: A global index of sustainable well-being*. London, United Kingdom: New Economics Foundation. Retrieved from http://www.happyplanetindex. org/assets/happy-planet-index-report.pdf

Abrahamse, W., & Steg, L. (2011). Factors related to household energy use and intention to reduce it: The role of psychological and socio-demographic variables. *Human Ecology Review, 18*(1), 30–40.

Abrahamse, W., Steg, L., Vlek, C., & Rothengatter, T. (2005). A review of intervention studies aimed at household energy conservation. *Journal of Environmental Psychology, 25*, 273–291. doi: 10.1016/j.jenvp.2005.08.002

Abrahamse, W., Steg, L., Vlek, C., & Rothengatter, T. (2007). The effect of tailored information, goal setting, and tailored feedback on household energy use, energy-related behaviors, and behavioral antecedents. *Journal of Environmental Psychology, 27*, 265–276. doi: 10.1016/j.jenvp.2007.08.002

Abram, D. (1996). *The spell of the sensuous: Perception and language in a more than human world*. New York, NY: Pantheon.

Adams, J.S. (1965). Inequity in social exchange. In L. Berkowitz (Ed.), *Advances in experimental social psychology: Vol. 2* (pp. 267–299). New York, NY: Academic Press.

Afzal, S., Bojesen, S.E., & Nordestgaard, B.G. (2014). Reduced 25-hydroxyvitamin D and risk of Alzheimer's disease and vascular dementia. *Alzheimer's & Dementia, 10*(3), 296–302. doi: 10.1016/j.jalz.2013.05.1765

Aguilar-Luzón, M.D.C., García-Martínez, J.M.Á., Calvo-Salguero, A., & Salinas, J.M. (2012). Comparative study between the theory of planned behavior and the value-belief-norm model regarding the environment, on Spanish housewives' recycling behavior. *Journal of Applied Social Psychology, 42*, 2797–2833. http://dx. doi.org/10.1111/j.1559-1816.2012.00962.x

Agyeman, J., Doppelt, B., Lynn, K., & Hatic, H. (2007). The climate-justice link: Communicating risk with low-income and minority audiences. In S.C. Moser, & L. Dilling (Eds.), *Creating a climate for change: Communicating climate change and facilitating social change* (pp. 119–138). New York, NY: Cambridge University Press.

Ahn, S.J., Bailenson, J.N., & Park, D. (2014). Short- and long-term effects of embodied experiences in immersive virtual environments on environmental locus of control and behavior. *Computers in Human Behavior, 39*, 235–245.

Aitken, C., McMahon, T., Wearing, A., & Finlayson, B. (1994). Residential water use: Predicting and reducing consumption. *Journal of Applied Social Psychology, 24*, 136–158. doi: 10.1111/j.1559-1816.1994.tb00562.x

Ajzen, I. (1991). The theory of planned behavior. *Organizational Behavior and Human Decision Processes, 50*(2), 179–211. doi: 10.1016/0749-5978(91)90020-T

Ajzen, I. (1998). Models of human social behavior and their application to health psychology. *Psychology & Health, 13*(4), 735–739. doi: 10.1080/08870449808407426

Ajzen, I., Rosenthal, L. H., & Brown, T. C. (2000). Effects of perceived fairness on willingness to pay. *Journal of Applied Social Psychology, 30,* 2439–2450. doi: 10.1111/j.1559-1816.2000.tb02444.x

Alarcon, G. M., Bowling, N. A., & Khazon, S. (2013). Great expectations: A meta-analytic examination of optimism and hope. *Personality and Individual Differences, 54,* 821–827.doi: 10.1016/j.paid.2012.12.004

Albrecht, G. (2005). Solastalgia, a new concept in human health and identity. *Philosophy Activism Nature, 3,* 41–44.

Albrecht, G. (2012). Psychoterratic conditions in a scientific and technological world. In P. H. Kahn, Jr., & P. H. Hasbach (Eds.), *Ecopsychology: Science, totems, and the technological species* (pp. 241–264). Cambridge, MA: MIT Press.

Albrecht, G., Sartore, G.-M., Connor, L., Higginbotham, N., Freeman, S., Kelly, B., . . . Pollard, G. (2007). Solastalgia: The distress caused by environmental change. *Australasian Psychiatry, 15*(1), S95–S98. doi: 10.1080/10398560701701288

Allcott, H. (2011). Social norms and energy conservation. *Journal of Public Economics, 95,* 1082–1095. doi:10.1016/j.jpubeco.2011.03.003

Allen, M., Wicks, R. B., & Schulte, S. (2013). Online environmental engagement among youth: Influence of parents, attitudes and demographics. *Mass Communication & Society, 16*(5), 661–686. doi: 10.1080/15205436.2013.770032

Alzahabi, R., & Becker, M. W. (2013). The association between media multitasking, task-switching, and dual-task performance. *Journal of Experimental Psychology: Human Perception and Performance, 39*(5), 1485–1495. doi: 10.1037/a0031208

Amel, E. L., & Manning, C. M. (2012). Exploring the effectiveness of ecological principles as a method for integrating environmental content into psychology courses. *Ecopsychology, 4*(2), 127–136. doi: 10.1089/eco.2012.0034

Amel, E. L., Manning, C. M., Forsman, J. W., & Scott, B. A. (2011, April). Goal specificity, goal difficulty and perceived need to develop skills in the context of environmental sustainability. Symposium for Society for Human Ecology, Las Vegas, NV.

Amel, E. L., Manning, C. M., & Scott, B. A. (2009). Mindfulness and sustainable behavior: Pondering attention and awareness as means for increasing green behavior. *Ecopsychology, 1,* 14–25. doi: 10.1089/eco.2008.0005

Amel, E. L., Manning, C. M., & Scott, B. A. (2013, September). Goal setting in context: The impact of psychological distance on the appeal of goals. 10th Biennial Conference on Environmental Psychology, Magdeburg, Germany.

Amel, E. L., Scott, B. A., Manning, C. M., & Forsman, J. W. (2009, April). Increasing employees' green behavior: The role of I/O constructs such as attitudes, norms, motivation, and leadership. In M. Dela Rosa (Chair), *What is I/O psychology's role in supporting green business initiatives?* Symposium for the annual meeting of the Society for Industrial and Organizational Psychology, New Orleans, Louisiana.

Amel, E. L., Scott, B. A., Manning, C. M., & Stinson, J. (2007, August). I'm not an environmentalist, I just behave like one. Presentation by first author in session on Conservation Psychology in the Field at the annual meeting of the American Psychological Association, San Francisco, CA.

American Academy of Pediatrics (2013). *Where we stand: TV viewing time.* Retrieved from http://www.healthychildren.org/English/family-life/Media/Pages/Where-We-Stand-TV-Viewing-Time.aspx

American Psychiatric Association (2013). *Diagnostic and statistical manual of mental disorders, 5th Edition: DSM-5.* Arlington, VA: American Psychiatric Publishing.

Anderson, R. C. (2005). *Mid-course correction: Toward a sustainable enterprise: The Interface model.* White River Junction, VT: Chelsea Green.

Andrews, L. W. (2013, March 12). Born to be junk food junkies. *Psychology Today.* Retrieved from http://www.psychologytoday.com/blog/minding-the-body/

Annerstedt, M., & Währborg, P. (2011). Nature-assisted therapy: Systematic review of controlled and observational studies. *Scandinavian Journal of Public Health, 39*(4), 371–388. doi: 10.1177/1403494810396400

Appleton, J. (1975). *The experience of landscape.* New York, NY: Wiley.

Arndt, J., Solomon, S., Kasser, T., & Sheldon, K. M. (2004). The urge to splurge: A terror management account of materialism and consumer behavior. *Journal of Consumer Psychology, 14*(3), 198–212. doi: 10.1207/s15327663jcp1403_2

Arnocky, S., Milfont, T. L., & Nicol, J. R. (2014). Time perspective and sustainable behavior: Evidence for the distinction between consideration of immediate and future consequences. *Environment and Behavior, 46*(5), 556–582. doi: 10.1177/0013916512474987

Arnocky, S., Stroink, M., & DeCicco, T. (2007). Self-construal predicts environmental concern, cooperation, and conservation. *Journal of Environmental Psychology 27,* 255–264. doi: 10.1016/j.jenvp.2007.06.005

Arnold, R. (1996). Overcoming Ideology. In P. D. Brick, & R. McGreggor Cawley (Eds.), *A wolf in the garden: The land rights movement and the new environmental debate* (pp. 15ff). Lanham, MD: Rowman & Littlefield Publishers.

Aronson, E. (1968). Dissonance theory: Progress and problems. In R. P. Abelson, E. Aronson, W. J. McGuire, T. M. Newcomb, M. J. Rosenberg, & P. H. Tannebaum (Eds.), *Theories of cognitive consistency: A sourcebook.* Chicago, IL: Rand McNally.

Aronson, E., & O'Leary, M. (1983). The relative effectiveness of models and prompts on energy conservation: A field experiment in a shower room. *Journal of Environmental Systems, 12,* 219–224.

Ashley, P., Nishioka, M., Wooton, M. A., Zewatsky, J., Gaitens, J., & Anderson, J. (2006). *Healthy homes issues: Pesticides in the home—Use, hazards, and integrated pest management.* U.S. Department of Housing and Urban Development (HUD). Retrieved from http://www.hud.gov/offices/lead/library/hhi/Pesticide_Final_Revised_04-26-06.pdf

Ashton, M. C., & Lee, K. (2005). Honesty-Humility, the Big Five, and the Five-Factor Model. *Journal of Personality, 73,* 1321–1353. doi: 10.1111/j.1467-6494.2005.00351.x

Ashton, M.C., & Lee, K. (2007). Empirical, theoretical, and practical advantages of the HEXACO model of personality structure. *Personality and Social Psychology Review, 11*(2), 150–166. doi: 10.1177/1088868306294907

Ashton, M.C., Lee, K., Perugini, M., Szarota, P., de Vries, R.E., DiBlas, L., . . . De Raad, B. (2004). A six-factor structure of personality-descriptive adjectives: Solutions from psycholexical studies in seven languages. *Journal of Personality and Social Psychology, 86*(2), 356–366. doi: 10.1037/0022-3514.86.2.356

Aspinwall, L.G., & Staudinger, U.M. (Eds.). (2003). *A psychology of human strengths: Fundamental questions and future directions for a positive psychology*. Washington, DC: American Psychological Association.

Attari, S.Z. (2014). Perceptions of water use. *Proceedings of the National Academy of Sciences, 111*(14), 5129–5134. doi: 10.1073/pnas.1316402111

Aubrey, A. (2014, February 21). How tracing the oil in your pop-tarts may help save rain forests. *National Public Radio*. Retrieved from http://www.npr.org/blogs/thesalt/2014/02/20/280257631/how-tracing-the-oil-in-your-pop-tarts-may-help-save-rainforests

Aubrun, A., Brown, A., & Grady, J. (2006). *Conceptualizing US food systems with simplifying models: Findings from the TalkBack testing*. Washington, DC: FrameWorks Institute Research Report.

Ayres, I., Raseman, S., & Shih, A. (2013). Evidence from two large field experiments that peer comparison feedback can reduce residential energy usage. *Journal of Law, Economics, and Organization, 29*(5), 992–1022. doi: 10.3386/w15386

Ayupan, L.B., & Oliveros, T.G. (1994). Filipino peasant women in defence of life. In V. Shiva (Ed.), *Close to home: Women reconnect ecology, health, and development* (pp. 113–120). London, United Kingdom: Earthscan Publications.

Backscheider, A.G., Shatz, M., & Gelman, S.A. (1993). Preschoolers' ability to distinguish living kinds as a function of regrowth. *Child Development, 64*, 1242–1257. doi: 10.1111/j.1467-8624.1993.tb04198.x

Bacon, F. (1620/1955). Preface of the great instauration. From H. G. Dick (Ed.), *Selected writings of Francis Bacon* (p. 447). New York, NY: The Modern Library.

Baier, M., Kals, E., & Müller, M.M. (2013). Ecological belief in a just world. *Social Justice Research, 26*(3), 272–300. doi: 10.1007/s11211-013-0192-0

Bailey, R. (1993). *Eco-scam: The false prophets of ecological apocalypse*. New York, NY: St. Martin's Press.

Bain, P.G., Hornsey, M.J., Bongiorno, R., & Jeffries, C. (2012). Promoting pro-environmental action in climate change deniers. *Nature Climate Change, 2*, 600–603. doi: 10.1038/nclimate1532

Baiocchi, G., & Lerner, J. (2007). Could participatory budgeting work in the United States? *The Good Society, 16*(1), 8–13. doi: 10.1353/gso.0.0009

Baltes, M.M., & Hayward, S.C. (1976). Application and evaluation of strategies to reduce pollution: Behavior control of littering in a football stadium. *Journal of Applied Psychology, 61*, 501–506. doi: 10.1037/0021-9010.61.4.501

Bamberg, S. (2013). Changing environmentally harmful behaviors: A stage model of self-regulated behavioral change. *Journal of Environmental Psychology, 34*, 151–159. doi: 10.1016/j.jenvp.2013.01.002

Bamberg, S., Ajzen, I., & Schmidt, P. (2003). Choice of travel mode in the theory of planned behavior: The roles of past behavior, habit, and reasoned action. *Basic and Applied Social Psychology, 25*(3), 175–187. doi: 10.1207/S15324834BASP2503_01

Bamberg, S., & Möser, G. (2007). Twenty years after Hines, Hungerford, and Tomera: A new meta-analysis of psycho-social determinants of pro-environmental behaviour. *Journal of Environmental Psychology, 27*, 14–25. doi: 10.1016/j.jenvp.2006.12.002

Bamberg, S., & Schmidt, P. (2001). Theory-driven subgroup-specific evaluation of an intervention to reduce private car use. *Journal of Applied Social Psychology, 31*(6), 1300–1329. doi: 10.1111/j.1559-1816.2001.tb02675.x

Bandura, A. (1977). *Social learning theory.* Englewood Cliffs, NJ: Prentice-Hall.

Bandura, A., & Cervone, D. (1983). Self-evaluative and self-efficacy mechanisms governing the motivational effects of goal systems. *Journal of Personality and Social Psychology, 45*, 1017–1028. doi: 10.1037/0022-3514.45.5.1017

Bansal, P. (2003). From issues to actions: The importance of individual concerns and organizational values in responding to natural environmental issues. *Organizational Science, 14*(5), 510–527. doi: 10.1287/orsc.14.5.510.16765

Bara ski, M., Srednicka-Tober, D., Volakakis, N., Seal, C., Sanderson, R., Stewart, G. B., . . . Leifert, C. (2014). Higher antioxidant and lower cadmium concentrations and lower incidence of pesticide residues in organically grown crops: a systematic literature review and meta-analyses. *British Journal of Nutrition, 112*, 794–811.

Barling, J., Christie, A., & Hoption, C. (2010). Leadership. In S. Zedeck (Ed.), *APA handbook of industrial and organizational psychology* (Vol. 1, pp. 183–240). Washington, DC: American Psychological Association.

Barona, E., Ramankutty, N., Hyman, G., & Coomes, O. T. (2010). The role of pasture and soybean in deforestation of the Brazilian Amazon. *Environmental Research Letters, 5*, 024002. doi:10.1088/1748-9326/5/2/024002

Barouki, R., Gluckman, P. D., Grandjean, P., Hanson, M., & Heindel, J. J. (2012). Developmental origins of non-communicable disease: Implications for research and public health. *Environmental Health, 11*, 42. doi: 10.1186/1476-069X-11-42

Barr, S. (2007). Factors influencing environmental attitudes and behaviors. *Environment and Behavior, 39*, 435–473. doi: 10.1177/0013916505283421

Bartels, J., & Onwezen, M. C. (2014). Consumers' willingness to buy products with environmental and ethical claims: the roles of social representations and social identity. *International Journal of Consumer Studies, 38*(1), 82–89. doi: 10.1111/ijcs.12067

Bashir, N. Y., Lockwood, P., Chasteen, A. L., Nadolny, D., & Noyes, I. (2013). The ironic impact of activists: Negative stereotype reduce social change influence. *European Journal of Social Psychology, 43*, 614–626. doi: 10.1002/ejsp.1983

Batson, C.D., Early, S., & Salvarani, G. (1997). Perspective taking: Imagining how another feels versus imagining how you would feel. *Personality and Social Psychology Bulletin, 23*(7), 751–758. doi: 10.1177/0146167297237008

Baum, W.M. (1994). *Understanding behaviorism: Science, behavior, and culture.* New York, NY: HarperCollins.

Baxter, K., Boisvert, A., Lindberg, C., & Mackrae, K. (2009). *Sustainability primer.* The Natural Step Canada. Retrieved from http://sustainabilitycommittee.uhh.hawaii.edu/documents/PrimerGuidebookNAT-LowRes_20090609.pdf

Bazerman, M.H. (2006). Climate change as a predictable surprise. *Climatic Change, 77,* 179–193. doi: 10.1007/s10584-006-9058-x

Bazerman, M.H., & Hoffman, A.J. (1999). Sources of environmentally destructive behavior: Individual, organizational, and institutional perspectives. *Research in Organizational Behavior, 2,* 39–79.

Beatley, T. (2011). *Biophilic cities: Integrating nature into urban design and planning.* Washington, DC: Island Press.

Beck, A.M., & Katcher, A.H. (2003). Future directions in human-animal bond research. *American Behavioral Scientist, 47,* 79–93. doi: 10.1177/0002764203255214

Beck, J.W., & Schmidt, A.M. (2013). State-level goal orientations as mediators of the relationship between time pressure and performance: A longitudinal study. *Journal of Applied Psychology, 98*(2), 354–363. doi: 10.1037/a0031145

Beck, L., & Madresh, E.A. (2008). Romantic partners and four-legged friends: An extension of attachment theory to relationships with pets. *Anthrozoos, 21,* 43–56. doi: 10.2752/089279308X274056

Becker, L.J. (1978). Joint effect of feedback and goal setting on performance: A field study of residential energy conservation. *Journal of Applied Psychology, 63,* 428–433.

Bedrosian, T.A., & Nelson, R.J. (2013). Influence of the modern light environment on mood. *Molecular Psychiatry, 18*(7), 751–757. doi:10.1038/mp.2013.70

Bekoff, M. (2007). *The emotional lives of animals: A leading scientist explores animal joy, sorrow, and empathy—And why they matter.* Novato, CA: New World Library.

Bellinger, D.C. (2012). A strategy for comparing the contributions of environmental chemicals and other risk factors to neurodevelopment of children. *Environmental Health Perspectives, 120,* 501–507. doi: 10.1289/ehp.1104170

Bellinger, D.C. (2013). Prenatal exposures to environmental chemicals and children's neurodevelopment: An update. *Safety and Health at Work, 4,* 1–11. doi: 10.5491/SHAW.2013.4.1.1

Bellinger, D.C., & Adams, H.F. (2001). Environmental pollutant exposures and children's cognitive abilities. In R.J. Sternberg, & E.L. Grigorenko (Eds.), *Environmental effects on cognitive abilities* (pp. 157–188). Mahway, NJ: Erlbaum.

Benyus, J. (2010). Mother Nature's school of design. In NW Earth Institute (Ed.), *Choices for Sustainable Living* (pp. 34–38). Portland, OR: NW Earth Institute.

Berenguer, J. (2007). The effect of empathy in proenvironmental attitudes and behavior. *Environment and Behavior, 39,* 269–283. doi: 10.1177/0013916506292937

Berenguer, J. (2010). The effect of empathy in environmental moral reasoning. *Environment and Behavior, 42*, 110–134. doi: 10.1177/0013916508325892

Bergman, A., Heindel, J. J., Jobling, S., Kidd, K. A., & Zoeller, R. T. (Eds.). (2013). *State of the science of endocrine disrupting chemicals—2012*. Retrieved from the United Nations Environment Programme and the World Health Organization website: http://unep.org/pdf/9789241505031_eng.pdf

Berman, M. G., Jonides, J., & Kaplan, S. (2008). The cognitive benefits of interacting with nature. *Psychological Science, 19*(12), 1207–1212. doi: 10.1111/j.1467-9280.2008.02225.x

Berman, M. G., Kross, E., Krpan, K. M., Askren, M. K., Burson, A., Deldin, P., . . . Jonides, J. (2012). Interacting with nature improves cognition and affect for individuals with depression. *Journal of Affective Disorders, 140*(3), 300–305. doi: 10.1016/j.jad.2012.03.012

Berto, R. (2005). Exposure to restorative environments helps restore attentional capacity. *Journal of Environmental Psychology, 25*(3), 249–259. doi: 10.1016/j.jenvp.2005.07.001

Beute, F., & de Kort, Y. A. W. (2013). Let the sun shine! Measuring explicit and implicit preferences for environments differing in naturalness, weather type, and brightness. *Journal of Environmental Psychology, 36*, 162–178. doi: 10.1016/j.jenvp.2013.07.016

Beyer, P. (2004). The global environment as a religious issue: A sociological analysis. *Religion, 22*, 1–19. doi: 10.1016/0048-721X(92)90034-2

Bhave, M. P. (2014). Our railway network is a huge dustbin. *The Hindu Business Line*. Retrieved from http://www.thehindubusinessline.com/opinion/our-railway-network-is-a-huge-dustbin/article4534553.ece

Biello, D. (2010, August 6). Where did the Carter White House's solar panels go? *Scientific American*. Retrieved from http://www.scientificamerican.com/

Bies, R. J. (2001). Interactional (in) justice: The sacred and the profane. In J. S. Greenberg, & R. Cropanzano (Eds.), *Advances in organizational justice* (pp. 89–118). Stanford, CA: Stanford University Press.

Bittman, M. (2008, January 27). Rethinking the meat-guzzler. *New York Times*. Retrieved from http://www.nytimes.com/2008/01/27/weekinreview/27bittman.html?pagewanted=all&_r=0

Bixler, R. D., & Floyd, M. F. (1997). Nature is scary, disgusting, and uncomfortable. *Environment and Behavior, 29*, 443–467. doi: 10.1177/001391659702900401

Blake, M. (2014, March/April). The scary new evidence on BPA-free plastics. *Mother Jones*. Retrieved from http://www.motherjones.com/

Blockstein, D. E. (2002). Passenger pigeon ectopistes migratorius. In A. Poole, & F. Gill (Eds.), *The birds of North America 611*. Philadelphia, PA: The Birds of North America, Inc.

Bloodhart, B., Swim, J. K., & Zawadzki, M. J. (2013). Spreading the eco-message: Using proactive coping to aid eco-rep behavior change programming. *Sustainability, 5*, 1661–1679. doi: 10.3390/su5041661

Bloom, A. (1987). *The closing of the American mind: How higher education has failed democracy and impoverished the souls of today's students*. New York, NY: Simon and Schuster.

Bloom, J. (2010). *American wasteland*. Cambridge, MA: DaCapo Press.

Blum, D. (2013, August 16). Is there danger lurking in your lipstick? *New York Times*. Retrieved from http://well.blogs.nytimes.com/2013/08/16/is-there-danger-lurking-in-your-lipstick/

Bodenhausen, G. V., & Gawronski, B. (2013). Attitude change. In D. Reisberg (Ed.), *The Oxford handbook of cognitive psychology* (pp. 957–969) New York, NY: Oxford University Press.

Bogard, P. (2013, August 19). Bringing back the night: A fight against light pollution. *Yale Environment 360*. Retrieved from http://e360.yale.edu/feature/bringing_back_the_night__a_fight_against_light_pollution/2681/

Boldemann, C., Blennow, M., Dal, H., Mårtensson, F., Raustorp, A., Yuen, K., & Wester, U. (2006). Impact of preschool environment upon children's physical activity and sun exposure. *Preventive Medicine, 42*(4), 301–308. doi: 10.1016/j.ypmed.2005.12.006

Boldero, J. (1995). The prediction of household recycling of newspapers: The role of attitudes, intentions, and situational factors. *Journal of Applied Social Psychology, 25*(5), 440–462. doi: 10.1111/j.1559-1816.1995.tb01598.x

Bor, D. (2012). *The ravenous brain: How the new science of consciousness explains our insatiable search for meaning*. New York, NY: Basic Books.

Borden, R. J., & Francis, J. L. (1978). Who cares about ecology? Personality and sex differences in environmental concern. *Journal of Personality, 46*, 190–203. doi: 10.1111/j.1467-6494.1978.tb00610.x

Borghese, M. M., Tremblay, M. S., Leduc, G., Boyer, C., LeBlanc, A. G., . . . Chaput, J. P. (2014). Independent and combined associations of total sedentary time and television viewing time with food intake patterns of 9- to 11-year-old Canadian children. *Applied Physiology, Nutrition, and Metabolism, 39*, 937–943. doi: 10.1139/apnm-2013–0551

Boseley, S. (2009). Climate change biggest threat to health, doctors say. *Guardian*. [On-line edition]. Retrieved from http://www.guardian.co.uk/environment/2009/may/13/climate-change-health-impact

Bostrom, A., & Lashof, D. (2007). Weather it's climate change? In S. C. Moser, & L. Dilling (Eds.), *Creating a climate for change: Communicating climate change and facilitating social change* (pp. 31–43). New York, NY: Cambridge University Press.

Boucher, O., Jacobson, S. W., Plusquellec, P., Dewailly, E., Avotte, P., Forget-Dubois, N., . . . Muckle, G. (2012). Prenatal methylmercury, postnatal lead exposure, and evidence of attention deficit/hyperactivity disorder among Inuit children in Artic Québec. *Environmental Health Perspectives, 120*(10), 1456–1461. doi: 10.1289/ehp.1204976

Boyle, C. A., Boulet, S., Schieve, L. A., Cohen, R. A., Blumberg, S. J., Yeargin-Allsopp, M., Visser, S., & Kogan, M. D. (2011). Trends in the prevalence of developmental disabilities in US children, 1997–2008. *Pediatrics, 127*(6), 1034–1042. doi: 10.1542/peds.2010-2989

Bragg, E. A. (1996). Towards ecological self: Deep ecology meets construction-ist self-theory. *Journal of Environmental Psychology, 16,* 93–108. doi: 10.1006/jevp.1996.0008

Braitman, L. (2014). *Animal madness: How anxious dogs, compulsive parrots, and ele-phants in recovery help us understand ourselves.* New York, NY: Simon & Schuster.

Brannan, D. B. (2011). *An experimental study of real-time feedback on driving behavior.* Behavior, Energy, & Climate Change Conference, Washington, DC.

Brehm, S. S., & Brehm, J. W. (1981). *Psychological reactance: A theory of freedom and control.* New York, NY: Academic Press.

Breton, M. J. (1998). *Women pioneers for the environment.* Boston, MA: Northeastern University Press.

Bright, A. D., & Tarrant, M. A. (2002). Effect of environment-based coursework on the nature of attitudes toward the Endangered Species Act. *The Journal of Environ-mental Education, 33*(4), 10–19. doi: 10.1080/00958960209599149

Bringslimark, T., Hartig, T., Patil, G. G. (2009). The psychological benefits of indoor plants: A critical review of the experimental literature. Journal of Environmental Psychology, 29, 422–433.

Bringslimark, T., Hartig, T., & Patil, G. G. (2011). Adaptation to windowlessness: Do office workers compensate for a lack of visual access to the outdoors? *Environment and Behavior, 43*(4), 469–487. doi: 10.1177/0013916510368351

Broadhurst, P. L. (1959). The interaction of task difficulty and motivation: The Yerkes-Dodson law revived. *Acta Psychologica, 16,* 321–338. doi: 10.1016/0001-6918(59)90105-2

Brod, C. (1984). *Technostress: The human cost of the computer revolution.* Reading, MA: Addison-Wesley.

Brodt, S., Kramer, K. J., Kendall, A., & Feenstra, G. (2013). Comparing environmen-tal impacts of regional and national-scale food supply chains: A case study of processed tomatoes. *Food Policy, 42,* 106–114. doi: 10.1016/j.foodpol.2013.07.004

Brook, A. (2011). Ecological footprint feedback: Motivating or discouraging? *Social Influence, 6*(2), 113–128. doi: 10.1080/15534510.2011.566801

Brown, K. W., & Kasser, T. (2005) Are psychological and ecological well-being com-patible? The role of values, mindfulness, and lifestyle. *Social Indicators Research, 74,* 349–368. doi: 10.1007/s11205-004-8207-8

Brown, S.-E. (2007). Companion animals as self-objects. *Anthrozoös, 20,* 329–343. doi: 10.2752/089279307X245654

Brown, W., & Brown, E. (2013). *Browsing nature's aisles: A year of foraging for wild food in the suburbs.* Gabriola Island, British Columbia, Canada: New Society Publishers.

Brown, W., & May, D. (2012). Organizational change and development: The efficacy of transformational leadership training. *Journal of Management Development, 31*(6), 520–536. doi: 10.1108/02621711211230830

Brügger, A., Kaiser, F. G., & Roczen, N. (2011). One for all? Connectedness to nature, inclusion of nature, environmental identity, and implicit associations with nature. *European Psychologist, 16,* 324–333. doi: 10.1027/1016-9040/a000032

Brundtland G. H., & World Commission on Environment and Development. (1987). *Our common future: Report of the World Commission on Environment and Development.* New York, NY: Oxford University Press.

Bulbeck, C. (2005). *Facing the wild: Ecotourism, conservation, and animal encounters.* New York, NY: Routledge.

Bureau of Labor Statistics. (2013, March 19). *News release: Employment in green goods and services—2011* (Publication No. USDL-13–0476). Retrieved from http://www. bls.gov/news.release/pdf/ggqcew.pdf

Bureau of Labor Statistics. (2014). Insulation workers. *Occupational Outlook Handbook.* Retrieved from http://www.bls.gov/ooh/construction-and-extraction/insulation-workers.htm#tab-6

Burger, J. M. (1999). The foot-in-the-door compliance procedure: A multiple-process analysis and review. *Personality and Social Psychology Review, 3*, 303–325. doi: 10.1207/s15327957pspr0304_2

Buzzell, L., & Chalquist, C. (2009). *Ecotherapy: Healing with nature in mind.* San Francisco, CA: Sierra Club Books.

Cabral, P. U., Canário, A. C., Spyrides, M. H., Uchôa, S. A., Eleutério, J. J., Giraldo, P. C., & Gonçalves, A. K. (2014). Physical activity and sexual function in middle-aged women. *Revista da Associação Médica Brasileira, 60*(1), 47–52. doi: 10.1590/1806-9282.60.01.011

Calderwood, C., Ackerman, P. L., & Conklin, E. M. (2014). What else do college students "do" while studying? An investigation of multitasking. *Computers & Education, 75*, 19–29. doi: 10.1016/j.compedu.2014.02.004

Campbell, N. M., Bevc, C. A., & Picou, J. S. (2013). Perceptions of toxic exposure: Considering "White male" and "Black female" effects. *Sociological Spectrum, 33*(4), 313–328. doi: 10.1080/02732173.2013.732882

Carrico, A. R., & Riemer, M. (2011). Motivating energy conservation in the workplace: An evaluation of the use of group-level feedback and peer education. *Journal of Environmental Psychology, 31*, 1–13. doi: 10.1016/j.jenvp.2010.11.004

Carson, R. (1962). *Silent spring.* New York, NY: Houghton Mifflin.

Catton, W. R. (1993). Carrying capacity and the death of a culture: A tale of two autopsies. *Sociological Inquiry, 63*, 202–222. doi: 10.1111/j.1475-682X.1993.tb00303.x

Center for Sustainable Systems, University of Michigan. (2013). *U.S. food system factsheet.* Pub. No. CSS01–06. Retrieved from http://css.snre.umich.edu/css_doc/CSS01-06.pdf

Center for Sustainable Systems, University of Michigan. (2013). *U.S. material use factsheet.* Pub. No. CSS05–18. Retrieved from http://css.snre.umich.edu/css_doc/CSS05–18.pdf

Centers for Disease Control and Prevention. (2011). *Developmental disabilities increasing in U.S.* Retrieved from http://www.cdc.gov/features/dsdev_disabilities/

Centers for Disease Control and Prevention (CDC). (2012). *Second national report on biochemical indicators of diet and nutrition in the U.S. population 2012: Executive summary.* Retrieved from http://www.cdc.gov/nutritionreport/pdf/ExeSummary_Web_032612.pdf

Centers for Disease Control and Prevention (CDC). (2014). *Overweight and obesity: Adult obesity facts.* Retrieved from www.cdc.gov/obesity/data/adult.html

Centers for Disease Control and Prevention, Advisory Committee on Childhood Lead Poisoning Prevention. (2012). *Low level lead exposure harms children: A renewed call for primary prevention.* Retrieved from http://www.cdc.gov/nceh/lead/ACCLPP/Final_Document_030712.pdf

Cervinka, R., Röderer, K., & Hefler, E. (2012). Are nature lovers happy? On various indicators of well-being and connectedness with nature. *Journal of Health Psychology, 17*(3), 379–388. doi: 10.1177/1359105311416873

Chaiken, S. (1987). The heuristic model of persuasion. In M. P. Zanna, J. M. Olson, & C. P. Herman (Eds.), *Social influence: The Ontario symposium.* (pp. 3–39). Hillsdale, NJ: Erlbaum.

Chalquist, C. (2009). A look at the ecotherapy research evidence. *Ecopsychology, 1*(2), 64–74. doi: 10.1089/eco.2009.0003

Chameides, B. (2011, June 14). The Toxic Substances Control Act's toxic baddies. *The Huffington Post.* Retrieved from http://www.huffingtonpost.com/green/

Charles, C., & Senauer, A. (2010). *Health benefits to children of contact with the outdoors and nature.* Retrieved from http://www.childrenandnature.org/downloads/CNNHealthBenefits2012.pdf

Chase, R. (2010, March 12). You asked: Does everyone in America own a car? *U.S. Department of State IIP Digital.* Retrieved from http://iipdigital.usembassy.gov

Chawla, L. (2006). Research methods to investigate significant life experiences: Review and recommendations. *Environmental Education Research, 12*(3–4), 359–374. doi: 10.1080/13504620600942840

Chawla, L. (2012). The development of conservation behaviors in childhood and youth. In S. Clayton (Ed.), *The Oxford handbook of environmental and conservation psychology* (pp. 527–555). New York, NY: Oxford University Press.

Chawla, L., & Derr, V. (2012). The development of conservation behaviors in childhood and youth. In S. Clayton (Ed.), *Handbook on environmental and conservation psychology.* New York, NY: Oxford University Press.

Chellappa, S. L., Steiner, R., Oelhafen, P., Lang, D., Götz, T., Krebs, J., & Cajochen, C. (2013). Acute exposure to evening blue-enriched light impacts on human sleep. *Journal of Sleep Research, 22*(5), 573–580. doi: 10.1111/jsr.12050

Chen, M. F., & Tung, P. J. (2010). The moderating effect of perceived lack of facilities on consumers' recycling intentions. *Environment and Behavior, 42*(6), 824–844. doi: 10.1177/0013916509352833

Chen, P. W. (2010, September 16). Teaching doctors about nutrition and diet. *The New York Times.* Retrieved from http://www.nytimes.com

Chess, C. & Johnson, B. B. (2007). Information is not enough. In S. C. Moser, & L. Dilling (Eds.), *Creating a climate for change: Communicating climate change and facilitating social change* (pp. 223–236). Cambridge, United Kingdom: Cambridge University Press.

Chudek, M., & Henrich, J. (2011). Culture–gene coevolution, norm-psychology and the emergence of human prosociality. *Trends in Cognitive Sciences, 15*(5), 218–226. doi: 10.1016/j.tics.2011.03.003

Cialdini, R.B. (2005). Basic social influence is underestimated. *Psychological Inquiry, 16*(4), 158–161. doi: 10.1207/s15327965pli1604_03

Cialdini, R.B. (2006). *Influence: The psychology of persuasion* (Revised edition). New York: Harper Business.

Cialdini, R.B., Kallgren, C.A., & Reno, R.R. (1991). A focus theory of normative conduct: A theoretical refinement and reevaluation of the role of norms in human behavior. *Advances in Experimental Social Psychology, 24,* 201–234. doi: 10.1016/S0065-2601(08)60330-5

Cialdini, R.B., Reno, R.R., & Kallgren, C.A. (1990). A focus theory of normative conduct: Recycling the concept of norms to reduce littering in public places. *Journal of Personality and Social Psychology, 58,* 1015–1026. doi: 10.1037/0022-3514.58.6.1015

Cinzano, P., Falchi, F., & Elvidge, C.D. (2001). The first World Atlas of the artificial night sky brightness. *Monthly Notices of the Royal Astronomical Society, 328,* 689–707. doi: 10.1046/j.1365-8711.2001.04882.x

Civilian Conservation Corps (CCC) Legacy. (2013). *Civilian Conservation Corps: Preserving America's natural resources, 1933–1942.* Retrieved from http://www.ccclegacy.org/

Clark, J.P., Marmol, L.M., Cooley, R., & Gathercoal, K. (2004). The effects of wilderness therapy on the clinical concerns (axes I, II, and IV) of troubled adolescents. *Journal of Experiential Education, 27,* 213–232. doi: 10.1177/105382590402700207

Clark, K.H., & Nicholas, K.A. (2013). Introducing urban food forestry: A multifunctional approach to increase food security and provide ecosystem services. *Landscape Ecology, 28*(9), 1649–1669. doi: 10.1007/s10980-013-9903-z

Clarke, M., & Lawn, P. (2008). A policy analysis of Victoria's Genuine Progress Indictor. *The Journal of Socio-Economics, 37,* 864–879.

Clarke, R. (1973). *Ellen Swallow: The woman who founded ecology.* Chicago, IL: Follett Publishing Company.

Clayton, S. (2000). Models of justice in the environmental debate. *Journal of Social Issues, 56,* 459–474. doi: 10.1111/0022-4537.00178

Clayton, S. (2003a). Environmental identity: A conceptual and an operational definition. In S. Clayton, & S. Opotow (Eds.), *Identity and the natural environment* (pp. 45–65). Cambridge, MA: MIT Press.

Clayton, S. (2003b). Justice and identity: Changing perspectives on what is fair. *Personality and Social Psychology Review, 7,* 298–310. doi: 10.1207/S15327957PSPR0704_03

Clayton, S., & Opotow, S. (2003). *Identity and the natural environment: The psychological significance of nature.* Cambridge, MA: Massachusetts Institute of Technology Press.

Clayton, S., Fraser, J., & Burgess, C. (2011). The role of zoos in fostering environmental identity. *Ecopsychology, 3*(2), 87–96. doi: 10.1089/eco.2010.0079

Clayton, S., Manning, C. M., & Hodge, C. (2014). *Beyond storms & droughts: The psychological impacts of climate change*. Washington, DC: American Psychological Association and ecoAmerica.

Clelland, J. D., Read, L. L., Drouet, V., Kaon, A., Kelly, A., Duff, K. E., . . . Clelland, C. L. (2014). Vitamin D insufficiency and schizophrenia risk: Evaluation of hyperprolinemia as a mediator of association. *Schizophrenia Research, 156*(1), 15–22. doi: 10.1016/j.schres.2014.03.017

Clements, R. (2004). An investigation of the status of outdoor play. *Contemporary Issues in Early Childhood, 5*(1), 68–80. doi: 10.2304/ciec.2004.5.1.10

Cleveland, M., Kalamas, M., & Laroche, M. (2012). "It's not easy being green": Exploring green creeds, green deeds, and internal environmental locus of control. *Psychology & Marketing, 29*(5), 293–305. doi: 10.1002/mar.20522

Clinebell, H. (1996). *Ecotherapy: Healing ourselves, healing the earth*. New York, NY: Haworth.

Cobern, M. K., Porter, B. E., Leeming, F. C., & Dwyer, W. O. (1995). The effect of commitment on adoption and diffusion of grass cycling. *Environment and Behavior, 27*(2), 213–232. doi: 10.1177/0013916595272006

Coffey, D. J., & Joseph, P. H. (2013). A polarized environment: The effect of partisanship and ideological values on individual recycling and conservation behavior. *American Behavioral Scientist, 57*(1), 116–139. doi: 10.1177/0002764212463362

Cohen, J. (1992). A power primer. *Psychological Bulletin, 112*, 155–159. doi: 10.1037/0033-2909.112.1.155

Cohen, J. (1995). *How many people can the earth support?* New York, NY: W. W. Norton & Co.

Cohen-Charash, Y., & Specter, P. E. (2001). The role of justice in organizations: A metaanalysis. *Organizational Behavior and Human Decision Processes, 86*(2), 278–321. doi: 10.1006/obhd.2001.2958

Colborn, T. (2004). Neurodevelopment and endocrine disruption. *Environmental Health Perspectives, 112*, 944–949.

Colborn, T., Dumanoski, D., & Myers, J. P. (1997). *Our stolen future: Are we threatening our fertility, intelligence, and survival? A scientific detective story*. New York, NY: Plume.

Cole, E., Erdman, E., & Rothblum, E. D. (Eds.). (1994). *Wilderness therapy for women: The power of adventure*. New York, NY: Routledge.

Cole, N. (with Watrous, S.). (2007). Across the great divide: Supporting scientists as effective messengers in the public sphere. In S. C. Moser, & L. Dilling (Eds.), *Creating a climate for change: Communicating climate change and facilitating social change* (pp. 180–198). Cambridge, United Kingdom: Cambridge University Press.

Cole, R. J. (2012). Transitioning from green to regenerative design. *Building Research & Information, 40*(1), 39–53.

Coleridge, S. T. (1798). The rime of the ancient mariner. In W. Wordsworth, & S. T. Coleridge, *Lyrical ballads, with a few other poems*. London, United Kingdom: J. & A. Arch.

Coley, J. D., Freeman, A. C., & Blaszczyk, K. (2003). Taxonomic and ecological relations in urban and rural children's folk biology. Retrieved from http://www.psych.neu.edu/faculty/j.coley/pub/CFB2003.pdf

Collado, S., Staats, H., & Corraliza, J. A. (2013). Experiencing nature in children's summer camps: Affective, cognitive and behavioural consequences. *Journal of Environmental Psychology, 33*, 37–44. doi: 10.1016/j.jenvp.2012.08.002

Common Sense Media. (2013). *Zero to eight: Children's media use in America, 2013.* Retrieved from https://www.commonsensemedia.org/research/zero-to-eight-childrens-media-use-in-america-2013

Container Recycling Institute. (2013). *Bottle bills promote recycling and reduce waste.* Retrieved from http://www.bottlebill.org/about/benefits/waste.htm

Cooper Marcus, C. (2006) Healing gardens in hospitals. In C. Wagenaar (Ed.), *The architecture of hospitals* (pp. 314–329). Rotterdam, The Netherlands: NAi Publishers.

Cordano, M., Welcomer, S. A., & Scherer, R. F. (2003). An analysis of the predictive validity of the new ecological paradigm scale. *Journal of Environmental Education, 34*(3), 22–28. doi: 10.1080/00958960309603490

Cordano, M., Welcomer, S., Scherer, R. F., Pradenas, L., and Parada, V. (2010). A cross-cultural assessment of three theories of pro-environmental behavior: A comparison between business students of Chile and the United States. *Environment and Behavior, 43*, 634–657. doi: 10.1177/0013916510378528

Costa, E. M., Spritzer, P. M., Hohl, A., & Bachega, T. A. (2014). Effects of endocrine disruptors in the development of the female reproductive tract. *Arquivos Brasileiros de Endocrinologia & Metabologia, 58*(2), 153–161. doi: 10.1590/0004-2730000003031

Costa, L. G., & Giordano, G. (2007). Developmental neurotoxicity of polybrominated diphenyl ether (PBDE) flame retardants. *NeuroToxicology, 28*, 1047–1067. doi: 10.1016/j.neuro.2007.08.007

Costa, P. T., Jr., & McCrae, R. R. (1992). *Revised NEO Personality Inventory (NEO-PI-R) and NEO Five-Factor Inventory (NEO-FFI) manual.* Odessa, FL: Psychological Assessment Resources.

Costanza, R., Kubiszewski, I., Giovannini, E., Lovins, H., McGlade, J., Pickett, K. E., . . . Wilkinson, R. (2014). Time to leave GDP behind. *Nature, 505*(7483), 283–285.

Coyle, K. (2005). *Environmental literacy in America: What ten years of NEETF/Roper research and related studies say about environmental literacy in the U.S.* Washington, DC: The National Environmental Education & Training Foundation. Retrieved from http://www.csu.edu/cerc/documents/EnvironmentalLiteracyInAmerica2005.pdf

Craig, C. S., & McCann, J. M. (1978). Assessing communication effects on energy conservation. *Journal of Consumer Research, 5*, 82–88. doi: 10.1086/208718

Crichton, G. E., & Alkerwi, A. (2014). Association of sedentary behavior time with ideal cardiovascular health: The ORISCAV-LUX study. *PLoS One, 9*(6), e99829. doi: 10.1371/journal.pone.0099829

Crompton, T. (2008). *Weathercocks & signposts: The environment movement at a crossroads.* World Wildlife Fund-United Kingdom Strategies for Change Project. Retrieved from wwf.org.uk/strategiesforchange

Crompton, T., & Kasser, T. (2009). *Identity campaigning: Bringing the person into the environmental movement.* Godalming, United Kingdom: World Wildlife Foundation.

Csikszentmihalyi, M. (2004). What we must accomplish in the coming decades. *Zygon, 39,* 359–366. doi: 10.1111/j.1467-9744.2004.00579.x

Cunsolo Willox, A., Harper, S.L., Ford, J.D., Landman, K., Houle, L., & Edge, V.L. (2012). "From this place and of this place:" Climate change, sense of place, and health in Nunatsiavut, Canada. *Social Science & Medicine, 75*(3), 538–547. doi: 10.1016/j.socscimed.2012.03.043

Daley, C.A., Abbott, A., Doyle, P.S., Nader, G.A., & Larson, S. (2010). A review of fatty acid profiles and antioxidant content in grass-fed and grain-fed beef. *Nutrition Journal, 9,* 10. doi: 10.1186/1475-2891-9-10

Darbi, W.P.K. (2012). Of mission and vision statements and their potential impact on employee behavior and attitudes: The case of a public but profit-oriented tertiary institution. *International Journal of Social Science, 3*(14), 95–109.

Darwin, C. (1859). *The origin of species.* Chicago, IL: Donahue Henbneberry & Co.

Dawes, R.M. (1980). Social dilemmas. *Annual Review of Psychology, 31,* 169–193. doi: 10.1146/annurev.ps.31.020180.001125

DeAngelis, T. (2013). Therapy gone wild. *Monitor on Psychology, 44*(8), 49–52.

Dearie, J., & Geduldig, C. (2013). *Where the jobs are: Entrepreneurship and the soul of the American economy.* New York, NY: John Wiley & Sons.

Dearing, J.W. (2009). Applying diffusion of innovation theory to intervention development. *Research on Social Work Practice, 19*(5), 503–518. doi: 10.1177/1049731509335569

Deci, E.L., & Ryan, R.M. (1985). *Intrinsic motivation and self-determination in human behavior.* New York, NY: Plenum.

DeLeon, I.G., & Fuqua, R.W. (1995). The effects of public commitment and group feedback on curbside recycling. *Environment and Behavior, 27*(2), 233–250. doi: 10.1177/0013916595272007

Denver Water. (2014). *2014 rules for outdoor water use.* Retrieved from http://www.denverwater.org/conservation/wateruserulesregulations/summerwateringrules

Deutsche Welle. (2014, January 9). *Alarm over soaring world meat consumption.* Retrieved from http://www.dw.de/p/1AoMG

de Vries, G., Terwel, B.W., & Ellemers, N. (2014). Spare the details, share the relevance: The dilution effect in communications about carbon dioxide capture and storage. *Journal of Environmental Psychology, 38,* 116–123. doi: 10.1016/j.jenvp.2014.01.003

De Wit, M.J. (2008). An exploration of modern consumer society and a guide towards mindful consuming. *Dissertation Abstracts International, 69*(3-B), 1948.

DeYoung, R. (2000). Expanding and evaluating motives for environmentally responsible behavior. *Journal of Social Issues, 56,* 509–526. doi: 10.1111/0022-4537.00181

Diamond, J. (2005). *Collapse: How societies choose to fail or survive.* London, United Kingdom: Allen Lane.

Dickerson, C.A., Thibodeau, R., Aronson, E., & Miller, D. (1992). Using cognitive dissonance to encourage water conservation. *Journal of Applied Social Psychology, 22,* 841–854. doi: 10.1111/j.1559-1816.1992.tb00928.x

Dickinson, J. L., Crain, R., Yalowitz, S., & Cherry, T. M. (2013). How framing climate change influences citizen scientists' intentions to do something about it. *The Journal of Environmental Education, 44*(3), 145–158, doi: 10.1080/00958964.2012.742032

Diekmann, A., & Preisendörfer, P. (1992). Persónliches umweltverhalten: Diskrepanzen zwischen anspruch und wirklichkeit. / Ecology in everyday life: Inconsistencies between environmental attitudes and behavior. *Kölner Zeitschrift für Soziologie und Sozialpsychologie, 44*(2), 226–251.

Diener, E., & Seligman, M.E.P. (2004). Beyond money: Toward an economy of well-being. *Psychological Science in the Public Interest, 5*, 1–31. doi: 10.1111/j.0963-7214.2004.00501001.x

Dierdorff, E. C., Norton, J. J., Drewes, D. W., Kroustalis, C. M., Rivkin, D., & Lewis, P. (2009). *Greening of the world of work: Implications for O*NET®-SOC and new and emerging occupations.* Raleigh, NC: National Center for O*Net Development.

Dietz, T. M., Fitzgerald, A., & Shwom, R. (2005). Environmental values. *Annual Review of Environment & Resources, 30*, 1–38. doi: 10.1146/annurev.energy.30.050504.144444

Dinan, T. G., Stanton, C., & Cryan, J. F. (2013). Psychobiotics: A novel class of psychotropic. *Biological Psychiatry, 74*(10), 720–726. doi: 10.1016/j.biopsych.2013.05.001

Doherty, T., & Clayton, S. (2011). The psychological impacts of global climate change. *American Psychologist, 66*, 265–276. doi: 10.1037/a0023141

Dolan, P., Hallsworth, M., Halpern, D., King, D., Metcalfe, R., & Vlaev, I. (2012). Influencing behavior: The mindspace way. *Journal of Economic Psychology, 33*, 264–277. doi: 10.1016/j.joep.2011.10.009

Dono, J., Webb, J., & Richardson, B. (2010). The relationship between environmental activism, pro-environmental behaviour and social identity. *Journal of Environmental Psychology, 30*(2), 178–186. doi: 10.1016/j.jenvp.2009.11.006

Doppelt, B. (2012). *From me to we: The five transformational commitments required to rescue the planet, your organization, and your life.* Sheffield, UK: Greenleaf Publishing.

Dr. Seuss. (1971). *The Lorax.* New York, NY: Random House.

Dreyer, S.J., & Walker, I. (2013). Acceptance and support for the Australian carbon policy. *Social Justice Research, 26*(3), 343–362. doi: 10.1007/s11211-013-0191-1

Dunlap, R. E., & Gale, R. P. (1974). Party membership and environmental politics: A legislative roll-call analysis. *Social Science Quarterly, 55*, 670–690.

Dunlap, R. E., & McCright, A. M. (2008a). A widening gap: Republican and Democratic views on climate change. *Environment: Science and Policy for Sustainable Development, 50*(5), 26–35. doi: 10.3200/ENVT.50.5.26-35

Dunlap, R. E., & McCright, A. M. (2008b). Social movement identity: Validating a measure of identification with the environmental movement. *Social Science Quarterly, 89*, 1045–1065. doi: 10.1111/j.1540-6237.2008.00573.x

Dunlap, R. E., Van Liere, K., Mertig, A., & Jones, R. E. (2000). New trends in measuring environmental attitudes: Measuring endorsement of the new ecological paradigm: A revised NEP scale. *Journal of Social Issues, 56*, 425–442. doi: 10.1111/0022-4537.00176

Dunn, A. D. (2010). Siting green infrastructure: Legal and policy solutions to alleviate urban poverty and promote healthy communities. *Environmental Affairs, 97,* 41–66.

Dunn, E. W., Gilbert, D. T., & Wilson, T. D. (2011). If money doesn't make you happy, then you probably aren't spending it right. *Journal of Consumer Psychology, 21*(2), 115–125. doi: 10.1016/j.jcps.2011.02.002

Dustin, D., Bricker, N., Arave, J., Wall, W., & Wendt, G. (2011). The promise of river running as a therapeutic medium for veterans coping with post-traumatic stress disorder. *Therapeutic Recreation Journal, 14,* 326–340.

Dutcher, D. D., Finley, J. C., Luloff, A. E., & Johnson, J. B. (2007). Connectivity with nature as a measure of environmental values. *Environment and Behavior, 39,* 474–493. doi: 10.1177/0013916506298794

Dyment, J. E., & Bell, A. C. (2008). Grounds for movement: Green school grounds as sites for promoting physical activity. *Health Education Research, 23*(6), 952–962. doi: 10.1093/her/cym059

Eagly, A. H., & Chaiken, S. (1998). *The psychology of attitudes.* Fort Worth, TX: Harcourt Brace.

Egerton, F. N. (2013). History of ecological sciences, part 47: Ernst Haeckel's ecology. *Bulletin of the Ecological Society of America, 94,* 222–244. doi.10.1890/0012-9623-94.3.222

Ehrhardt-Martinez, K., Donnelly, K. A., & Laitner, J. A. (2010). *Advanced metering initiatives and residential feedback programs: A meta-review for economy-wide electricity-saving opportunities* (Report No. E105). Washington, DC: American Council for an Energy-Efficient Economy.

Ehrlich, P. R., & Ehrlich, A. H. (1991). *Healing the planet: Strategies for resolving the environmental crisis* (pp. 7–10). Reading, MA: Addison-Wesley.

Ehrlich, P. R., & Ehrlich, A. H. (2008). *The dominant animal: Human evolution and the environment.* Washington, DC: Island Press.

Ehrlich, P. R., & Holdren, J. (1971). The impact of population growth. *Science, 171,* 1212–1217. doi: 10.1126/science.171.3977.1212

Einstein, E. (2005). A. Calaprice (Ed.), *The new quotable Einstein.* Princeton, NJ: Princeton University Press.

Eisler, A. D., Eisler, H., & Yoshida, M. (2003). Perception of human ecology: Cross-cultural and gender comparisons. *Journal of Environmental Psychology, 23*(1), 89–101. doi: 10.1016/S0272-4944(02)00083-X

Electronics Take Back Coalition. (n.d.). State legislation: States are passing e-waste legislation. Retrieved from http://www.electronicstakeback.com/promote-good-laws/state-legislation/

Elgin, D. (2000). *The garden of simplicity.* Retrieved from http://www.simpleliving.net/webofsimplicity/the_garden_of_simplicity.asp

Eriksen, M., Mason, S., Wilson, S., Box, C., Zellers, A., Edwards, W., . . . Amato, S. (2013). Microplastic pollution in the surface waters of the Laurentian Great Lakes. *Marine Pollution Bulletin, 77*(1–2). 177–182.

Evans, G.W., & Stecker, R. (2004). Motivational consequences of environmental stress. *Journal of Environmental Psychology, 24,* 143–165. doi: 10.1016/S0272-4944(03)00076-8

Everett, P.B., Hayward, S.C., & Meyers, A.W. (1974). The effects of a token reinforcement procedure on bus ridership. *Journal of Applied Behavior Analysis, 7,* 1–9. doi: 10.1901/jaba.1974.7-1

Faber Taylor, A., & Kuo, F.E. (2006). Is contact with nature important for healthy child development? State of the evidence. In C. Spencer, & M. Blades (Eds.), *Children and their environments: Learning, using and designing spaces* (pp. 124–140). New York, NY: Cambridge University Press.

Faber Taylor, A., & Kuo, F.E. (2009). Children with attention deficits concentrate better after walk in the park. *Journal of Attention Disorders, 12,* 402–409. doi: 10.1177/1087054708323000

Faber Taylor, A., & Kuo, F.E. (2011). Could exposure to everyday green spaces help treat ADHD? Evidence from children's play settings. *Applied Psychology: Health and Well-Being, 3,* 281–303.

Faber Taylor, A., Kuo, F.E., & Sullivan, W.C. (2001). Coping with ADD: The surprising connection to green play settings. *Environment and Behavior, 33,* 54–77. doi: 10.1177/00139160121972864

Faber Taylor, A., Wiley, A., Kuo, F.E., & Sullivan, W.C. (1998). Growing up in the inner city: Green spaces as places to grow. *Environment and Behavior, 30,* 3–27. doi: 10.1177/0013916598301001

Federal Communications Commission. (2005). *Communications history: Golden age, 1930 through 1950s.* Retrieved from http://transition.fcc.gov/omd/history/tv/1930–1959.html

Festinger, L. (1954). A theory of social comparison processes. *Human Relations, 7*(2), 117–140.

Festinger, L. (1957). *A theory of cognitive dissonance.* Stanford, CA: Stanford University Press.

Feygina, I., Jost, J.T., & Goldsmith, R.E. (2010). System justification, the denial of global warming, and the possibility of "system-sanctioned change." *Personality and Social Psychology Bulletin, 36*(3), 326–338. doi: 10.1177/0146167209351435

Fielding, K.S., McDonald, R., & Louis, W. (2008). Theory of planned behavior, identity and intentions to engage in environmental activism. *Journal of Environmental Psychology, 28*(4), 318–326. doi: 10.1016/j.jenvp.2008.03.003

Fillmore, R. (2009). *The growth of environmentalism, 1900–1949.* Gale, Cengage. Retrieved from http://www.sciencescribe.net/articles/Developments_in_Ecology.pdf

Finley, J.R., Benjamin, A.S., & McCarley, J.S. (2014). Metacognition of multitasking: How well do we predict the costs of divided attention? *Journal of Experimental Psychology: Applied, 20*(2), 158–165. doi: 10.1037/xap0000010

Fiore, V., Marci, M., Poggi, A., Giagulli, V.A., Licchelli, B., Iacoviello, M., . . . Triggiani, V. (2015). The association between diabetes and depression: A very disabling condition. *Endocrine, 48,* 14–24. doi: 10.1007/s12020–014–0323-x

Fischer, B. (2013, August 22). The tradition continues: The United States wastes more energy than it uses [Web log post]. Retrieved from http://blog.opower.com/2013/08/the-tradition-continues-the-united-states-wastes-more-energy-than-it-uses/

Fischer, J., Dyball, R., Fazey, I., Gross, C., Dovers, S., Ehrlich, P.R. . . . Borden, R.J. (2012). Human behavior and sustainability. *Frontiers in Ecology and the Environment, 10*(3), 153–160. doi:10.1890/110079

Fisher, A. (2002). *Radical eco-psychology: Psychology in the service of life.* Albany, NY: State University of New York Press.

Fisman, L. (2005). The effects of local learning on environmental awareness in children: An empirical investigation. *The Journal of Environmental Education, 36*(3), 39–50. doi: 10.3200/JOEE.36.3.39-50

Fitzherbert, E. B., Struebig, M. J., Morel, A., Danielsen, F., Brühl, C. A., Donald, P. F., & Phalan, B. (2008). How will oil palm expansion affect biodiversity? *Trends in Ecology & Evolution, 23*(10), 538–545. doi: 10.1016/j.tree.2008.06.012

Fitzmaurice, A. G., & Bronstein, J. M. (2011). Pesticides and Parkinson's disease. In M. Stoytcheva (Ed.), *Pesticides in the modern world—Effects of pesticide exposure* (pp. 307–22). Rijeka, Croatia: Intech Europe.

Fjørtoft, I. (2004). Landscape as playscape: The effects of natural environments on children's play and motor development. *Children, Youth and Environments, 14*(2), 21–44.

Flint, H. J., Scott, K. P., Louis, P., & Duncan, S. H. (2012). The role of the gut microbiota in nutrition and health. *Nature Reviews Gastroenterology and Hepatology, 9,* 577–589. doi: 10.1038/nrgastro.2012.156

Flynn, J., Slovic, P., & Mertz, C. K. (1994). Gender, race, and perception of environmental health risks. *Risk Analysis, 14,* 1101–1108. doi: 10.1111/j.1539-6924.1994.tb00082.x

Food and Agriculture Organization of the United Nations (FAO). (2013). *FAO statistical yearbook 2013: World food and agriculture.* Rome: FAO. Retrieved from http://www.fao.org /docrep/018/i3107e/i3107e.PDF

Food and Agriculture Organization of the United Nations. (2013a). *Food wastage footprint: Impact on natural resources summary report.* Retrieved from http://www.fao.org/docrep /018/i3347e/i3347e.pdf

Food and Agriculture Organization of the United Nations. (2013b). *The state of food insecurity in the world. Executive summary.* Retrieved from http://www.fao.org/docrep/018/i3458e/i3458e.pdf

Foster, J. A. (2014). Gut feelings: Bacteria and the brain. In B. Glovin (Ed.), *Cerebrum 2013: Emerging ideas in brain science. The Dana Foundation's Cerebrum* (pp. 71–80). Washington, DC: Dana Press.

Fowler, C. (2009). Cary Fowler: One seed at a time, protecting the future of food [Video file]. Retrieved from http://www.ted.com/talks/cary_fowler_one_seed_at_a_time_protecting _the_future_of_food

Frantz, C. M., & Mayer, F. (2009). The emergency of climate change: Why are we failing to take action. *Analyses of Social Issues and Public Policy (ASAP), 9*(1), 205–222. doi:10.1111/j.1530-2415.2009.01180.x

Fredrickson, B.L. (1998). What good are positive emotions? *Review of General Psychology, 2*, 300–319. doi: 10.1037/1089-2680.2.3.300

French, J.R.P., & Raven, B.H. (1959). The bases of social power. In D. Cartwright (Ed.) *Studies in social power* (pp. 150–167). Ann Arbor, MI: Institute for Social Research.

Freud, S. (1938/1964). Splitting of the ego in the process of defense. In J. Strachey, & A. Freud (Eds.), *The standard edition of the complete psychological works of Sigmund Freud* (pp. 275–276). London, United Kingdom: Hogarth Press.

Frick, J., Kaiser, F.G., & Wilson, M. (2004). Environmental knowledge and conservation behavior: exploring prevalence and structure in a representative sample. *Personality and Individual Differences, 37*(8), 1597–1613. doi: 10.1016/j.paid.2004.02.015

Friedman, R.S., & Förster, J. (2010). Implicit affective cues and attentional tuning: An integrative review. *Psychological Bulletin, 136*(5), 875–893. doi: 10.1037/a0020495

Friese, G., Hendee, J., & Kinziger, M. (1998). The wilderness experience program industry in the United States: Characteristics and dynamics. *Journal of Experiential Education, 21*(1) 40–45. doi: 10.1177/105382599802100109

Frijda, N.H. (2007). *The laws of emotion*. Mahwah, NJ: Erlbaum.

Fröhlich, G., Sellmann, D., & Bogner, F.X. (2013). The influence of situational emotions on the intention for sustainable consumer behaviour in a student-centred intervention. *Environmental Education Research, 19*(6), 747–764. doi:10.1080/13504622.2012.749977

Fuhrman, J., Sarter, B., Glaser, D., & Acocella, S. (2010). Changing perceptions of hunger on a high nutrition density diet. *Nutrition Journal, 9*, 51. doi:10.1186/1475-2891-9-51

Fuller, T. (1732). *Gnomologia, adagies and proverbs, wise sentences and witty sayings, ancient and modern, foreign and British*. Retrieved from http://dictionary.reference.com/browse/stitch+in+time,+a

Fulton, L.V., Ivanitskaya, L.V., Bastian, N.D., Erofeev, D.A., & Mendez, F.A. (2013). Frequent deadlines: Evaluating the effect of learner control on healthcare executives' performance in online learning. *Learning and Instruction, 23*, 24–32. doi: 10.1016/j.learninstruc.2012.09.001

Galli, A., Wackernagel, M., Iha, K., & Lazarus, E. (2014). Ecological footprint: Implications for biodiversity. *Biological Conservation, 173*, 121–132. doi: 10.1016/j.biocon.2013.10.019

Gallup. (2014a). *In U.S., most do not see global warming as serious threat*. Retrieved from http://www.gallup.com/poll/167879/not-global-warming-serious-threat.aspx

Gallup. (2014b). *Americans again pick environment over economic growth*. Retrieved from http://www.gallup.com/poll/168017/americans-again-pick-environment-economic-growth.aspx

Gangestad, S.W., & Simpson, J.A. (Eds.). (2007). *The evolution of mind: Fundamental questions and controversies*. New York, NY: Guilford Press.

Gardner, G.T., & Stern, P.C. (2008). The short list: The most effective actions U.S. households can take to curb climate change. *Environment: Science and Policy for Sustainable Development, 50*, 12–23. doi: 10.3200/ENVT.50.5.12-25

Gärling, T., & Schuitema, G. (2007). Effectiveness, public acceptability, and political feasibility of policy measures to change demand for private car use. *Journal of Social Issues, 63*, 139–153. doi: 10.1111/j.1540-4560.2007.00500.x

Gaspari, L., Sampino, D. R., Paris, F., Audran, F., Orsini, M., Neto, J. B., & Sultan, C. (2012). High prevalence of micropenis in 2710 male newborns from an intensive-use pesticide area of Northeastern Brazil. *International Journal of Andrology, 35*(3), 253–264. doi: 10.1111/j.1365-2605.2011.01241.x

Gatersleben, B., & Andrews, M. (2013). When walking in nature is not restorative—The role of prospect and refuge. *Health & Place, 20,* 91–101. doi: 10.1016/j.healthplace.2013.01.001

Gatersleben, B., Steg, L., & Vlek, C. (2002). Measurement and determinants of environmentally significant consumer behavior. *Environment and Behavior, 34*(3), 335–362. doi: 10.1177/0013916502034003004

Gaulin, S.J.C., & McBurney, D.H. (2003). *Evolutionary psychology* (2nd ed.). Upper Saddle River, NJ: Pearson.

Gawronski, B., & Strack, F. (Eds.). (2012). *Cognitive consistency: A fundamental principle in social cognition.* New York, NY: Guilford Press.

Geller, E.S. (1987). Applied behavior analysis and environmental psychology: From strange bedfellows to a productive marriage. In D.S. Stokols, & I. Altman (Eds.), *Handbook of environmental psychology* (Vol. 1, pp. 361–387). New York, NY: John Wiley & Sons.

Geller, E.S. (1992). It takes more than information to save energy. *American Psychologist, 47*, 814–815. doi: 10.1037/0003-066X.45.10.1109

Geller, E.S. (2002). The challenge of increasing proenvironmental behavior. In R.B. Bechtel, & A. Churchman (Eds.), *Handbook of environmental psychology* (pp. 525–540), New York: Wiley.

Georgia Southern University (GSU) researchers study wilderness therapy for PTSD. (2014, June 27). *GSU Newsroom.* Retrieved from http://news.georgiasouthern.edu/2014/06/27/georgia-southern-university-researchers-study-wilderness-therapy-for-ptsd/

Giddens, J.L., Schermer, J.A., & Vernon, P.A. (2009). Material values are largely in the family: A twin study of genetic and environmental contributions to materialism. *Personality and Individual Differences, 46*(4), 428–431. doi: 10.1016/j.paid.2008.11.008

Giddings, B., Hopwood, B., & O'Brien, G. (2002), Environment, economy and society: Fitting them together into sustainable development. *Sustainable Development, 10*, 187–196. doi: 10.1002/sd.199

Gifford, R. (2011). The dragons of inaction: Psychological barriers that limit climate change mitigation and adaptation. *American Psychologist, 66*(4), May-Jun 2011, 290–302. doi: 10.1037/a0023566 Gifford 2013

Gifford, R. (2014). Environmental psychology matters. *Annual Review of Psychology, 65*, 541–79. doi: 10.1146/annurev-psych-010213-115048

Gifford, R., & Comeau, L.A. (2011). Message framing influences perceived climate change competence, engagement, and behavioral intentions. *Global Environmental Change, 21*, 1301–1307. doi: 10.1016/j.gloenvcha.2011.06.004

Gifford, R., & Nilsson, A. (2014). Personal and social factors that influence pro-environmental concern and behavior: A review. *International Journal of Psychology, 49*(3), 141–157. doi: 10.1002/ijop.12034

Gifford, R., & Sussman, R. (2012). Environmental attitudes. In S.D. Clayton (Ed.), *The Oxford handbook of conservation and environmental psychology* (pp. 65–80). New York, NY: Oxford University Press.

Ginde, A.A., Liu, M.C., & Camargo, C.A. (2009). Demographic differences and trends of vitamin D insufficiency in the US population, 1988–2004. *Archives of Internal Medicine, 169*(6), 626–632. doi:10.1001/archinternmed.2008.604

Giordano, G., & Costa, L.G. (2012). Developmental neurotoxicity: Some old and new issues. *International Scholarly Research Network Toxicology, 2012*, np. doi: 10.5402/2012/814795

Giuliani, M.V., & Scopelliti, M. (2009). Empirical research in environmental psychology: Past, present, and future. *Journal of Environmental Psychology, 29*, 375–386. doi: 10.1016/j.jenvp.2008.11.008

Glendinning, C. (1994). *My name is Chellis & I'm in recovery from western civilization.* Boston, MA: Shambhala.

Gluckman, P., & Hanson, M. (2008). *Mismatch: The lifestyle diseases timebomb.* New York, NY: Oxford University Press.

Goldenberg, S. (2009, January 16). The worst of times: Bush's environmental legacy examined. *The Guardian.* Retrieved from http://www.theguardian.com/politics/

Goldstein, M.C., Titmus, A.J., & Ford, M. (2013). Scales of spatial heterogeneity of plastic marine debris in the northeast Pacific Ocean. *PLoS ONE 8*(11), e80020. doi: 10.1371/journal.pone.0080020

Goldstein, N.J., Cialdini, R.B., & Griskevicius, V. (2008). A room with a viewpoint: Using social norms to motivate environmental conservation in hotels. *Journal of Consumer Research, 35*, 472–482. doi: 10.1086/586910

Goleman, D. (2009). *Ecological intelligence: How knowing the hidden impacts of what we buy can change everything.* New York, NY: Doubleday.

Gonzalez, M.T., Hartig, T., Patil, G.G., Martinsen, E.W., & Kirkevold, M. (2010). Therapeutic horticulture in clinical depression: A prospective study of active components. *Journal of Advanced Nursing, 66*(9), 2002–2013. doi: 10.1111/j.1365-2648.2010.05383.x

Gorini, F., Muratori, F., & Morales, M.A. (2014). The role of heavy metal pollution in neurobehavioral disorders: A focus on Autism. *Review Journal of Autism and Developmental Disorders (online edition).* doi: 10.1007/s40489-014-0028-3

Gottlieb, B.H., & Bergen, A.E. (2010). Social support concepts and measures. *Journal of Psychosomatic Research, 69*(5), 511–520. doi: 10.1016/j.jpsychores.2009.10.001

Gottlieb, R. (2005). *Forcing the spring: The transformation of the American environmental movement* (revised and updated edition). Washington, DC: Island Press.

Graber, A., & Junge, R. (2009). Aquaponic systems: Nutrient recycling from fish wastewater by vegetable production. *Desalination, 246*, 147–156.

Grandjean, P., & Landrigan, P. J. (2006). Developmental neurotoxicity of industrial chemicals. *Lancet, 368,* 2167–2178. doi: 10.1016/S0140-6736(06)69665-7

Grandjean, P., & Landrigan, P. J. (2014). Neurobehavioural effects of developmental toxicity. *The Lancet Neurology, 13*(3), 330–338. doi: 10.1016/S1474-4422(13)70278-3

Greaves, M., Zibarras, L. D., & Stride, C. (2013). Using the theory of planned behavior to explore environmental behavioral intentions in the workplace. *Journal of Environmental Psychology, 34,* 109–120. doi: 10.1016/j.jenvp.2013.02.003

Green-Demers, I., Pelletier, L. G., & Ménard, S. (1997). The impact of behavioural difficulty on the saliency of the association between self-determined motivation and environmental behaviours. *Canadian Journal of Behavioural Science, 29,* 157–166.

Greenwald, A., & Banaji, M. (1995). Implicit social cognition: Attitudes, self-esteem, and stereotypes. *Psychological Review, 102,* 4–27. doi: 10.1037/0033-295X.102.1.4

Greenwald, A., McGhee, D., & Schwartz, J. (1998). Measuring individual differences in implicit cognition: The Implicit Association Test. *Journal of Personality and Social Psychology, 74,* 1464–1480. doi: 10.1037/0022-3514.74.6.1464

Greenwald, A. G., Poehlman, T. A., Uhlmann, E. L., & Banaji, M. R. (2009). Understanding and using the Impliciat Association Test: III. Meta-analysis of predictive validity. *Journal of Personality and Social Psychology, 97*(1), 17–41. doi: 10.1037/a0015575

Greenway, R. (1995). The wilderness effect and ecopsychology. In T. Roszak, M. E. Gomes, & A. D. Kanner (Eds.), *Eco-psychology: Restoring the earth, healing the mind* (pp. 122–135). San Francisco, CA: Sierra Club Books.

Griskevicius, V., Cantú, S. M., & van Vugt, M. (2012). The evolutionary bases for sustainable behavior: Implications for marketing, policy, and social entrepreneurship. *Journal of Public Policy & Marketing, 31,* 115–128. doi: 10.1509/jppm.11.040

Griskevicius, V., Tybur, J. M., & Van den Bergh, B. (2010). Going green to be seen: Status, reputation, and conspicuous conservation. *Journal of Personality and Social Psychology, 98,* 392–404. doi: 10.1037/a0017346

Groce, L. (1975). *Junk food junkie* [audio recording]. Los Angeles, CA: Peaceable Records.

Grooten, M. (Ed.). (2012). World Wildlife Fund living planet report 2012: Biodiversity, biocapacity, and better choices. *World Wildlife Fund.* Retrieved from http://awsassets.panda.org/downloads/1_lpr_2012_online_full_size_single_pages_final_120516.pdf.

Grotto, D., & Zied, E. (2010). The Standard American Diet and its relationship to the health status of Americans. *Nutrition in Clinical Practice, 25*(6), 603–612. doi: 10.1177/088453361038623

Grover, E. O. (1909). *The book of good cheer: A little bundle of cheery thoughts.* Chicago, IL: P. F. Volland & Company.

Guadagno, R. E., & Cialdini, R. B. (2010). Preference for consistency and social influence: A review of current research findings. *Social Influence, 5*(3), 152–163. doi: 10.1080/15534510903332378

Guéguen, N. (2012). Dead indoor plants strengthen belief in global warming. *Journal of Environmental Psychology 32*, 173–177. doi:10.1016/j.jenvp.2011.12.002

Guilarte, T. R., Opler, M., & Pletnikov, M. (2012). Is lead exposure in early life an environmental risk factor for Schizophrenia? Neurobiological connections and testable hypotheses. *NeuroToxicology, 33,* 560–574. doi: 10.1016/j.neuro.2011.11.008

Guillette, E. A., Meza, M. M., Aquilar, M. G., Soto, A. D., & Garcia, I. E. (1998). An anthropological approach to the evaluation of preschool children exposed to pesticides in Mexico. *Environmental Health Perspectives, 106,* 347–353.

Gunders, D. (2012). *Wasted: How America is losing up to 40 percent of its food from farm to fork to landfill.* Washington, DC: Natural Resources Defense Council. Retrieved from http://www.nrdc.org/food/files/wasted-food-ip.pdf

Guney, M., & Zagury, G. J. (2012). Heavy metals in toys and low-cost jewelry: Critical review of U.S. and Canadian legislations and recommendations for testing. *Environmental Science & Technology, 46*(8), 4265–4274. doi: 10.1021/es203470x

Haidt, J. (2001). The emotional dog and its rational tail: A social intuitionist approach to moral judgment. *Psychological Review, 108,* 814–834. doi: 10.1037//0033-295X.108.4.814

Haim, A., & Portnov, B. A. (2013). Light pollution as a new risk factor for human breast and prostate cancers. New York, NY: Springer.

Han, K.-T. (2010). An exploration of relationships among the responses to natural scenes: Scenic beauty, preference, and restoration. *Environment and Behavior, 42*(2), 243–270. doi: 10.1177/0013916509333875

Hancock, A. (2002). *Report on the Sonoma county Ecological Footprint project.* Sebastopol, CA: Sustainable Sonoma County.

Hardin, G. (1968, December 13). The tragedy of the commons. *Science, 162,* 1234–1248. doi:10.1126/science.162.3859.1243

Hargreaves, T., Nye, M., & Burgess, J. (2010). Making energy visible: A qualitative field study of how householders interact with feedback from smart energy monitors. *Energy Policy, 38,* 6111–6119. doi: 10.1016/j.enpol.2010.05.068

Harper, N., Russell, K., Cooley, R., & Cupples, J. (2007). Catherine Freer wilderness therapy expeditions: An exploratory case study of adolescent wilderness therapy, family functioning, and the maintenance of change. *Child & Youth Care Forum, 36*(2/3), 111–129. doi:10.1007/s10566-007-9035-1

Harrigan, N., Achananuparp, P., & Lim, E. P. (2012). Influentials, novelty, and social contagion: The viral power of average friends, close communities, and old news. *Social Networks, 34,* 470–480. doi: 10.1016/j.socnet.2012.02.005

Hart, R. A. (1997). *Children's participation: The theory and practice of involving young citizens in community development and environmental care.* New York, NY: UNICEF.

Hartig, T., Böök, A., Garvill, J., Olsson, T., & Gärling, T. (1996). Environmental influences on psychological restoration. *Scandinavian Journal of Psychology, 37*(4), 378–393. doi: 10.1111/j.1467-9450.1996.tb00670.x

Hartig, T., & Cooper Marcus, C. (2006). Essay: Healing gardens—Places for nature in health care. *Lancet, 368,* S36–S37. doi: 10.1016/S0140-6736(06)69920-0

Hartig, T., Evans, G.W., Jamner, L.D., Davis, D.S., & Gärling, T. (2003). Tracking restoration in natural and urban field settings. *Journal of Environmental Psychology, 23,* 109–123. doi: 10.1016/S0272-4944(02)00109-3

Hartig, T., & Staats, H. (2006). The need for psychological restoration as a determinant of environmental preferences. *Journal of Environmental Psychology, 26,* 215–226. doi: 10.1016/j.jenvp.2006.07.007

Haub, C., & Kaneda, T. (2013). *2013 world population data sheet.* Retrieved from Population Reference Bureau website: http://www.prb.org/pdf13/2013-population-data-sheet_eng.pdf

Hawcroft, L.J., & Milfont, T.L. (2010). The use (and abuse) of the new environmental paradigm scale over the last 30 years: A meta-analysis. *Journal of Environmental Psychology 30,* 143–158. doi:10.1016/j.jenvp.2009.10.003

Hawken, P. (1993, September/October). A declaration of sustainability: 12 steps society can take to save the whole enchilada. *Utne Reader,* 54–61.

Hawken, P. (2007). *Blessed unrest: How the largest movement in the world came into being, and why no one saw it coming.* New York, NY: Viking Press.

Hawken, P., Lovins, A., & Lovins, L.H. (1999). *Natural capitalism: Creating the next industrial revolution.* New York, NY: Little, Brown and Company.

Hayes, D. (1990). Earth day 1990: Threshold of the green decade. *World Policy Journal, 7*(2), 289–304.

Heft, H. (1988). Affordances of children's environments: A functional approach to environmental description. *Children's Environments Quarterly, 5,* 29–37.

Helvarg, D. (2004). *The war against the greens: The "Wise-Use" movement, the New Right, and the browning of America.* Boulder, CO: Johnson Books.

Hendricks, S. (2005). *Divine destruction: Dominion theology and American environmental policy.* New York, NY: Melville House.

Herring, H., & Sorrell, S. (2008). *Energy efficiency and sustainable consumption: The rebound effect.* New York, NY: Palgrave.

Herweg, F., & Müller, D. (2011). Performance of procrastinators: On the value of deadlines. *Theory and Decision, 70*(3), 329–366. doi: 10.1007/s11238-010-9195-6

Hickling, A.K., & Gelman, S.A. (1995). How does your garden grow?: Early conceptualization of seeds and their place in the plant growth cycle. *Child Development, 66,* 856–876. doi: 10.1111/j.1467-8624.1995.tb00910.x

Hilbig, B.E., Zettler, I., Moshagen, M., & Heydasch, T. (2013). Tracing the path from personality—Via cooperativeness—To conservation. *European Journal of Personality, 27*(4), 319–327. doi: 10.1002/per.1856

Hines, J.M., Hungerford, H.R., & Tomera, A.N. (1986). Analysis and synthesis of research on responsible environmental behavior: A meta-analysis. *Journal of Environmental Education, 18*(2), 1–8. doi: 10.1080/00958964.1987.9943482

Hines, M. (2011). Gender development and the human brain. *Annual Review of Neuroscience, 34,* 69–88. doi: 10.1146/annurev-neuro-061010-113654

Hipp, J.A., & Ogunseitan, O.A. (2011). Effect of environmental conditions on perceived psychological restorativeness of coastal parks. *Journal of Environmental Psychology, 31*(4), 421–429. doi: 10.1016/j.jenvp.2011.08.008

Hirsh, J. B. (2010). Personality and environmental concern. *Journal of Environmental Psychology, 30*(2), 245–248. doi: 10.1016/j.jenvp.2010.01.004

Hirsh, J. B., & Dolderman, D. (2007). Personality predictors of consumerism and environmentalism: A preliminary study. *Personality and Individual Differences, 43*(6), 1583–1593. doi: 10.1016/j.paid.2007.04.015

Hobbes, T. (1651/1962). *Leviathan: Or the matter, forme, and power of commonwealth ecclesiastical and civil.* New York, NY: Collier.

Hochman, D. (2013, May 3). Urban gardening: An appleseed with attitude. *The New York Times.* Retrieved from http://www.nytimes.com/2013/05/05/fashion/urban-gardening-an-appleseed-with-attitude.html?pagewanted=all&_r=2& Hoffman, 2013

Holland, L., & Carter, J. S. (2005). Words v. deeds: A comparison of religious belief and environmental action. *Sociological Spectrum, 25*(6), 739–753. doi: 10.1080/02732170500260908

Holmgren, D. (2002). *Permaculture: Principles and pathways beyond sustainability.* Hepburn Springs, Vic, Australia: Holmgren Design Services.

Homonoff, T. A. (2013). Essays in Behavioral Economics and Public Policy. Unpublished Doctoral Dissertation, Princeton University, Princeton, NJ.

Hood, E. (2005). Are EDCs blurring issues of gender? *Environmental Health Perspectives, 113,* A670–677. doi: 10.1289/ehp.113-a670

Hopkins, R. (2008). *The transition handbook: From oil dependency to local resilience.* White River Junction, VT: Chelsea Green Publishing.

Houde, S., Todd, A., Sudarshan, A., Flora, J. A., & Armel, K. C. (2013). Real-time feedback and electricity consumption: A field experiment assessing the potential for savings and persistence. *The Energy Journal, 34*(1), 87–102.

Hovland, C. I., Janis, I. L., & Kelley, H. H. (1953). *Communication and persuasion: Psychological studies of opinion change.* New Haven, CT: Yale University Press.

Howell, A. J., Dopko, R. L., Passmore, H.-A., & Buro, K. (2011). Nature connectedness: Associations with well-being and mindfulness. *Personality and Individual Differences, 51*(2), 166–171. doi: 10.1016/j.paid.2011.03.037

Hurst, M., Dittmar, H., Bond, R., & Kasser, T. (2013). The relationship between materialistic values and environmental attitudes and behaviors: A meta-analysis. *Journal of Environmental Psychology, 36,* 257–269. doi: 10.1016/j.jenvp.2013.09.003

Hussar, K. M., & Horvath, J. C. (2011). Do children play fair with mother nature? Understanding children's judgments of environmentally harmful actions. *Journal of Environmental Psychology, 31,* 309–313. doi: 10.1016/j.jenvp.2011.05.001

Inagaki, K., & Hatano, G. (2004). Vitalistic causality in young children's naïve biology. *Trends in Cognitive Sciences, 8*(8), 356–362. doi: 10.1016/j.tics.2004.06.004

Intergovernmental Panel on Climate Change (IPCC). (2013). Summary for policy makers. In T. F. Stocker, D. Qin, G.-K. Plattner, M. Tignor, S. K. Allen, J. Boschung, . . . P. M. Midgley (Eds.), *Climate change 2013: The physical science basis. Contribution of Working Group I to the fifth assessment report of the Intergovernmental Panel on Climate Change.* New York, NY: Cambridge University Press. Retrieved from http://www.climatechange2013.org/images/report/WG1AR5_SPM_FINAL.pdf

Intergovernmental Panel on Climate Change (IPCC). (2014). *Climate change 2014: Impacts, adaptation, and vulnerability*. New York, NY: Cambridge University Press. Retrieved from http://www.ipcc.ch/report/ar5/wg2/

Jacka, F. N., Pasco, J. A., Mykletun, A., Williams, L. J., Hodge, A. M., O'Reilly, S. L., . . . Berk, M. (2010). Association of western and traditional diets with depression and anxiety in women. *American Journal of Psychiatry, 167*, 305–311. doi:10.1176/appi.ajp.2009.09060881

Jackson, T. (2005). *Motivating sustainable consumption: A review of evidence on consumer behaviour and behavioural change*. London, United Kingdom: SDRN.

Jain, R. K., Taylor, J. E., & Culligan, P. J. (2013). Investigating the impact eco-feedback information representation has on building occupant energy consumption behaviors and savings. *Energy and Buildings, 64*, 408–414. doi: 10.1016/j.enbuild.2013.05.011

James, W. T. (1890). *The principles of psychology*. New York, NY: Holt, Rinehart & Winston.

James, W. (1899/2008). *Talks to teachers on psychology and to students on some of life's ideals*. Rockville, MD: Arc Manor.

Janagan, K., Sathish, V., & Vijayakumar, A. (2003). A sustainable system for solid waste treatment vermiculture. In M. J. Bunch, V. M. Suresh, & T. V. Kumaran (Eds.), *Proceedings of the third international conference on environment and health* (pp. 175–185). Chennai, India: Department of Geography, University of Madras and Faculty of Environmental Studies, York University.

Jeng, H. A. (2014). Exposure to endocrine disrupting chemicals and male reproductive health. *Frontiers in Public Health, 2,* Article 55. doi: 10.3389/fpubh.2014.00055

Jenkins, J. (1999). *Humanure handbook: A guide to composting human manure* (2nd ed.). Grove City, PA: Joseph Jenkins. Retrieved from http://humanurehandbook.com/downloads/H2.pdf

Jensen, D. (2006, May/June). Beyond hope. *Orion Magazine*, Retrieved from http://www.orionmagazine.org/index.php/articles/article/170/

Joireman, J., Truelove, H. B., & Duell, B. (2010). Effect of outdoor temperature, heat primes and anchoring on belief in global warming. *Journal of Environmental Psychology, 30*, 358–367. doi:10.1016/j.jenvp.2010.03.004

Jost, J. T., Liviatan, I., van der Toorn, J., Ledgerwood, A., Mandisodza, A., & Nosek, B. A. (2010). System justification: How do we know it's motivated? In D. R. Bobocel, A. C. Kay, M. P. Zanna, & J. M. Olson (Eds.), *The psychology of justice and legitimacy* (pp. 173–203). New York, NY: Psychology Press.

Joye, Y. (2007). Architectural lessons from environmental psychology: The case of biophilic architecture. *Review of General Psychology, 11,* 305–328. doi: 10.1037/1089-2680.11.4.305

Joye, Y. (2012). Can architecture become second nature? An emotion-based approach to nature-oriented architecture. In P. H. Kahn, Jr., & P. H. Hasbach (Eds.), *Ecopsychology: Science, totems, and the technological species* (pp. 195–217). Cambridge, MA: MIT Press.

Julius, H., Beetz, A., Kotrschal, K., Turner, D., & Uvnäs-Moberg, K. (2013). *Attachment to pets: An integrative view of the human-animal relationship with implications for therapeutic practice.* Cambridge, MA: Hogrefe.

Kabat-Zinn, J. (2013). *Full catastrophe living (Revised edition): Using the wisdom of your body and mind to face stress, pain, and illness.* New York, NY: Bantam Books.

Kahan, D. M., Braman, D., Gastil, J., Slovic, P., & Mertz, C. K. (2007). Culture and identity-protective cognition: Explaining the white male effect in risk perception. *Journal of Empirical Legal Studies, 4*(3), 465–505. doi: 10.1111/j.1740-1461.2007.00097.x

Kahan, D. M., Peters, E., Wittlin, M., Slovic, P., Ouellette, L. L., Braman, D., & Mandel, G. (2012). The polarizing impact of science literacy and numeracy on perceived climate change risks. *Nature Climate Change, 2*, 732–735. doi: 10.1038/nclimate1547

Kahn, P. H., Jr. (1997). Developmental psychology and the biophilia hypothesis: Children's affiliation with nature. *Developmental Review, 17*, 1–61. doi: 10.1006/drev.1996.0430

Kahn, P. H., Jr. (1999). *The human relationship with nature: Development and culture.* Cambridge, MA: Massachusetts Institute of Technology Press.

Kahn, P. H., Jr. (2002). Children's affiliations with nature: Structure, development, and the problem of environmental generational amnesia. In P. H. Kahn, Jr., & S. R. Kellert (Eds.), *Children and nature: Psychological, sociocultural, and evolutionary investigations* (pp. 93–116). Cambridge, MA: Massachusetts Institute of Technology Press.

Kahn, P. H., Jr. (2011). *Technological nature: Adaptation and the future of human life.* Cambridge, MA: MIT Press.

Kahn, P. H., Jr., Friedman, B., Pérez-Granados, D. R., & Freier, N. G. (2006). Robotic pets in the lives of preschool children. *Interaction Studies, 7*, 405–436. doi:10.1.1.130.3260

Kahn, P. H., Jr., & Hasbach, P. H. (2012). *Ecopsychology: Science, totems, and the technological species.* Cambridge, MA: MIT Press.

Kahn, P. H., Jr., & Hasbach, P. H. (2013). *The rediscovery of the wild.* Cambridge, MA: MIT Press.

Kahn, P. H., Jr., & Lourenco, O. (2002). Water, air, fire, and earth: A developmental study in Portugal of environmental moral reasoning. *Environment and Behavior, 34*(4), 405–430. doi: 10.1177/00116502034004001

Kahn, P. H., Jr., Saunders, C. D., Severson, R. L., Myers, O. E., Jr., & Gill, B. T. (2008). Moral and fearful affiliations with the animal world: Children's conceptions of bats. *Anthrozoös, 21*(4), 375–385. doi: 10.2752/175303708X371591

Kahneman, D. (2011). *Thinking fast and slow.* New York, NY: Farrar, Straus, and Giroux.

Kaiser, F. G. (1998). A general measure of ecological behavior. *Journal of Applied Social Psychology, 28*(5), 395–422. doi: 10.111/j.1559-1816.1998.tb01712.x

Kaiser, F. G., & Biel, A. (2000). Assessing general ecological behavior: A cross-cultural comparison between Switzerland and Sweden. *European Journal of Psychological Assessment, 16*(1), 44–52. doi: 10.1027//1015-5759.16.1.44

Kaiser, F. G., & Byrka, K. (2011). Environmentalism as a trait: Gauging people's prosocial personality in terms of environmental engagement. *International Journal of Psychology, 46*(1), 71–79. doi: 10.1080/00207594.2010.516830

Kaiser, F.G., & Fuhrer, U. (2003). Ecological behavior's dependency on different forms of knowledge. *Applied Psychology: An International Review, 52*(4), 598–613. doi: 10.1111/1464–0597.00153

Kaiser, F.G., & Schultz, P.W. (2009). The attitude-behavior relationship: A test of three models of the moderating role of behavioral difficulty. *Journal of Applied Social Psychology, 39*, 186–207. doi: 10.1111/j.1559-1816.2008.00435.x

Kaiser, F.G., Wölfing, S., & Fuhrer, U. (1999). Environmental attitude and ecological behaviour. *Journal of Environmental Psychology, 19*(1), 1–19. doi: 10.1006/jevp.1998.0107

Kals, E., & Ittner, H. (2003). Children's environmental identity: Indicators and behavioral impacts. In S. Clayton, & S. Opotow (Eds.), *Identity and the natural environment: The psychological significance of nature* (pp. 135–157). Cambridge, MA: Massachusetts Institute of Technology Press.

Kanner, A.D., & Gomes, M.E. (1995). The all-consuming self. In T. Roszak, M.E. Gomes, & A.D. Kanner (Eds.), *Eco-psychology: Restoring the earth, healing the mind* (pp. 77–91). San Francisco, CA: Sierra Club.

Kantola, S., Syme, G.J., & Campbell, N.A. (1984). Cognitive dissonance and energy conservation. *Journal of Applied Psychology, 69*(3), 416–421. doi: 10.1037/0021-9010.69.3.416

Kaplan, R., & Austin, M.E. (2004). Out in the country: Sprawl and the quest for nature nearby. *Landscape and Urban Planning, 69*, 235–243. doi: 10.1016/j.landurbplan.2003.09.006

Kaplan, R., & Kaplan, S. (1989). *The experience of nature.* Cambridge, MA: Cambridge University Press.

Kaplan, S. (1992). Environmental preference in a knowledge-seeking, knowledge-using organism. In J.H. Barkow, L. Cosmides, & J. Tooby (Eds.), *The adapted mind: Evolutionary psychology and the generation of culture.* New York, NY: Oxford University Press.

Kaplan, S. (1995). The restorative benefits of nature: Toward an integrative framework. *Journal of Environmental Psychology, 15*, 169–182. doi: 10.1016/0272-4944(95)90001-2

Kaplan, S., & Talbot, J.F. (1983). Psychological benefits of a wilderness experience. In I. Altman, & J.F. Wohlwill (Eds.), *Behavior and the natural environment* (pp. 163–203). New York, NY: Plenum Press.

Kardefelt-Winther, D. (2014). A conceptual and methodological critique of internet addiction research: Towards a model of compensatory internet use. *Computers in Human Behavior, 31*, 351–354. doi: 10.1016/j.chb.2013.10.059

Karjalainen, S. (2011). Consumer preferences for feedback on household electricity consumption. *Energy and Buildings, 43*, 458–467. doi: 10.1016/j.enbuild.2010.10.010

Karniol, R., & Ross, M. (1996). The motivational impact of temporal focus: Thinking about the future and the past. *Annual Review of Psychology, 47*, 593–620. doi: 10.1146/annurev.psych.47.1.593

Karp, D.G. (1996). Values and their effect on pro-environmental behavior. *Environment and Behavior, 28*(1), 111–133. doi: 10.1177/0013916596281006

Kasser, T. (2009). Shifting values in response to climate change. In R. Engleman, M. Renner, & J. Sawan (Eds.), *2009 state of the world: Into a warming world* (pp. 122–125). New York, NY: W. W. Norton & Co.

Kasser, T. (2011). Ecological challenges, materialistic values, and social change. In R. Biswas- Diener (Ed.), *Positive psychology as social change.* (pp. 89–108). New York, NY: Springer Science+Business Media B. V.

Katcher, A., & Wilkins, G. (1998). Animal-assisted therapy in the treatment of disruptive behavior disorders. In A. Lundberg (Ed.), *The environment and mental health* (pp. 193–204). Mahwah, NJ: Erlbaum.

Kates, R. W. (1976). Experiencing the environment as hazard. In S. Wapner, S. Cohen, & B. Kaplan (Eds.), *Experiencing the environment* (pp. 133–156). New York, NY: Plenum Press.

Katzev, R., & Johnson, T. (1983). A social-psychological analysis of residential electricity consumption: The impact of minimal justification techniques. *Journal of Economic Psychology, 3*, 267–284. doi: 10.1016/0167-4870(83)90006-5

Katzmarzyk, P. T., Church, T. S., Craig, C. L., & Bouchard, C. (2009). Sitting time and mortality from all causes, cardiovascular disease, and cancer. *Medicine & Science in Sports & Exercise, 41*(5), 998–1005. doi: 10.1249/MSS.0b013e3181930355

Keating, D. P. (Ed.). (2011). *Nature and nurture in early child development.* New York, NY: Cambridge University Press.

Keep America Beautiful. (2013). *Recycling facts and stats.* Retrieved at http://www.kab.org/site/PageServer?pagename=recycling_facts_and_stats

Keil, D. (2011). An ounce of prevention is worth a pound of cure: Reframing the debate about law school affirmative action. *Denver University Law Review, 88*(4), 791–806.

Kellert, S. R. (1997). *Kinship to mastery: Biophilia in human evolution and development.* Washington, DC: Island Press.

Kellert, S. R. (2005). *Building for life: Designing and understanding the human-nature connection.* Covelo, CA: Island Press.

Kellert, S. R., Heerwagen, J., & Mador, M. (Eds.). (2008). *Biophilic design: The theory, science, and practice of bringing buildings to life.* Hoboken, NJ: Wiley.

Kellert, S. T., & Wilson, E. O. (Eds.). (1993). *The biophilia hypothesis.* Washington, DC: Island Press.

Kenis, A., & Mathijs, E. (2012). Beyond individual behaviour change: The role of power, knowledge and strategy in tackling climate change. *Environmental Education Research, 18*(1), 45–65. doi: 10.1080/13504622.2011.576315

Kennedy, K. A., & Pronin, E. (2008). When disagreement gets ugly: Perceptions of bias and the escalation of conflict. *Personality and Social Psychology Bulletin, 34*(6), 833–848. doi: 10.1177/0146167208315158

Kidner, D. W. (2001). *Nature and psyche: Radical environmentalism and the politics of subjectivity.* Albany, NY: State University of New York Press.

Kiesling, F. M., & Manning, C. M. (2010). How green is your thumb? Environmental gardening identity and ecological gardening practices. *Journal of Environmental Psychology, 30*(3), 315–327. doi: 10.1016/j.jenvp.2010.02.004

King, C. A. (2008). Community resilience and contemporary agri-ecological systems: Reconnecting people and food, and people with people. *Systems Research and Behavioral Science, 25*(1), 111–124. doi: 10.1002/sres.854

Kirby, E. D., Muroy, S. E., Sun, W. G., Covarrubias, D., Leong, M. J., Barchas, L. A., & Kaufer, D. (2013). Acute stress enhances adult rat hippocampal neurogenesis and activation of newborn neurons via secreted astrocytic FGF2. *eLife, 2013*(2), e00362. doi: 10.7554/eLife.00362

Kirkby, M. (1989). Nature as refuge in children's environments. *Children's Environments Quarterly, 6*, 7–12.

Klein, H. J., Wesson, M. J., Hollenbeck, J. R., Alge, B. J. (1999). Goal commitment and the goal-setting process: Conceptual clarification and empirical synthesis. *Journal of Applied Psychology, 84*(6), 885–896. doi: 10.1037/0021-9010.84.6.885

Klingberg, T. (2008). *The overflowing brain: Information overload and the limits of working memory*. New York, NY: Oxford University Press.

Klinkenborg, V. (2008, November). Light pollution. *National Geographic*. Retrieved from http://ngm.nationalgeographic.com/2008/11/light-pollution/klinkenborg-text

Klöckner, C. A., & Friedrichsmeier, T. (2011). A multi-level approach to travel mode choice- How person characteristics and situation specific aspects determine car use in a student sample. *Transportation Research Part F: Traffic Psychology and Behaviour, 14*, 261–277. doi: 10.1016/j.trf.2011.01.006

Kluger, A. N., & DeNisi, A. (1996). The effects of feedback interventions on performance: A historical review, a meta-analysis, and a preliminary feedback intervention theory. *Psychological Bulletin, 119*(2), 254–284. doi: 10.1037/0033-2909.119.2.254

Knippenberg, S., Damoiseaux, J., Bol, Y., Hupperts, R., Taylor, B. V., Ponsonby, A. L., . . . van der Mei, I. A. F. (2014). Higher levels of reported sun exposure, and not vitamin D status, are associated with less depressive symptoms and fatigue in multiple sclerosis. *Acta Neurologica Scandinavica, 129*(2), 123–131. doi: 10.1111/ane.12155

Knussen, C., Yule, F., MacKenzie, J., & Wells, M. (2004). An analysis of intentions to recycle household waste: The roles of past behaviour, perceived habit, and perceived lack of facilities. *Journal of Environmental Psychology, 24*, 237–246. doi: 10.1016/j.jenvp.2003.12.001

Ko, C.-H., Yen, J.-Y., Chen, S.-H., Wang, P.-H., Chen, C.-S., & Yen, C.-F. (2014). Evaluation of the diagnostic criteria of internet gaming disorder in the DSM-5 among young adults in Taiwan. *Journal of Psychiatric Research, 53*, 103–110. doi: 10.1016/j. jpsychires. 2014.02.008

Koger, S. M., Schettler, T., & Weiss, B. (2005). Environmental toxicants and developmental disabilities: A challenge for psychologists. *American Psychologist, 60*, 243–255. doi: 10.1037/0003-066X.60.3.243

Koger, S. M., & Scott, B. A. (2007). Psychology and environmental sustainability: A call for integration. *Teaching of Psychology, 34*, 10–18. doi: 10.1080/00986280709336642

Kollmuss, A., & Agyeman, J. (2002). Mind the gap: Why do people act environmentally and what are the barriers to pro-environmental behavior? *Environmental Education Research, 8,* 239–260. doi: 10.1080/13504620220145401

Korpela, K.M. (2012). Place attachment. In S. Clayton (Ed.), *The Oxford handbook of environmental and conservation psychology* (pp. 148–163). New York, NY: Oxford University Press.

Korpela, K.M. (2013). Perceived restorativeness of urban and natural scenes—Photographic illustrations. *Journal of Architectural and Planning Research, 30*(1), 23–38.

Korpela, K., Borodulin, K., Neuvonen, M., Paronen, O., & Tyrväinen, L. (2014). Analyzing the mediators between nature-based outdoor recreation and emotional well-being. *Journal of Environmental Psychology, 37,* 1–7. doi: 10.1016/j.jenvp.2013.11.003

Krakauer, J. (1991, December). Brown fellas. *Outside,* 67–69.

Krakauer, J. (1995, October). Loving them to death. *Outside Magazine, 20,* 72–80, 82, 142–143.

Kruger, R.F., South, S., Johnson, W., & Iacono, W. (2008). The heritability of personality is not always 50%: Gene-environment interactions and correlations between personality and parenting. *Journal of Personality, 76*(6), 1485–1522. doi: 10.1111/j.1467-6494.2008.00529.x

Krupp, F., & Horn, M. (2008). *Earth, the sequel: The race to reinvent energy and stop global warming.* New York, NY: W.W. Norton & Co.

Kubiszewski, I., Costanza, R., Franco, C., Lawn, P., Talberth, J., Jackson, T., & Aylmer, C. (2013). Beyond GDP: Measuring and achieving global genuine progress. *Ecological Economics, 93,* 57–68. doi: 10.1016/j.ecolecon.2013.04.019

Kuhn, M.H. (1960). Self-attitudes by age, sex and professional training. *Sociological Quarterly, 1*(1), 39–56. doi: 10.1111/j.1533-8525.1960.tb01459.x

Kuhn, M.H., & McPartland, T.S. (1954). An empirical investigation of self-attitudes. *American Sociological Review, 19*(1), 68–76. doi: 10.2307/2088175

Kuntsler, J.H. (2005). *The long emergency: Surviving the end of oil, climate change, and other converging catastrophes of the twenty-first century.* New York, NY: Grove Press.

Kuo, F.E., & Faber Taylor, A. (2004). A potential natural treatment for attention-deficit/hyperactivity disorder: Evidence from a national study. *American Journal of Public Health, 94,* 1580–1586. doi: 10.2105/AJPH.94.9.1580

Kutz, G.D., & O'Connell, A. (2007). Residential treatment programs: Concerns regarding abuse and death in certain programs for troubled youth. United States Government Accountability Office. Retrieved from http://www.gao.gov/new.items/d08146t.pdf

LaDuke, W. (1999). *All my relations.* Cambridge, MA: South End Press.

Lambrinidou, Y., Triantafyllidou, S., & Edwards, M. (2010). Failing our children: Lead in U.S. school drinking water. *New Solutions, 20*(1), 25–47. doi: 10.2190/NS.022010eov

Lariviere, M., Couture, R., Ritchie, S.D., Cote, D., & Oddson, B. (2012). Behavioural assessment of wilderness therapy participants: Exploring the consistency of observational data. *The Journal of Experiential Education, 35*(1), 290–302. doi: 10.1177/105382591203500106

Larson, M.E., Houlihan, D., & Goernert, P.N. (1995). Effects of informational feedback on aluminum can recycling. *Behavioral Interventions, 10*, 111–117. doi: 10.1002/bin.2360100207

Latane, B., & Darley, J.M. (1968). Group inhibition of bystander intervention in emergencies. *Journal of Personality and Social Psychology, 10*, 215–221. doi: 10.1037/h0026570

Latane, B., & Darley, J.M. (1970). *The unresponsive bystander: Why doesn't he help?* New York, NY: Appleton-CenturyCrofts.

Latham, G.P., & Locke, E.A. (2006). Enhancing the benefits and overcoming the pitfalls of goal setting. *Organizational Dynamics, 35*(4), 332–340. doi: 10.1016/j.orgdyn.2006.08.008

Laumann, K., Gärling, T., & Stormark, K.M. (2003). Selective attention and heart rate responses to natural and urban environments. *Journal of Environmental Psychology, 23*, 125–134. doi: 10.1016/S0272-4944(02)00110-X

Lavergne, K.L., Sharp, E.C., Pelletier, L.G., & Holtby, A. (2010). The role of perceived government style in the facilitation of self-determined and non self-determined motivation for pro-environmental behavior. *Journal of Environmental Psychology, 30*, 169–177. doi: 10.1016/j.jenvp.2009.11.002

Lavoie, R.A., Jardine, T.D., Chumchal, M.M., Kidd, K.A., & Campbell, L.M. (2013). Biomagnification of mercury in aquatic food webs: A worldwide meta-analysis. *Environmental Science & Technology, 47*, 13385–13394. doi: 10.1021/es403103t

Lazarus, R.S., & Folkman, S. (1984). *Stress, appraisal, and coping.* New York, NY: Springer Publishing Company.

Lee, K., & Ashton, M.C. (2005). Psychopathology, Machiavellianism, and narcissism in the Five-Factor Model and the HEXACO model of personality structure. *Personality and Individual Differences, 38*(7), 1571–1582. doi: 10.1016/j.paid.2004.09.016

Lee, Y.-K., Chang, C.-T., Lin, Y., & Cheng, Z.-H. (2014). The dark side of smartphone usage: Psychological traits, compulsive behavior and technostress. *Computers in Human Behavior, 31*, 373–383. doi: 10.1016/j.chb.2013.10.047

Leeper, T.J., & Slothuus, R. (2014). Political parties, motivated reasoning, and public opinion formation. *Advances in Political Psychology, 35*(Suppl. 1), 129–156. doi: 10.1111/pops.12164

Lehman, P.K., & Geller, E.S. (2004). Behavior analysis and environmental protection: Accomplishments and potential for more. *Behavior and Social Issues, 13*, 13–32. doi: 10.5210/bsi.v13i1.33

Leiserowitz, A. (2005). American risk perceptions: Is climate change dangerous? *Risk Analysis, 25*, 1433–1442. doi: 10.1111/j.1540-6261.2005.00690.x

Leiserowitz, A. (2007). Communicating the risks of global warming: American risk perceptions, affective images, and interpretive communities. In S.C. Moser, & L. Dilling (Eds.), *Creating a climate for change: Communicating climate change and facilitating social change* (pp. 44–63). New York, NY: Cambridge University Press.

Leiserowitz, A., Maibach, E., & Roser-Renouf, C. (2009). *Saving energy at home and on the road: A survey of Americans' energy saving behaviors, intentions, motivations, and barriers.* Yale Project on Climate Change. George Mason University, Center for Climate Change Communication, Fairfax County, VA. Retrieved from http://environment.yale.edu/uploads/SavingEnergy.pdf

Leiserowitz, A., Maibach, E., Roser-Renouf, C., Feinberg, G., & Rosenthal, S. (2014). *Politics & Global Warming, Spring 2014.* Yale University and George Mason University. New Haven, CT: Yale Project on Climate Change Communication. Retrieved from http://environment.yale.edu/climate-communication/article/politics-and-global-warming-spring-2014/

Leiserowitz, A., Maibach, E., Roser-Renouf, C., Feinberg, G., Rosenthal, S., & Marlon, J. (2014). *Climate change in the American mind: Americans' global warming beliefs and attitudes in November 2013.* Yale University and George Mason University. New Haven, CT: Yale Project on Climate Change Communication.

Leiter, M.P., Laschinger, H.K.S., Day, A., & Gilin-Oore, D. (2011). The impact of civility interventions on employee social behavior, distress, and attitudes. *Journal of Applied Psychology, 96*(6), 1258–1274. doi:10.1037/a0024442

Leonard-Barton, D. (1981). The diffusion of active-residential solar energy equipment in California. In A. Shama (Ed.), *Marketing solar energy innovations* (pp. 243–257). New York, NY: Praeger.

León-Latre, M., Moreno-Franco, B., Andrés-Esteban, E.M., Ledesma, M., Laclaustra, M., Alcalde, V., . . . Aragon Workers' Health Study investigators. (2014). Sedentary lifestyle and its relation to cardiovascular risk factors, insulin resistance and inflammatory profile. *Revista Española de Cardiologia, 67*(6), 449–455. doi: 10.1016/j.rec.2013.10.015

Leopold, A. (1949). *A Sand County almanac: And sketches here and there.* New York, NY: Oxford University Press.

Lerner, M.J. (1980). *The belief in a just world: A fundamental delusion.* New York, NY: Plenum Press.

Leslie, K.E., & Koger, S.M. (2011). A significant factor in autism: Methyl mercury induced oxidative stress in genetically susceptible individuals. *Journal of Physical and Developmental Disabilities, 23,* 313–324. doi: 10.1007/s10882-011-9230-8

Levine, D.S., & Strube, M.J. (2012). Environmental attitudes, knowledge, intentions and behaviors among college students. *The Journal of Social Psychology, 152*(3), 308–326. doi: 10.1080/00224545.2011.604363

Lewandowski, S. (1987). Diohe'ko, the Three Sisters in Seneca life: Implications for a native agriculture in the Finger Lakes region of New York State. *Journal of Agriculture and Human Values, 4*(2–3), 76–93. doi: 10.1007/BF01530644

Lewicka, M. (2011). Place attachment: How far have we come in the last 40 years? *Journal of Environmental Psychology, 31*(3), 207–230. doi: 10.1016/j.jenvp.2010.10.001

Lewis, A.L., & Eves, F.F. (2012). Prompts to increase stair climbing in stations: The effect of message complexity. *Journal of Physical Activity and Health, 9*(7), 954–961.

Lind, C. (1992). *The Wright style: Re-creating the spirit of Frank Lloyd Wright.* New York, NY: Simon & Schuster.

Lindberg, C. (2000). Eschatology and fanaticism in the reformation era: Luther and the Anabaptists. *Concordia Theological Quarterly, 64,* 259–278.

Locke, E. A. (2000). Motivation by goal setting. In R. Golembiewski (Ed.), *Handbook of Organizational Behavior* (Vol. 2, pp. 43–56). New York, NY: Marcel Dekker.

Locke, E. A., & Latham, G. P. (1990). *A theory of goal setting and task performance.* Englewood Cliffs, NJ: Prentice-Hall.

Locke, E. A., & Latham, G. P. (2002). Building a practically useful theory of goal setting and task motivation: A 35-year odyssey. *American Psychologist, 57,* 705–717. doi: 10.1037/0003-066X.57.9.705

Locke, J. (1690/1939) An essay concerning the true original extent and end of civil government, Sections 40–41. Reproduced in Burtt (Ed.), *The English philosophers from Bacon to Mill* (pp. 403–503). New York, NY: The Modern Library.

Lockley, S. W., & Foster, R. G. (2012). *Sleep: A very short introduction.* New York, NY: Oxford University Press.

Logsdon-Conradsen, S. C., & Allred, S. L. (2010). Motherhood and environmental activism: A developmental framework. *Ecopsychology, 2*(3), 141–146. doi: 10.1089/eco.2010.0027

Lomborg, B. (2001). *The skeptical environmentalist: Measuring the real state of the world.* Cambridge, NY: Cambridge University Press.

Lopez, S. J. (Ed.). (2009). *The encyclopedia of positive psychology.* New York, NY: Blackwell Publishing.

Louv, R. (2005/2008). *Last child in the woods: Saving our children from nature-deficit disorder.* Chapel Hill, NC: Algonquin Books of Chapel Hill.

Lövdén, M., Xu, W., & Wang, H. X. (2013). Lifestyle change and the prevention of cognitive decline and dementia: What is the evidence? *Current Opinion in Psychiatry, 26,* 239–243. doi: 10.1097/YCO.0b013e32835f4135

Lucas, A. J., & Dyment, J. E. (2010). Where do children choose to play on the school ground? The influence of green design. *Education 3–13: International Journal of Primary, Elementary, and Early Years Education. 38*(2), 177–189. doi: 10.1080/03004270903130812

Lv, Z., Qi, H., Wang, L., Fan, X., Han, F., Wang, H., & Bi, S. (2014). Vitamin D status and Parkinson's disease: A systematic review and meta-analysis. *Neurological Sciences, 2014,* n. p. doi: 10.1007/s10072-014-1821-6

Lynne, G. D., Casey, C. F., Hodges, A., & Rahmani, M. (1995). Conservation technology adoption and the theory of planned behaviour. *Journal of Economic Psychology, 16,* 581–598. doi: 10.1016/0167-4870(95)00031-6

Maathai, W. (2003). *The Green Belt Movement: Sharing the approach and the experience.* Brooklyn, NY: Lantern Books.

Macon, M. B., & Fenton, S. E. (2013). Endocrine disruptors and the breast: Early life effects and later life disease. *Journal of Mammary Gland Biology and Neoplasia, 18*(1), 43–61. doi: 10.1007/s10911-013-9275-7

MacPherson, D.B. (1962). *The political theory of possessive individualism: Hobbes to Locke*. Oxford, United Kingdom: Clarendon Press.

Macy, J. (1995). Working through environmental despair. In T. Roszak, M.E. Gomes, & A.D. Kanner (Eds.), *Ecopsychology: Restoring the earth, healing the mind* (pp. 240–259). San Francisco, CA: Sierra Club Books.

Macy, J. (2012). The greening of the self. In L. Buzzell, & C. Chalquist (Eds.), *Ecotherapy: Healing with nature in mind*. San Francisco, CA: Sierra Club Books.

Macy, J., & Brown, M.Y. (1998). *Coming back to life: Practices to reconnect our lives, our world*. Gabriola Island, British Columbia: New Society Publishers.

Maibach, E.W., Nisbet, M., Baldwin, P., Akerlof, K., & Diao, G. (2010). Reframing climate change as a public health issue: An exploratory study of public reactions. *BMC Public Health, 10*, 299. doi: 10.1186/1471-2458-10-299

Main, D. (2014, July 30). Gulf of Mexico dead zone is the size of Connecticut. *NBC News*. Retrieved from http://www.nbcnews.com/science/science-news/gulf-mexico-dead-zone- size-connecticut-f6C10798946

Manes, C. (1996). Nature and silence. In C. Glotfelty, & H. Fromm (Eds.), *The ecocriticism reader* (pp. 15–29). Athens, GA and London, United Kingdom: The University of Georgia Press.

Mannetti, L., Pierro, A., & Livi, S. (2004). Recycling: Planned and self-expressive behaviour. *Journal of Environmental Psychology, 24*, 227–236. doi: 10.1016/j.jenvp.2004.01.002

Manning, C.M., Amel, E.L., Forsman, J.W., & Scott, B.A. (2009). Framing climate change solutions: The importance of getting the numbers right. *International Journal of Climate Change Strategies and Management, 1*(4), 326–339. doi: 10.1108/17568690911002861

March, J.G., Gual, M., & Orozco, F. (2004). Experiences on greywater re-use for toilet flushing in a hotel (Mallarca Island, Spain). *Desalination, 164*, 241–247.

Marino, L. (2012). Construct validity of animal assisted therapy and activities: How important is the animal in AAT? *Anthrozoös, 25*, S139-S151. doi: 10.2752/175303712X13353430377219

Markowitz, E.M., Goldberg, L.R., Ashton, M.C., & Lee, K. (2012). Profiling the "proenvironmental individual": A personality perspective. *Journal of Personality, 80*(1), 81–111. doi: 10.1111/j.1467-6494.2011.00721.x

Marques-Pinto, A., & Carvalho, D. (2013). Human infertility: Are endocrine disruptors to blame? *Endocrine Connections*. Retrieved from http://www.endocrineconnections.com/content/2/3/R15.full.pdf

Marshall, B. (1930). The problem of the wilderness. *Scientific Monthly, 30*, 141–148.

Marshall, B.K. (2004). Gender, race, and perceived environmental risk: The "white male" effect in Cancer Alley, LA. *Sociological Spectrum, 24*(4), 453–478. doi: 10.1080/02732170490459485

Martin, G., & Pear, J. (2009). *Behavior modification: What it is and how to do it* (8th ed.). Englewood Cliffs, NJ: Prentice-Hall.

Maslach, C., & Jackson, S.E. (1981). The measurement of experienced burnout. *Journal of Occupational Behavior, 2*, 99–113. doi: 10.1002/job.4030020205

Maslach, C., & Leiter, M. P. (2008). Early predictors of job burnout and engagement. *Journal of Applied Psychology, 93*(3), 498–512. doi: 10.1037/0021-9010.93.3.498

Maslow, A. H. (1943). A theory of human motivation. *Psychological Review, 50,* 370–396. doi: 10.1037/h0054346

MassAudubon. (2014). *History of Mass Audubon.* Retrieved from http://www.massaudubon.org/about-us/history

Masters, R. D. (2003). The social implications of evolutionary psychology: Linking brain biochemistry, toxins, and violent crime. In R. W. Bloom, & N. K. Dess (Eds.), *Evolutionary psychology and violence: A primer for policymakers and public policy advocates* (pp. 23–56). Westport, CT: Praeger.

Mayer, F. S., & Frantz, C. M. (2004). The connectedness to nature scale: A measure of individuals' feeling in community with nature. *Journal of Environmental Psychology, 24,* 503–515. doi: 10.1016/j.jenvp.2004.10.001

Mazar, N., & Zhong, C.-B. (2010). Do green products make us better people? *Psychological Science, 21,* 494–498. doi: 10.1177/0956797610363538

McBride, D. L., & Korell, G. (2005). Wilderness therapy for abused women. *Canadian Journal of Counselling, 39,* 3–14.

McCalley, L. T. (2006). From motivation and cognition theories to everyday applications and back again: The case of product-integrated information and feedback. *Energy Policy, 34,* 129–137. doi: 10.1016/j.enpol.2004.08.024

McCalley, L. T., & Midden, C.J.H. (2002). Energy conservation through product-integrated feedback: The roles of goal-setting and social orientation. *Journal of Economic Psychology, 23*(5), 589–603. doi: 10.1016/S0167-4870(02)00119-8

McCrae, R. R., & Costa, P. T. (1987). Validation of the five-factor model of personality across instruments and observers. *Journal of Personality and Social Psychology, 52*(1), 81–90. doi: 10.1037/0022-3514.52.1.81

McCright, A. M., & Dunlap, R. E. (2011). The politicization of climate change and polarization in the American public's views of global warming, 2001–2010. *The Sociological Quarterly, 52*(2), 155–194. doi: 10.1111/j.1533-8525.2011.01198.x

McCright, A. M., & Dunlap, R. E. (2013). Bringing ideology in: The conservative white male effect on worry about environmental problems in the USA. *Journal of Risk Research, 16*(2), 211–226. doi: 10.1080/13669877.2012.726242

McDonough, W., & Braungart, M. (1998, October 1). The NEXT industrial revolution. *The Atlantic.* Retrieved from http://www.theatlantic.com/issues/98oct/industry.htm

McDonough, W., & Braungart, M. (2002). *Cradle to cradle: Remaking the way we make things.* New York, NY: North Point Press.

McFarlane, R. (2004, October 29). Where the wild things are. *The Guardian.* Retrieved from http://www.theguardian.com/books/2004/oct/30/featuresreviews.guardianreview35

McGinn, A. P. (2002). Reducing our toxic burden. In C. Flavin, H. French, & G. Gardner (Eds.), *State of the world, 2002: A worldwatch institute report on progress toward a sustainable society* (pp. 75–100). New York, NY: W. W. Norton & Co.

McKenzie-Mohr, D. (2000a). Fostering sustainable behavior through community-based social marketing. *American Psychologist, 55*(5), 531–537. doi: 10.1037/0003-066X.55.5.531

McKenzie-Mohr, D. (2000b). Promoting sustainable behavior: An introduction to community- based social marketing. *Journal of Social Issues, 56*(3), 543–555. doi: 10.1111/0022-4537.00183

McKenzie-Mohr, D. (2011). *Fostering sustainable behavior: An introduction to community-based social marketing* (3rd ed.). Gabriola Island, Canada: New Society.

McKenzie-Mohr, D., Lee, N., Schultz, P. W., & Kotler, P. (2012). *Social marketing to protect the environment: What works.* Thousand Oaks, CA: Sage.

McKenzie-Mohr, D., & Smith, W. (1999). *Fostering sustainable behavior: An introduction to community-based social marketing.* Gabriola Island, British Columbia: New Society Publishers.

McKibben, B. (2014). A call to arms: An invitation to demand action on climate change. *Rolling Stone.* Retrieved from http://www.rollingstone.com/politics/news/a-call-to-arms-an- invitation-to-demand-action-on-climate-change-20140521

McLain, R., Poe, M., Hurley, P. T., Lecompte-Mastenbrook, J., & Emery, M. R. (2012). Producing edible landscapes in Seattle's urban forest. *Urban Forestry & Urban Greening, 11*(2), 187–194. doi: 10.1016/j.ufug.2011.12.002

Meadows, D. (2008). *Thinking in systems: A primer.* D. Wright (Ed.). White River Jct., VT: Chelsea Green Publishing.

Meadows, D., Randers, J., & Meadows, D. (2004). *Limits to growth: The 30-year update.* White River Junction, VT: Chelsea Green Publishing.

Medin, D. L., & Atran, S. (Eds.). (1999). *Folkbiology.* Cambridge, MA: MIT Press.

Medin, D. L., & Atran, S. (2004). The native mind: Biological categorization and reasoning in development and across cultures. *Psychological Review, 111*, 960–983. doi: 10.1037/0033-295X.111.4.960

Melson, G. F. (2001). *Why the wild things are: Animals in the lives of children.* Cambridge, MA: Harvard University Press.

Melson, G. F. (2013a). Children and wild animals. In P. H. Kahn, Jr., & P. H. Hasbach (Eds.), *The rediscovery of the wild* (pp. 93–117). Cambridge, MA: MIT Press.

Melson, G. F. (2013b). Children's ideas about the moral standing and social welfare of non- human species. *Journal of Sociology and Social Welfare, 40*(4), 81–106.

Melson, G. F., Kahn, J. H., Beck, A., & Friedman, B. (2009). Robotic pets in human lives: Implications for the human–animal bond and for human relationships with personified technologies. *Journal of Social Issues, 65*(3), 545–567. doi:10.1111/j.1540-4560.2009.01613.x

Melson, G. F., Kahn, P. H., Jr., Beck, A. M., Friedman, B., Roberts, T., Garrett, E., & Gill, B. T. (2009). Children's behavior toward and understanding of robotic and living dogs. *Journal of Applied Developmental Psychology, 30*, 92–102. doi: 10.1016/j.appdev.2008.10.011

Merchant, C. (1980). *The death of nature: Women, ecology and the scientific revolution.* San Francisco, CA: Harper.

Merchant, C. (1995). *Earthcare: Women and the environment.* New York, NY: Routledge.

Mesmer-Magnus, J., Glew, D.J., & Viswesvaran, C. (2012). A meta-analysis of positive humor in the workplace. *Journal of Managerial Psychology, 27*(2), 155–190. doi: 10.1108/02683941211199554

Messaoudi, M., Violle, N., Bisson, J.F., Desor, D., Javelot, H., & Rougeot, C. (2011). Beneficial psychological effects of a probiotic formulation (*Lactobacillus helveticus* R0052 and *Bifidobacterium longum* R0175) in healthy human volunteers. *Gut Microbes, 2*(4), 256–261. doi: 10.4161/gmic.2.4.16108

Messer Diehl, E.R. (2009). Gardens that heal. In L. Buzzell, & C. Chalquist (Eds.), *Ecotherapy: Healing with nature in mind* (pp. 166–173). San Francisco, CA: Sierra Club Books.

Meyer, D.S. (2007). Building social movements. In S.C. Moser, & L. Dilling (Eds.), *Creating a climate for change: Communicating climate change and facilitating social change* (pp. 451–461). Cambridge, United Kingdom: Cambridge University Press.

Meyer, P.J. (2003). *Attitude is everything: If you want to succeed above and beyond.* Meyer Resource Group.

Milfont, T.L. (2010). Global warming, climate change and human psychology. In V. Corral- Verdugo, C.H. Garcia-Cadena, & M. Frias-Arment (Eds.), *Psychological approaches to sustainability: Current trends in theory, research and practice.* New York, NY: Nova Science Publishers.

Milfont, T.L., & Duckitt, J. (2010). The environmental attitudes inventory: A valid and reliable measure to assess the structure of environmental attitudes. *Journal of Environmental Psychology, 30,* 80–94. doi:10.1016/j.jenvp.2009.09.001

Milfont, T.L., Richter, I., Sibley, C.G., Wilson, M.S., & Fischer, R. (2013). Environmental consequences of the desire to dominate and be superior. *Personality and Social Psychology Bulletin, 39*(9), 1127–1138. doi: 10.1177/0146167213490805

Milfont, T.L., & Sibley, C.G. (2012). The big five personality traits and environmental engagement: Associations at the individual and societal level. *Journal of Environmental Psychology, 32*(2), 187–195. doi: 10.1016/j.jenvp.2011.12.006

Milfont, T.L., Wilson, J., & Diniz, P. (2012). Time perspective and environmental engagement: A meta-analysis. *International Journal of Psychology, 47*(5), 325–334. doi: 10.1080/00207594.2011.647029

Milinski, M., Sommerfeld, R.D., Krambeck, H.J., Reed, F.A., & Marotzke, J. (2008). The collective-risk social dilemma and the prevention of simulated dangerous climate change. *Proceedings of the National Academy of Sciences, 105,* 2291–2294. doi: 10.1073/pnas.0709546105

Miller, G.T. (2002). *Living in the environment: Principles, connections and solutions* (12th ed.). Belmont, CA: Wadsworth/Thompson Learning.

Miller, G.T. (2007). *Living in the environment: Principles, connections and solutions* (14th ed.). Belmont, CA: Wadsworth/Thompson Learning.

Miller, G.T., & Spoolman, S. (2012). *Living in the environment: Principles, connections and solutions* (17th ed.). Belmont, CA: Brooks/Cole, Cengage Learning.

Millichap, J.G., & Yee, M.M. (2012). The diet factor in attention-deficit/hyperactivity disorder. *Pediatrics, 129*(2), 330–337. doi: 10.1542/peds.2011-2199

Mishra, A., & Tripathi, S. (1978). *The chipko movement*. New Delhi, India: People's Action/Gandhi Peace Foundation.

Mitchell, T. R., & Daniels, D. (2003). Motivation. In W. C. Borman, D. R. Ilgen, & R. J. Klimoski (Eds.), *Handbook of psychology: Industrial and organizational psychology* (Vol. 12, pp. 225–254). Hoboken, NJ: John Wiley & Sons.

Mitkus, R. J., King, D. B., Walderhaug, M. O., & Forshee, R. A. (2014). A comparative pharmacokinetic estimate of mercury in U.S. Infants following yearly exposures to inactivated influenza vaccines containing thimerosal. *Risk Analysis, 34*(4), 735–750. doi: 10.1111/risa.12124

Mohai, P. (1992). Men, women, and the environment: An examination of the gender gap in environmental concern and activism. *Society and Natural Resources, 5,* 1–19.

Moller, A. C., Ryan, R. M., & Deci, E. L. (2006). Self-determination theory and public policy: Improving the quality of consumer decisions without using coercion. *Journal of Public Policy & Marketing, 25,* 104–116. doi: 10.1509/jppm.25.1.104

Monroe, M. C. (2003). Two avenues for encouraging conservation behaviors. *Human Ecology Review, 10,* 113–125.

Montiel-Castro, A. J., González-Cervantes, R. M., Bravo-Ruiseco, G., & Pacheco-López, G. (2013). The microbiota-gut-brain axis: Neurobehavioral correlates, health, and sociality. *Frontiers in Integrative Neuroscience, 7,* Article 70. doi: 10.3389/fnint.2013.00070

Morris, D. (1969). *The human zoo*. London, UK: Jonathan Cape.

Morrison, D., Lin, Q., Wiehe, S., Liu, G., Rosenman, M., Fuller, T., . . . Filippelli, G. (2013). Spatial relationships between lead sources and children's blood lead levels in the urban center of Indianapolis (USA). *Environmental Geochemistry and Health, 35*(2), 171–183. doi: 10.1007/s10653-012-9474-y

Moschis, G. P. (2007). Life course perspectives on consumer behavior. *Journal of the Academy of Marketing Science, 35*(2), 295–307. doi: 10.1007/s11747-007-0027-3

Moser, S. C. (2007). More bad news: The risk of neglecting emotional responses to climate change information. In S. C. Moser, & L. Dilling (Eds.), *Creating a climate for change: Communicating climate change and facilitating social change* (pp. 64–80). Cambridge, UK: Cambridge University Press.

Mostafa, M. M. (2007). Gender differences in Egyptian consumers' green purchase behaviour: The effects of environmental knowledge, concern and attitude. *International Journal of Consumer Studies, 31*(3), 220–229. doi: 10.1111/j.1470-6431.2006.00523.x

Mostafa, M. M. (2012). Does globalization affect consumers' pro-environmental intentions? A multilevel analysis across 25 countries. *International Journal of Sustainable Development and World Ecology, 19,* 229–237. doi: 10.1080/13504509.2011.614289

Muchinsky, P. M. (2012). *Psychology applied to work* (10th ed.). Summerfield, NC: Hypergraphic Press.

Muir, J. (1911/1988). *My first summer in the Sierra*. San Francisco, CA: Sierra Club Books.

Mulkern, A. C., & ClimateWire. (2013). If you know how a cow feels, will you eat less meat? *Scientific American*. Retrieved from http://www.scientificamerican.com/article/if-you-know-how-cow-feels-will-you-eat-less-meat/

Murphy, N. (2014). Improving water quality in 19th century Massachusetts. *MIT Libraries News*. Retrieved from http://libraries.mit.edu/news/improving-water-quality/14244/

Murray, R. (Director). (2011). *Meet the climate sceptics* [Film]. London, UK: BBC.

Murtagh, N., Gatersleben, B., & Uzzell, D. (2012). Self-identity threat and resistance to change: Evidence from regular travel behaviour. *Journal of Environmental Psychology, 32,* 318–326. doi: 10.1016/j.jenvp.2012.05.008

Myers, G. (2007). *The significance of children and animals: Social development and our connections to other species.* West Lafayette, IN: Purdue University Press.

Myers, G.J., & Davidson, P.W. (2000). Does methylmercury have a role in causing developmental disabilities in children? *Environmental Health Perspectives, 108*(3), 413–420.

Myers, J. (2014, January 17). California governor declares drought emergency. *USA Today.* Retrieved from http://www.usatoday.com/story/weather/2014/01/17/california-drought- emergency/4581761/

Myers, N., Mittermeier, R.A., Mittermeier, C.G., daFonseca, G.A.B., & Kent, J. (2000). Biodiversity hotspots for conservation priorities. *Nature, 403,* 853–858. doi: 10.1038/35002501

Myers, O.E., Jr. (2012). Children in nature. In S. Clayton (Ed.), *The Oxford handbook of environmental and conservation psychology* (pp. 113–127). New York, NY: Oxford University Press.

Myers, O.E., & Saunders, C.D. (2002). Animals as links toward developing caring relationships with the natural world. In P.H. Kahn, Jr., & S.R. Kellert (Eds.), *Children and nature: Psychological, sociocultural, and evolutionary investigations* (pp. 153–178). Cambridge, MA: Massachusetts Institute of Technology Press.

Myers, O.E., Jr., Saunders, C.D., & Garrett, E. (2004). What do children think animals need? Developmental trends. *Environmental Education Research, 10,* 545–562. doi: 10.1080/13504620303461

Naess, A. (1985). Identification as a source of deep ecological attitudes. In M. Tobias (Ed.), *Deep ecology* (pp. 256–270). San Diego, CA: Avant Books.

National Audubon Society. (2014). *Timeline of accomplishments.* Retrieved from http://www.audubon.org/timeline-accomplishments

National Corn Growers Association. (2013). *National Corn Growers Association 2013 report.* Retrieved from http://www.ncga.com/upload/files/documents/pdf/WOC%202013.pdf

National Environmental Education and Training Foundation (NEETF). (2005). *Environmental literacy in America: What ten years of NEETF/roper research studies say about environmental literacy in the US.* Washington, DC: Author. Retrieved from http://www.neefusa.org/pdf/ELR2005.pdf

National Institute of Mental Health, U.S. Department of Health and Human Services. (2007). *Global use of ADHD medications rises dramatically.* Retrieved from http://www.nimh.nih.gov/science-news/2007/global-use-of-adhd-medications-rises-dramatically.shtml

National Oceanic and Atmospheric Administration. (2010). What we know about: The "garbage patches." Retrieved from http://marinedebris.noaa.gov/sites/default/files/Gen_GP-hi_7–18–11_0.pdf

National Oceanic and Atmospheric Administration. (2014, January). *State of the climate report.* Retrieved from http://www.ncdc.noaa.gov/sotc/

National Park Service. (2006). *Gerald R. Ford: Park ranger, 38th President of the United States.* Retrieved from http://home.nps.gov/applications/release/print.cfm?id=717

National Park Service. (2013). *Theodore Roosevelt and conservation.* Retrieved from http://www.nps.gov/thro/historyculture/theodore-roosevelt-and-conservation.htm

Neal, D. T., Wood, W., & Quinn, J. M. (2006). Habits—A repeat performance. *Current Directions in Psychological Science, 15*(4), 198–202. doi: 10.1111/j.1467-8721.2006.00435.x

Nedovic, S., & Morrissey, A.-M. (2013). Calm active and focused: Children's responses to an organic outdoor learning environment. *Learning Environments Research, 16*(2), 281–295. doi: 10.1007/s10984-013-9127-9

Needleman, H. L. (2004). Lead poisoning. *Annual Review of Medicine, 55*(1), 209–222. doi: 10.1146/annurev.med.55.091902.103653

Needleman, H. L., McFarland, C., Ness, R. B., Fienberg, S. E., & Tobin, M. J. (2002). Bone lead levels in adjudicated delinquents: A case control study. *NeuroToxicology and Teratology, 24*(6), 711–717. doi: 10.1016/S0892-0362(02)00269-6

Nevin, R. (2007). Understanding international crime trends: The legacy of preschool lead exposure. *Environmental Research, 104,* 315–336. doi: 10.1016/j.envres.2007.02.008

Ng, M., Fleming, T., Robinson, M., Thomson, B., Graetz, N., Margono, C., . . . Gakidou, E. (2014). Global, regional, and national prevalence of overweight and obesity in children and adults during 1980–2013: A systematic analysis for the Global Burden of Disease Study 2013. *The Lancet, 384,* 766–781. doi: 10.1016/S0140-6736(14)60460-8

Ng, Y. K. (2008). Environmentally responsible happy nation index: Toward an internationally acceptable national success indicator. *Social Indices Research, 85,* 425–446. doi: 10.1007/s11205-007-9135-1

Nicks, D., & Stout, D. (2014, January 10). Government declares state of emergency after West Virginia chemical spill. *Time.* Retrieved from http://nation.time.com/2014/01/10/west-virginia-governor-declares-state-of-emergency-after-chemical-spill/

Nisbet, E. K., & Zelenski, J. M. (2011). Underestimating nearby nature: Affective forecasting errors obscure the happy path to sustainability. *Psychological Science, 22*(9), 1101–1106. doi: 10.1177/0956797611418527

Nisbet, E. K., Zelenski, J. M., & Murphy, S. A. (2009). The Nature Relatedness Scale: Linking individuals' connection with nature to environmental concern and behavior. *Environment and Behavior, 41,* 715–740. doi: 10.1177/0013916508318748

Nolan, J. M., Schultz, P. W., Cialdini, R. B., Goldstein, N. J., & Griskevicius, V. (2008). Normative social influence in underdetected. *Personality and Social Psychology Bulletin, 34,* 913–923. doi: 10.1177/0146167208316691

Nordgren, L., & Engström, G. (2014). Animal-assisted intervention in dementia: Effects on quality of life. *Clinical Nursing Research, 23*(1), 7–19. doi: 10.1177/1054773813492546

Norgaard, K.M. (2011). *Living in denial: Climate change, emotions, and everyday life.* Cambridge, MA: MIT University Press.

Nuclear Energy Institute. (2014). World statistics: Nuclear energy around the world. Retrieved from http://www.nei.org/Knowledge-Center/Nuclear-Statistics/World-Statistics

Oddy, W.H., Robinson, M., Ambrosini, G.L., O'Sullivan, T.A., de Klerk, N.H., Beilin, L.J,. . . . Stanley, F.J. (2009). The association between dietary patterns and mental health in early adolescence. *Preventive Medicine, 49*, 39–44. doi: 10.1016/j.ypmed.2009.05.009

Ohtomo, S., & Hirose, Y. (2007). The dual-process of reactive and intentional decision-making involved in eco-friendly behavior. *Journal of Environmental Psychology, 27*, 117–125. doi: 10.1016/j.jenvp.2007.01.005

Ojala, M. (2012). Hope and climate change: The importance of hope for environmental engagement among young people. *Environmental Education Research, 18*(5), 625–642. doi: 10.1080/13504622.2011.637157

O'Neill, S., & Nicholson-Cole, S. (2009). "Fear won't do it": Promoting positive engagement with climate change through visual and iconic representations. *Science Communication, 30*(3), 355–379. doi: 10.1177/1075547008329201

Opotow, S., & Brook, A. (2003). Identity and exclusion in rangeland conflict. In S. Clayton, & S. Opotow (Eds.), *Identity and the natural environment: The psychological significance of nature* (pp. 249–272). Cambridge, MA: Massachusetts Institute of Technology Press.

Ornstein, R.E., & Ehrlich, P.R. (2000). *New world new mind: moving toward conscious evolution.* Cambridge, MA: Malor Books, ISHK.

Orr, D.W. (1992). *Ecological literacy: Education and the transition to a postmodern world.* Albany, NY: State University of New York Press.

Orr, D.W. (1993). Love it or lose it: The coming biophilia revolution. In S.R. Kellert, & E.O. Wilson (Eds.), *The biophilia hypothesis* (pp. 415–440). Washington, DC: Island Press.

Orzeł-Gryglewska, J. (2010). Consequences of sleep deprivation. *International Journal of Occupational Medicine and Environmental Health, 23*(1), 95–114. doi: 10.2478/v10001-010-0004-9

Osbaldiston, R., & Schott, J.P. (2012). Environmental sustainability and behavioral science: Meta-analysis of proenvironmental behavior experiments. *Environmental Behavior, 44*, 257–299. doi: 10.1177/0013916511402673

Oskamp, S., Harrington, M.J., Edwards, T.C., Sherwood, D.L., Okuda, S.M., & Swason, D.C. (1991). Factors influencing household recycling behavior. *Environment and Behavior, 23*, 494–519. doi: 10.1177/0013916591234005

Ostrom, E., Burger, J., Field, C.B., Norgaard, R.B., & Policansky, D. (2007). Revisiting the commons: Local lessons, global challenges. In D.J. Penn, & I. Mysterud (Eds.), *Evolutionary perspectives on environmental problems* (pp. 129–140). New Brunswick, NJ: Transaction Publishers.

O'Sullivan, J. (1845, December 27). [Untitled editorial] *New York Morning News.*

Our Children's Trust (2014). *Federal lawsuit.* Retrieved from http://ourchildrenstrust. org/US/Federal-Lawsuit

Packard, V. (1960). *The waste makers.* New York, NY: D. McKay Co.

Pahl, S., & Bauer, J. (2013). Overcoming the distance: Perspective taking with future humans improves environmental engagement. *Environment and Behavior, 45,* 155–169. doi: 10.1177/0013916511417618

Partap, U., & Ya, T. (2012). The human pollinators of fruit crops in Maoxian County, Sichuan, China: A case study of the failure of pollination services and farmers' adaptation strategies. *Mountain Research and Development, 32*(2), 176–186. doi: http://dx.doi.org/10.1659/MRD-JOURNAL-D-11-00108.1

Patel, A. V., Bernstein, L., Deka, A., Feigelson, H. S., Campbell, P. T., Gapstur, S. M., . . . Thun, M. J. (2010). Leisure time spent sitting in relation to total mortality in a prospective cohort of US adults. *American Journal of Epidemiology, 172*(4), 419–429. doi: 10.1093/aje/kwq155

Payne, S. R. (2011). Soundscapes within urban parks: Their restorative value. In M. Bonaiuto, M. Bonnes, A. M. Nenci, & G. Carrus (Eds.), *Urban diversities— Environmental and social issues. Advances in people-environment studies* (Vol. 2, pp. 147–158). Cambridge, MA: Hogrefe Publishing.

PBS. (2014). *Silence of the bees. How can you help the bees?* Retrieved from http://www.pbs. org/wnet/nature/episodes/silence-of-the-bees/how-can-you-help-the-bees/36/

Peck, J. (2008). Carbon offsetting: Forgive my carbon sin? *The Ecologist.* Retrieved from http://www.theecologist.org/investigations/climate_change/268713/carbon_ offsetting_for give_my_carbon_sin.html

Pelletier, L. G., Tuson, K. M., Green-Demers, I., Noels, K., & Beaton, A. M. (1998). Why are you doing things for the environment? The motivation toward the environment scale (MTES). *Journal of Applied Social Psychology, 28,* 437–468. doi: 10.1111/ j.1559-1816.1998.tb01714.x

Perera, F., & Herbstman, J. (2011). Prenatal environmental exposures, epigenetics, and disease. *Reproductive Toxicology, 31,* 363–373. doi: 10.1016/j.reprotox.2010.12.055

Perkins, H. (2010). Measuring love and care for nature. *Journal of Environmental Psychology, 30,* 455–463. doi: 10.1016/j.jenvp.2010.05.004

Perrin, J. L., & Benassi, V. A. (2009). The connectedness to nature scale: A measure of emotional connection to nature? *Journal of Environmental Psychology, 29,* 434–440. doi: 10.1016/j.jenvp.2009.03.003

Perugini, M. (2005). Predictive models of implicit and explicit attitudes. *British Journal of Social Psychology, 44,* 29–45. doi: 10.1348/014466604X23491

Petersen, J. E., Shunturov, V., Janda, K., Platt, G., & Weinberger, K. (2007). Dormitory residents reduce electricity consumption when exposed to real-time visual feedback and incentives. *International Journal of Sustainability in Higher Education, 8*(1), 16–33. doi: 10.1108/14676370710717562

Petrova, P. K., Cialdini, R. B., & Sills, S. J. (2007). Consistency-based compliance across cultures. *Journal of Experimental Social Psychology, 43,* 104–111. doi: 10.1016/j. jesp.2005.04.002

Pettus, A.M., & Giles, M.B. (1987). Personality characteristics and environmental attitudes. *Population and Environment, 9*(3), 127–137. doi: 10.1007/BF01259303

Pierro, A., Raven, B.H., Amato, C., & Bélanger, J.J. (2013). Bases of social power, leadership styles, and organizational commitment. *International Journal of Psychology, 48*(6), 1122–1134. doi: 10.1080/00207594.2012.733398

Pimentel, D., & Pimentel, M. (2003). Sustainability of meat-based and plant-based diets and the environment. *American Journal of Clinical Nutrition, 78*(suppl), 660S–663S.

Pinchot, G. (1947/1998). *Breaking new ground.* Original copyright Estate of Gifford Pinchot. Reprinted by Washington, DC: Island Press.

Pirages, D.C., & Ehrlich, P.R. (1974). *Ark II: Social response to environmental imperatives.* San Francisco, CA: W.H. Freeman.

Pirog, R., & Benjamin, A. (2003). Checking the food odometer: Comparing food miles for local versus conventional produce sales to Iowa institutions. Leopold Center for Sustainable Agriculture. Retrieved from http://www.leopold.iastate.edu/ sites/default/files/pubs-and- papers/2003–07-checking-food-odometer-comparing-food-miles-local-versus- conventional-produce-sales-iowa-institution.pdf

Polańska, K., Jurewicz, J., & Hanke, W. (2013). Review of current evidence on the impact of pesticides, polychlorinated biphenyls and selected metals on attention deficit/hyperactivity disorder in children. *International Journal of Occupational Medicine and Environmental Health, 26*(1), 16–38. doi: 10.2478/s13382-013-0073-7

Poliquin, R. (2012). *The breathless zoo: Taxidermy and the cultures of longing.* University Park, PA: Penn State Press.

Pollan, M. (2006). *The omnivore's dilemma: A natural history of four meals.* New York, NY: Penguin.

Ponting, C. (1991). *A green history of the world: The environment and collapse of great civilizations.* New York, NY: St. Martin's Press.

Poortinga, W., Whitmarsh, L.E., and Suffolk, C. (2013). The introduction of a single-use carrier bag charge in Wales: Attitude change and behavioural spillover effects. *Journal of Environmental Psychology, 36*, 240–247. doi: 10.1016/j.jenvp.2013.09. 001

Postma, J.A., & Lynch, J.P. (2012). Complementarity in root architecture for nutrient uptake in ancient maize/bean and maize/bean/squash polycultures. *Annals of Botany, 110*(2), 521–534. doi: 10.1093/aob/mcs082

Potts, S.G., Roberts, S.P.M., Dean, R., Marris, G., Brown, M.A., Jones, R., . . . Settele, J. (2010). Declines of managed honey bees and beekeepers in Europe. *Journal of Apicultural Research, 49*(1), 15–22. doi: 10.3896/IBRA.1.49.1.02

Pratto, F., Sidanius, J., Stallworth, L.M., & Malle, B.F. (1994). Social dominance orientation: A personality variable predicting social and political attitudes. *Journal of Personality and Social Psychology, 67*(4), 741–763. doi:10.1037/0022-3514.67.4. 741

Price, W.A. (2009). *Nutrition and physical degeneration* (8th ed.). Lemon Grove, CA: Price- Pottenger Nutrition Foundation.

Prichard, J.R., & Hartmann, M. (2014). What is the cost of poor sleep for college students? Calculating the contribution to academic failures using a large national sample. *Sleep, 37* (Abstract supplement), A375.

Puk, T.G., & Stibbards, A. (2012). Systemic ecological illiteracy? Shedding light on meaning as an act of thought in higher learning. *Environmental Education Research, 18*(3), 353–373. doi: 10.1080/13504622.2011.622840

Pyle, R.M. (1993). *The thunder tree: Lessons from an urban wildland.* Boston, MA: Houghton Mifflin.

Rabinovich, A., & Morton, T.A. (2012). Unquestioned answers or unanswered questions: Beliefs about science guide responses to uncertainty in climate change risk communication. *Risk Analysis, 32*(6), 992–1002. doi: 10.1111/j.1539-6924.2012.01771.x

Rabinovich, A., Morton, T.A., & Birney, M.E. (2012). Communicating climate science: The role of perceived communicator's motives. *Journal of Environmental Psychology, 32,* 11–18. doi: http://dx.doi.org.ezproxy.stthomas.edu/10.1016/j.jenvp.2011.09.002

Rainforest Alliance. (2012, June). Where's the sustainable beef? *The Canopy.* Retrieved from http://www.rainforest-alliance.org/publications/newsletter/canopy-june-2012

Raison, C. L, Lowry, C.A., & Rook, G.A. (2010). Inflammation, sanitation, and consternation: loss of contact with coevolved, tolerogenic microorganisms and the pathophysiology and treatment of major depression. *Archives of General Psychiatry, 67*(12), 1211–1224. doi: 10.1001/archgenpsychiatry.2010.161

Rana, S.V. (2014). Perspectives in endocrine toxicity of heavy metals—A review. *Biological Trace Elements Research, 160*(1), 1–14. doi: 10.1007/s12011-014-0023-7

Ratcliffe, E., Gatersleben, B., & Sowden, P.T. (2013). Bird sounds and their contributions to perceived attention restoration and stress recovery. *Journal of Environmental Psychology, 36,* 221–228. doi: 10.1016/j.jenvp.2013.08.004

Ratey, J. (2013). *Spark: The revolutionary new science of exercise and the brain.* New York, NY: Little Brown and Company.

Raymund, J.F. (1995). From barnyards to backyards: An exploration through adult memories and children's narratives in search of an ideal playscape. *Children's Environments, 12*(3), 362–380.

Redman, C.L. (1999). *Human impact on ancient environments.* Tucson, AZ: University of Arizona Press.

Reeder, G., Pryor, J.B., Wohl, M.J.A., & Griswell, M.L. (2005). On attributing negative motives to others who disagree with our opinions. *Personality and Social Psychology Bulletin, 31,* 1498–1510. doi: 10.1177/0146167205277093

Reese, R.F., & Myers, J.E. (2012). EcoWellness: The missing factor in holistic wellness models. *Journal of Counseling & Development, 90*(4), 400–406. doi: 10.1002/j.1556-6676.2012.00050.x

Reid, K. (2004). Happy days—For petroleum marketers, the 1950s lived up to the nostalgia. *National Petroleum News, 96,* 24–25.

Reno, R.R., Cialdini, R.B., & Kallgren, C.A. (1993). The transsituational influence of social norms. *Journal of Personality and Social Psychology, 64,* 104–112.

Reser, J. (2003). Thinking through "conservation psychology": Prospects and challenges. *Human Ecology Review, 10*, 167–174.

Reser, J. P., & Swim, J. K. (2011). Adapting to and coping with the threat and impacts of climate change. *American Psychologist, 66*, 277–289. doi: 10.1037/a0023412

Rice, K. M., Walker, E. M., Jr., Wu, G. C., Gillette, C., & Blough, E. R. (2014). Environmental mercury and its toxic effects. *Journal of Preventative Medicine and Public Health, 47*(2), 74–83. doi: 10.3961/jpmph.2014.47.2.74

Rich, C., & Longcore, T. (Eds.). (2005). *Ecological consequences of artificial night lighting.* Washington, DC: Island Press.

Risen, J. L., & Critcher, C. R. (2011). Visceral fit: While in a visceral state, associated sates of the world seem more likely. *Journal of Personality and Social Psychology, 100*(5) 777–793. doi: 10.1037/a0022460

Robelia, B., & Murphy, T. (2012). What do people know about key environmental issues? A review of environmental knowledge surveys. *Environmental Education Research, 18*(3), 299–321. doi: 10.1080/13504622.2011.618288

Roberts, D. (2011, November 29). Renewables in the U.S.: Growing fast, but not fast enough. *Grist.* Retrieved from http://grist.org/renewable-energy/2011–11–28-renewables-in-the-u-s-growing-fast-but-not-fast-enough/

Roberts, E. M., English, P. B., Grether, J. K., Windham, G. C., Somberg, L., & Wolff, C. (2007). Maternal residence near agricultural pesticide applications and autism spectrum disorders among children in the California central valley. *Environmental Health Perspectives, 115,* 1482–1489.

Robertson, J. L., & Barling, J. (2013). Greening organizations through leaders' influence on employees' pro-environmental behaviors. *Journal of Organizational Behavior, 34*, 176–194. doi: 10.1002/job.1820

Rogers, E. (2003). *Diffusion of innovations.* New York, NY: Free Press.

Rogers, E. M. (1962/2010). *Diffusion of innovations* (4th ed.). New York, NY: Simon & Schuster.

Rogers, H. (2005). *Gone tomorrow: The hidden life of garbage.* New York, NY: The New Press.

Rogers, S. (2011, March 18). Nuclear power plant accidents, listed and ranked since 1952. *The Guardian.* Retrieved from http://www.theguardian.com/news/datablog/2011/mar/14 /nuclear-power-plant-accidents-list-rank

Rogers, T. B., Kuiper, N. A., & Kirker, W. S. (1977). Self-reference and the encoding of personal information. *Journal of Personality and Social Psychology, 35*, 677–688. doi: 10.1037/0022-3514.35.9.677

Rogers, Z., & Bragg, E. (2012). The power of connection: Sustainable lifestyles and sense of place. *Ecopsychology, 4*(4), 307–318. doi: 10.1089/eco.2012.0079

Roe, J., & Aspinall, P. (2011). The restorative benefits of walking in urban and rural settings in adults with good and poor mental health. *Health & Place, 17*, 103–113. doi: 10.1016/j.healthplace.2010.09.003

Rolston, K. (2012). *Virtual Bus simulator uses real-time driving data to boost fuel efficiency.* Retrieved from https://inlportal.inl.gov/portal/server.pt/community/newsroom/257/feature_story_details/1269?featurestory=DA_589410

Rosen, L. (2012). *iDisorder: Understanding our obsession with technology and overcoming its hold on us*. New York, NY: Palgrave MacMillan.

Rosen, L.D., Lim, A.F., Felt, J., Carrier, L.M., Cheever, N.A., Lara-Ruiz, J.M., Mendoza, J., . . . Rokkum, J. (2014). Media and technology use predicts ill-being among children, preteens and teenagers independent of the negative health impacts of exercise and eating habits. *Computers in Human Behavior, 35*, 364–375. doi: 10.1016/j.chb.2014.01.036

Ross, L. (1977). The intuitive psychologist and his shortcomings: Distortions in the attribution process. In L. Berkowitz (Ed.), *Advances in experimental social psychology* (Vol. 10, pp. 173–220). New York, NY: Academic Press.

Ross, N., Medin, D., Coley, J.D., & Atran, S. (2003). Cultural and experimental differences in the development of folkbiological induction. *Cognitive Development, 18*, 25–47. doi: 10.1016/S0885-2014(02)00142-9

Rossignol, D.A., Genuis, S.J., & Frye, R.E. (2014). Environmental toxicants and autism spectrum disorders: A systematic review. *Translational Psychiatry, 4*(2), e360. doi: 10.1038/tp.2014.4

Roszak, T. (1992). *The voice of the earth: An exploration of eco-psychology*. New York, NY: Simon & Schuster.

Roszak, T., Gomes, M.E., & Kanner, A.D. (Eds.). (1995). *Ecopsychology: Restoring the earth, healing the mind*. San Francisco, CA: Sierra Club Books.

Rotter, J.B. (1966). Generalized expectancies for internal versus external control of reinforcement. *Psychological Monographs: General and Applied, 80*, 1–28. doi: 10.1037/h0092976

Royte, E. (2006). *Garbage land: On the secret trail of trash*. New York, NY: Back Bay Books.

Rubin, K.H., Fein, G.G., & Vandenberg, B. (1983). Play. In E.M. Hetherington (Ed.), *Handbook of child psychology: socialization, personality, and social development* (4th ed., Vol. 4, pp. 693–774). New York, NY: John Wiley & Sons.

Ruckart, P.Z., Kakolewski, K., Bove, F.J., & Kaye, W.E. (2004). Long-term neurobehavioral health effects of methyl parathion exposure in children in Mississippi and Ohio. *Environmental Health Perspectives, 112*, 46–51.

Russell, K.C. (2005). Two years later: a qualitative assessment of youth well-being and the role of aftercare in outdoor behavioral healthcare treatment. *Child and Youth Care Forum, 34*, 209–239. doi: 10.1007/s10566-005-3470-7

Russell, K.C. (2012). Therapeutic uses of nature. In S.D. Clayton (Ed.), *The Oxford handbook of environmental and conservation psychology* (pp. 428–444). New York, NY: Oxford University Press.

Rust, M. (2009). Why and how to therapists become ecotherapists? In L. Buzzell, & C. Chalquist (Eds.), *Ecotherapy: Healing with nature in mind* (pp. 37–45). San Francisco, CA: Sierra Club Books.

Ryan, R.M., & Deci, E.L. (2000). Self determination theory and the facilitation of intrinsic motivation, social development, and well-being. *American Psychologist, 55*, 68–78. doi: 10.1037/0003-066X.55.1.68

Saad, L. (2013). *Americans' concerns about global warming on the rise: Majority believe global warming is happening, but many still say it's exaggerated.* Retrieved from http://www.gallup.com/poll/161645/americans-concerns-global-warming-rise.aspx.

Sabini, M. (Ed.) (2002). *The Earth has a soul: The nature writings of C. G. Jung.* Berkeley, CA: North Atlantic Books.

Sabloff, A. (2001). *Reordering the natural world: Human and animals in the city.* Toronto, Canada: University of Toronto Press.

Sadalla, E., & Krull, J. (1995). Self-presentational barriers to resource conservation. *Environment and Behavior, 27,* 328–353. doi: 10.1177/0013916595273004

Satija, N., & Root, J. (2013, November 9). Concerns as Austin residents drill new wells. *The New York Times.* Retrieved from http://www.nytimes.com/2013/11/10/us/concerns-as-residents-drill-new-wells.html?_r=0 January 22, 2014

Saunders, C. D. (2003). The emerging field of conservation psychology. *Human Ecology Review, 10,* 137–149.

Saunders, M.N.K., & Thornhill, A. (2003). Organisational justice, trust and the management of change: An exploration. *Personnel Review, 32*(3), 360–375. doi: 10.1108/00483480310467660

Savitz, A. W., & Weber, K. (2006). *The triple bottom line: How today's best-run companies are achieving economic, social, and environmental success—And how you can too.* San Francisco, CA: Jossey-Bass.

Schettler, T., Stein, J., Reich, F., & Valenti, M. (2000). *In harm's way: Toxic threats to child development.* Cambridge, MA: Greater Boston Physicians for Social Responsibility.

Schlenker, B. R. (2012). Self-presentation. In M. R. Leary, & J. P. Tangney (Eds.), *Handbook of self and identity* (2nd ed., pp. 542–570). New York: Guilford Press.

Schlosser, E. (2012). *Fast food nation: The dark side of the all-American meal.* New York, NY: Mariner Books; Reprint edition.

Schmeichel, B.J., & Vohs, K. (2009). Self-affirmation and self-control: Affirming core values counteracts ego depletion. *Journal of Personality and Social Psychology, 96*(4), 770–782. doi: 10.1037/a0014635

Schmid, D., & Leitzmann, M. F. (2014). Television viewing and time spent sedentary in relation to cancer risk: A meta-analysis. *Journal of the National Cancer Institute, 106,* dju098. doi: 10.1093/jnci/dju098

Schmuck, P., & Schultz, P. W. (2002). Sustainable development as a challenge for psychology. In. P. Schmuck, & W. P. Schultz (Eds.), *Psychology of sustainable development* (pp. 3–17). Boston, MA: Kluwer Academic Publishers.

Schultz, P. W. (2000). Empathizing with nature: The effects of perspective taking on concern for environmental issues. *Journal of Social Issues, 56,* 391–406. doi: 10.1111/0022-4537.00174

Schultz, P. W. (2001). The structure of environmental concern: Concern for self, other people, and the biosphere. *Journal of Environmental Psychology, 21,* 1–13. doi: 10.1006/jevp.2001.0227

Schultz, P. W. (2002a). Environmental attitudes and behaviors across cultures. In W. J. Lonner, D. L. Dinnel, S. A. Hayes, & D. N. Sattler (Eds.), *OnLine readings in psychology and culture.* Bellingham, WA: Western Washington University, Department of Psychology, Center for Cross-Cultural Research. Retrieved from http://www.wwu.edu/~culture

Schultz, P. W. (2002b). Inclusion with nature: Understanding the psychology of human-nature interactions. In P. Schmuck, & P. W. Schultz (Eds.), *The psychology of sustainable development* (pp. 61–78). New York, NY: Kluwer.

Schultz, P. W. (2014). Strategies for promoting proenvironmental behavior: Lots of tools but few instructions. *European Psychologist, 19,* 107–117. doi:10.1027/1016-9040/a000163

Schultz, P. W., Gouveia, V., Cameron, L., Tankha, G., Schmuck, P., & Franek, M. (2005). Values and their relationship to environmental concern and conservation behavior. *Journal of Cross-Cultural Psychology, 36,* 457–475. doi: 10.1177/0022022105275962

Schultz, P. W., & Kaiser, F. G. (2012). Promoting pro-environmental behavior. In S. D. Clayton (Ed.), *The Oxford handbook of environmental and conservation psychology.* Oxford, UK: Oxford University Press.

Schultz, P. W., Khazian, A., & Zaleski, A. (2008). Using normative social influence to promote conservation among hotel guests. *Social Influence, 3,* 4–23. doi: 10.1080/15534510701755614

Schultz, P. W., Nolan, J. M., Cialdini, R. B., Goldstein, N. J., & Griskevicius, V. (2007). The constructive, destructive and reconstructive power of social norms. *Psychological Science, 18*(5), 429–434. doi: 10.1111/j.1467-9280.2007.01917.x

Schultz, P. W., Shriver, C., Tabanico, J., & Khazian, A. (2004). Implicit connections with nature. *Journal of Environmental Psychology, 24,* 31–42. doi: 10.1016/S0272-4944(03)00022-7

Schultz, P. W., & Zelezny, L. (2003). Reframing environmental messages to be congruent with American values. *Human Ecology Review, 10,* 126–136.

Schwartz, S. H. (1968). Words, deeds and the perception of consequences and responsibility in action situations. *Journal of Personality and Social Psychology, 10,* 232–242. doi: 10.1037/h0026569

Schwartz, S. H. (1977). Normative influences on altruism. In L. Berkowitz (Ed.), *Advances in experimental social psychology* (Vol. 19, pp. 221–279). New York, NY: Academic Press.

Schwartz, S. H. (1992). Universals in the content and structure of values: Theory and empirical tests in 20 countries. In M. Zanna (Ed.), *Advances in experimental social psychology* (Vol. 25, pp. 1–65). New York, NY: Academic Press.

Schwartz, S. H. (1994). Are there universal aspects in the structure and contents of human values? *Journal of Social Issues, 50,* 19–45. doi: 10.1111/j.1540-4560.1994.tb01196.x

Schwartz, S. H. (2012). An overview of the Schwartz theory of basic values. *Online Readings in Psychology and Culture, 2.* Retrieved from http://scholarworks.gvsu.edu/orpc/vol2/iss1/11/

Scopelliti, M., Carrus, G., & Bonnes, M. (2012). Natural landscapes. In S. Clayton (Ed.), *The Oxford handbook of environmental and conservation psychology* (pp. 332–347). New York, NY: Oxford University Press.

Scott, B. A., Amel, E. L., & Manning, C. M. (2011, April). Why I'm a concerned citizen, but not an environmental activist. Symposium for Society for Human Ecology, Las Vegas, NV.

Scott, B. A., Amel, E. L., & Manning, C. M. (2014). In and of the wilderness: Ecological connection through participation in nature. *Ecopsychology, 6*(2), 81–91. doi:10.1089/eco.2013.0104.

Scott, B. A., Manning, C. M., & Amel, E. L. (2008, September). Dangerous boys and daring girls: Participating in nature vs. protecting it. In B. A. Scott (Chair), *Shades of Green: Implications of Human Diversity for Environmental Education and Advocacy.* Symposium for Society for Human Ecology, Bellingham, WA.

Seguin, R., Buchner, D. M., Lui, J., Allison, M., Manini, T., Wang, C. Y., . . . LaCroix, A. Z. (2014). Sedentary behavior and mortality in older women. *American Journal of Preventative Medicine, 46*(2), 122–135. doi: http://dx.doi.org/10.1016/j.amepre.2013.10. 021

Seligman, M.E.P. (1971). Phobias and preparedness. *Behavior Therapy, 2,* 307–320. doi: 10.1016/S0005-7894(71)80064-3

Seligman, M.E.P. (1975). *Helplessness: On depression, development and death.* San Francisco, CA: Freeman.

Seligman, M.E.P. (2012). *Flourish: A visionary new understanding of happiness and well being.* New York, NY: FreePress.

Senbel, M., Ngo, V. D., & Blair, E. (2014). Social mobilization of climate change: University students conserving energy through multiple pathways for peer engagement. *Journal of Environmental Psychology, 38,* 84–93. doi: 10.1016/j.jenvp.2014.01.001

Sevillano, V., Aragones, J. I., Schultz, P. W. (2007). Perspective taking, environmental concern, and the moderating role of dispositional empathy. *Environment and Behavior, 39,* 685–705. doi: 10.1177/0013916506292334

Sewell, L. (1999). *Sight and sensibility: The eco-psychology of perception.* New York, NY: Tarcher/Putnam.

Shader, M. (2012, January 24). Recall: Super Luchamania action figures—lead paint. *Consumer Reports News.* Retrieved from http://www.consumerreports.org/cro/news/

Shaheen, S. A., & Cohen, A. P. (2013). Carsharing and personal vehicle services: Worldwide market developments and emerging trends. *International Journal of Sustainable Transportation, 7*(1), 5–34. doi: 10.1080/15568318.2012.660103

Shapiro, K., Randour, M. L., Krinsk, S., & Wolf, J. L. (2014). *The assessment and treatment of children who abuse animals: The AniCare child approach.* New York, NY: Springer.

Sharp, R., & Pestano, J. P. (2013). Water treatment contaminants: Forgotten toxics in American water. *Environmental Working Group.* Retrieved from http://www.ewg.org/research/water-treatment-contaminants

Sheer, R., & Moss, D. (Eds.). (2011, April 27). Dirt poor: Have fruits and vegetables become less nutritious? *Scientific American.* Retrieved from http://www.scientificamerican.com/ article/soil-depletion-and-nutrition-loss/

Sheldon, K.M., Nichols, C.P., & Kasser, T. (2011). Americans recommend smaller ecological footprints when reminded of intrinsic American values of self-expression, family, and generosity. *Ecopsychology, 3*(2), 97–104. doi: 10.1089/eco.2010.0078

Shellenberger, M., & Nordhaus, T. (2004). *The death of environmentalism: Global warming politics in a post-environmental world.* Retrieved from http://thebreakthrough.org/images/Death_of_Environmentalism.pdf

Shepard, P. (1998). *Nature and madness.* Athens, GA: University of Georgia Press.

Sheppard, S.R.J. (2012). *Visualizing climate change: A guide to visual communication of climate change and developing local solutions.* New York, NY: Routledge.

Shields, T., & Zeng, K. (2012). The reverse environmental gender gap in China: Evidence from "The China Survey". *Social Science Quarterly, 93*(1), 1–20. doi: 10.1111/j.1540-6237.2011.00802.x

Shiva, V. (1988). *Staying alive: Women, ecology and survival in India,* New Delhi, India: Zed Press.

Shiva, V. (1994). *Close to home: Women reconnect ecology, health and development worldwide.* London, United Kingdom: Earthscan.

Shor, J.B. (2002). Cleaning the closet: Toward a new fashion ethic. In J.B. Schor, & B. Taylor (Eds.), *Sustainable planet: Solutions for the 21st century* (pp. 45–60). Boston, MA: Beacon Press.

Sibley, C.G., & Kurz, T. (2013). A model of climate belief profiles: How much does it matter if people question human causation? *Analyses of Social Issues and Public Policy (ASAP), 13*(1), 245–261. doi: 10.1111/asap.12008

Sierio, F.W., Bakker, A.B., Dekker, G.B., & Van Den Burg, M.T.C. (1996). Changing organizational energy consumption behaviour through comparative feedback. *Journal of Environmental Psychology, 16*, 235–246. doi: 10.1006/jevp.1996.0019

Sierra Club. (2008). *Hetch Hetchy history.* Retrieved from http://www.sierraclub.org/ca/ hetchhetchy/history.asp

Simon, J.L. (1981). *The ultimate resource.* Princeton, NJ: Princeton University Press.

Singer, D.G., & Singer, J.L. (1990). *The house of make-believe.* Cambridge, MA: Harvard University Press.

Skibins, J.C., & Powell, R.B. (2013). Conservation caring: Measuring the influence of zoo visitors' connection to wildlife on pro-conservation behaviors. *Zoo Biology, 32*(5), 528–540. doi: 10.1002/zoo.21086

Skinner, B.F. (1948). *Walden two.* New York, NY: Macmillan.

Skinner, B.F. (1953). *Science and human behavior.* New York, NY: The Free Press, Macmillan Publishing Co., Inc.

Skinner, B.F. (1971). *Beyond freedom and dignity.* New York, NY: Alfred A. Knopf, Inc.

Skinner, B.F. (1987). Why we are not acting to save the world. In B.F. Skinner (Ed.), *Upon further reflection* (pp. 1–14). Englewood Cliffs, NJ: Prentice Hall.

Skinner, B.F. (1991). Why we are not acting to save the world. In W. Ishaq (Ed.), *Human behavior in today's world* (pp. 19–29). New York, NY: Praeger.

Sloman, S.A. (1996). The empirical case for two systems of reasoning. *Psychological Bulletin, 119*, 3–22. doi: 10.1037/0033-2909.119.1.3

Slovic, P. (1987). Perception of risk. *Science, New Series, 236*(4799), 280–285.

Slovic, P., Finucane, M., Peters, E., & MacGregor, D. G. (2002). The affect heuristic. In T. Gilovich, D. Griffin, & D. Kahneman (Eds.), *Heuristics and biases: The psychology of intuitive judgment* (pp. 397–420). New York, NY: Cambridge University Press.

Slovic, P., & Peters, E. (2006). Risk perception and affect. *Current Directions in Psychological Science, 15*(6), 322–325. doi: 10.1111/j.1467-8721.2006.00461.x

Smith, D., Miles-Richardson, S., LeConte, D., & Archie-Booker, E. (2013) Interventions to improve access to fresh food in vulnerable communities: A review of the literature. *International Journal on Disability and Human Development, 12*(4), 409–417. doi: 10.1515/ijdhd-2013-0203

Smith, D. B. (2010, January 27). Is there an ecological unconscious? *New York Times.* Retrieved from http://www.nytimes.com/2010/01/31/magazine/31ecopsych-t.html?pagewanted=all&_r=0

Smith, N., & Leiserowitz, A. (2013). American evangelicals and global warming. *Global Environmental Change, 23*(5), 1009–1017. doi: 10.1016/j.gloenvcha.2013.04.001

Sobel, D. (1996). *Beyond ecophobia: Reclaiming the heart in nature education.* Great Barrington, MA: Orion Society.

Sohr, S. (2001). Eco-activism and well-being: Between flow and burnout. In P. Schmuck, & K. M. Sheldon (Eds.), *Life goals and well-being: Towards a positive psychology of human striving* (pp. 202–215). Ashland, OH: Hogrefe & Huber Publishers.

Solomon, S., Greenberg, J., & Pyszczynski, T. (2004). The cultural animal: Twenty years of terror management theory and research. In J. Greenberg, S. L. Koole, & T. Pyszczynski (Eds.), *Handbook of experimental existential psychology* (pp. 13–34). New York, NY: Guilford.

Spence, A., Poortinga, W., & Pidgeon, N. (2012). The psychological distance of climate change. *Risk Analysis, 32*(6), 957–972. doi: 10.1111/j.1539-6924.2011.01695.x

St. John, D., & MacDonald, D. A. (2007). Development and initial validation of a measure of ecopsychological self. *Journal of Transpersonal Psychology, 39,* 48–67.

Staats, H. (2012). Restorative environments. In S. Clayton (Ed.), *The Oxford handbook of environmental and conservation psychology* (pp. 445–458). New York: Oxford University Press.

Stansfield, S. A., Clark, C., & Crombie, R. (2012). Noise. In S. Clayton (Ed.), *The Oxford handbook of environmental and conservation psychology* (pp. 375–390). New York, NY: Oxford University Press.

Steg, L., & De Groot, J.I.M. (2012). Environmental values. In S. D. Clayton (Ed.), *The Oxford handbook of conservation and environmental psychology* (pp. 81–92). New York, NY: Oxford University Press.

Steg, L., Dreijerink, L., & Abrahamse, W. (2005). Factors influencing the acceptability of energy policies: A test of VBN theory. *Journal of Environmental Psychology, 25,* 415–425. doi: 10.1016/j.jenvp.2005.08.003

Steg, L., & Vlek, C. (2009). Encouraging pro-environmental behaviour: An integrative review and research agenda. *Journal of Environmental Psychology, 29,* 309–317. doi: 10.1016/j.jenvp.2008.10.004

Stein, J., Schettler, T., Rohrer, B., & Valenti, M. (2008). Environmental threats to healthy aging. *Greater Boston Physician's for Social Responsibility and Science and Environmental Health Network*. Retrieved from http://www.psr.org/site/DocServer/GBPSRSEHN_HealthyAging1017.pdf?docID=5930

Stern, P. C. (1992). Psychological dimensions of global environmental change. *Annual Review of Psychology, 43*, 269–302. doi: 10.1146/annurev.ps.43.020192.001413

Stern, P. C. (2000). Psychology and the science of human-environment interactions. *American Psychologist, 55*, 523–530. doi: 10.1037/0003-066X.55.5.523

Stern, P. C. (2005). Understanding individuals' environmentally significant behavior. *Environmental Law Reporter: News and Analysis, 35*, 10785–10790.

Stern, P. C., Dietz, T., & Kalof, L. (1993). Value orientations, gender, and environmental concern. *Environment and Behavior, 25*, 322–348. doi: 10.1177/0013916593255002

Stokstad, E. (2009, November 25). Americans' eating habits more wasteful than ever. *American Association for the Advancement of Science*. Retrieved from http://news.sciencemag.org/2009/11/americans-eating-habits-more-wasteful-ever

Stone, J. (2012). Consistency as a basis for behavioral interventions: Using hypocrisy and cognitive dissonance to motivate behavior change. In B. Gawronski, & F. Strack (Eds.), *Cognitive consistency: A fundamental principle in social cognition* (pp. 326–347). New York, NY: Guilford Press.

Stone, J., & Fernandez, N. C. (2008). To practice what we preach: The use of hypocrisy and cognitive dissonance to motivate behavior change. *Social and Personality Psychology Compass, 2*(2), 1024–1051. doi: 10.1111/j.1751-9004.2008.00088.x

Stragier, J., Hauttekeete, L., De Marez, L., & Brondeel, R. (2012). Measuring energy-efficiency behavior in households: The development of a standardized measure. *Ecopsychology, 4*(1), 64–71. doi: 10.1089/eco.2012.0026

Suedfeld, P. (2009). Integrative complexity. In M. R. Leary, & R. H. Hoyle (Eds.), *Handbook of individual differences in social behavior* (pp. 354–366). New York, NY: Guilford Press.

Sunstein, C. R., & Reisch, L. A. (2013). Automatically green: Behavioral economics and environmental protection. *Harvard Environmental Law Review, 38*, 127–158. doi: 10.2139/ssrn.2245657

Sussman, R., & Gifford, R. (2012). Please turn off the lights: The effectiveness of visual prompts. *Applied Ergonomics, 43*(3), 596-603. doi:10.1016/j.apergo.2011.09.008

Sussman, R., Greeno, M., Gifford, R., & Scannell, L. (2013). The effectiveness of models and prompts on waste diversion: A field experiment on composting by cafeteria patrons. *Journal of Applied Social Psychology, 43*, 24–34. doi: 10.1111/j.1559-1816.2012.00978.x

Swami, V., Chamorro-Premuzic, T., Snelgar, R., & Furnham, A. (2010). Egoistic, altruistic, and biospheric environmental concerns: A path analytic investigation of their determinants. *Scandinavian Journal of Psychology, 51*(2), 139–145. doi: 10.1111/j.1467-9450.2009.00760.x

Swider, B. W., & Zimmerman, R. D. (2010). Born to burnout: A meta-analytic path model of personality, job burnout and work outcomes. *Journal of Vocational Behavior, 76*(3), 487–506. doi: 10.1016/j.jvb.2010.01.003

Swim, J.K., & Bloodhart, B. (2013). Admonishment and praise: Interpersonal mechanisms for promoting proenvironmental behavior. *Ecopsychology, 5*, 24–35. doi: 10.1089/eco.2012.0065

Swim, J.K., Clayton, S., Doherty, T., Gifford, R., Howard, G., Reser, J., . . . Weber, E. (2009). *Psychology and global climate change: Addressing a multi-faceted phenomenon and set of challenges.* Retrieved from http://www.apa.org/science/about/publications/climate-change.pdf

Syme, G.L., & Nancarrow, B.E. (2012). Justice and the allocation of natural resources: Current concepts and future directions. In S. Clayton (Ed.), *The Oxford handbook of environmental and conservation psychology* (pp. 93–112). New York, NY: Oxford University Press.

Tainter, J.A. (2012). Regenerative design in science and society. *Building Research & Information, 40*(3), 369–372.

Tam, K. (2013). Dispositional empathy with nature. *Journal of Environmental Psychology, 35*, 92–104.

Tam, K.-P. (2014). Are anthropomorphic persuasive appeals effective? The role of the recipient's motivations. *British Journal of Social Psychology, 54*, 187–200. doi: 10.1111/bjso.12076

Tanaka, C., Reilly, J.J., & Huang, W.Y. (2014). Longitudinal changes in objectively measured sedentary behaviour and their relationship with adiposity in children and adolescents: systematic review and evidence appraisal. *Obesity Reviews, 15*, 791–803. doi: 10.1111/obr.12195

Tanner, C. (1999). Constraints on environmental behaviour. *Journal of Environmental Psychology, 19*, 145–157. doi: 10.1006/jevp.1999.0121

Tanner, T. (1980). Significant life experiences: A new research area in environmental education. *Journal of Environmental Education, 11*(4), 20–24. doi:10.1080/00958964.1980.9941386

Tarakeshwar, N., Swank, A.B., Pargament, K.I., & Mahoney, A. (2001). The sanctification of nature and theological conservatism: A study of opposing religious correlates of environmentalism. *Review of Religious Research, 42*, 387–404.

Tassi, P., Rohmer, O., Bonnefond, A., Margiocchi, F., Poisson, F., & Schimchowitsch, S. (2013). Long term exposure to nocturnal railway noise produces chronic signs of cognitive deficits and diurnal sleepiness. *Journal of Environmental Psychology, 33*, 45–52. doi: 10.1016/j.jenvp.2012.10.003

Taubes, G. (2011, April 13). Is sugar toxic? *The New York Times Magazine.* Retrieved from http://www.nytimes.com/2011/04/17/magazine/mag-17Sugar t.html?pagewanted=all&_r=0

Tavris, C., & Aronson, E. (2007). *Mistakes were made (but not by me): Why we justify foolish beliefs, bad decisions, and hurtful acts.* New York, NY: Houghton Mifflin Harcourt.

Taylor, B., & Tilford, D. (2000). Why consumption matters. In J.B. Schor, & D.B. Holt (Eds.), *The consumer society reader.* New York, NY: The New Press.

Taylor, D.M., Segal, D., & Harper, N.J. (2010). The ecology of adventure therapy: An integral systems approach to therapeutic change. *Ecopsychology, 2*(2), 77–83. doi: 10.1089/eco.2010.0002

Taylor, S., & Todd, P. (1995). An integrated model of waste management behaviour: A test of household recycling and composting intentions. *Environment and Behaviour, 27*, 603–630. doi: 10.1177/0013916595275001

Temin, P. (2000). The great depression. In S.L. Engelman, & R.E. Gallman (Eds.), *The Cambridge economic history of the United States, volume III: The twentieth century* (pp. 301–328). Cambridge, UK: Cambridge University Press. doi: 10.1017/CHOL9780521553087

Tennessen, C.M., & Cimprich, B. (1995). Views to nature: Effects on attention. *Journal of Enviornmental Psychology, 15*(1), 77–85. doi: 10.1016/0272-4944(95)90016-0

Thaler, R.H., & Sunstein, C.R. (2009). *Nudge: Improving decisions about health, wealth, and happiness.* New York, NY: Penguin Group.

Thayer, S. (2006*). Forager's harvest: A guide to identifying, harvesting, and preparing wild plants.* Birchwood, WI: Forager's Harvest Press.

The National Academies. (2002). *Coal Waste Impoundments: Risks, Responses, and Alternatives.* Washington, DC: The National Academies Press.

The Outdoor Foundation. (2010). *Special report on youth: The next generation of outdoor champions.* Retrieved from http://www.outdoorfoundation.org/pdf/Research Youth.pdf

The World Bank. (2014). *Road sector gasoline fuel consumption per capita.* Retrieved from http://data.worldbank.org/indicator/IS.ROD.SGAS.PC

Thibaut, J.W., & Walker, L. (1975). *Procedural justice: A psychological analysis.* Hillsdale, NJ: Erlbaum.

Thøgersen, J., & Grønhøj, A. (2010). Electricity saving in households—A social cognitive approach. *Energy Policy, 38*, 7732–7743. doi: 10.1016/j.enpol.2010.08.025

Thomashow, M. (1995). *Ecological identity: Becoming a reflective environmentalist.* Cambridge, MA: Massachusetts Institute of Technology Press.

Thompson, K.R. (1998, Winter). Confronting the paradoxes in a total quality environment. *Organizational Dynamics*, 62–74.

Thoreau, H.D. (1854/1995a). Solitude. In *Walden; Or, life in the woods* (p. 90). Mineola, NY: Dover Publications.

Thoreau, H.D. (1854/1995b). Economy. In *Walden; Or, life in the woods* (p. 8). Mineola, NY: Dover Publications.

Thoreau, H.D. (1856/2009). In D. Searles (Ed.), *The journal of Henry David Thoreau, 1837–1861* (p. 356). New York, NY: The New York Review of Books.

Thoreau, H.D. (1862/2015). Walking. *The Atlantic.* Retrieved from http://www.theatlantic.com/magazine/archive/1862/06/walking/304674/

Toates, F. (2009). *Burrhus F. Skinner.* Basingstoke, United Kingdom: Macmillan Palgrave.

Tobler, C., Visschers, V.H.M., & Siegrist, M. (2011). Eating green. Consumers' willingness to adopt ecological food consumption behaviors. *Appetite, 57*(3), 674–682. doi: 10.1016/j.appet.2011.08.010

Tobler, C., Visschers, V.H.M., & Siegrist, M. (2012). Addressing climate change: Determinants of consumers' willingness to act and to support policy measures. *Journal of Environmental Psychology, 32*, 197–207. doi: 10.1016/j.jenvp.2012.02.001

Tolba, A. H., & Mourad, M. (2011). Individual and cultural factors affecting diffusion of innovation. *Journal of International Business & Cultural Studies, 5*. Retrieved from http://www.aabri.com/manuscripts/11806.pdf

Toner, K., Gan, M., & Leary, M. R. (2014). The impact of individual and group feedback on environmental intentions and self-beliefs. *Environment and Behavior, 46*(1), 24–45. doi: 10.1177/0013916512451902

Trope, Y., & Liberman, N. (2010). Construal-level theory of psychological distance. *Psychological Review, 117*(2), 440–463. doi: 10.1037/a0018963

Tuckey, M., Brewer, N., & Williamson, P. (2002). The influence of motives and goal orientation on feedback seeking. *Journal of Occupational and Organizational Psychology, 75*, 195–216. doi: 10.1348/09631790260098677

Tudor, T., Adam, E., & Bates, M. (2007). Drivers and limitations for the successful development and functioning of EIPs (eco-industrial parks): A literature review. *Ecological Economics, 61*, 199–207. doi: 10.1016/j.ecolecon.2006.10.010

Tuomisto, H. L., Hodge, I. D., Riordan, P., & Macdonald, D. W. (2012). Comparing energy balances, greenhouse gas balances and biodiversity impacts of contrasting farming systems with alternative land uses. *Agricultural Systems, 108*, 42–49. doi:10.1016/j.agsy.2012.01.004

Tversky, A., & Kahneman, D. (1973). Availability: A heuristic for judging frequency and probability. *Cognitive Psychology, 5*(2), 207–233. doi: 10.1016/0010-0285(73)90033-9

Twenge, J. M., & Kasser, T. (2013). Generational changes in materialism and work centrality, 1976–2007: Associations with temporal changes in societal insecurity and materialistic role modeling. *Personality and Social Psychology Bulletin, 39*(7), 883–897. doi: 10.1177/0146167213484586

Tyrväinen, L., Ojala, A., Korpela, K., Lanki, T., Tsunetsugu, Y., & Kagawa, T. (2014). The influence of urban green environments on stress relief measures: A field experiment. *Journal of Environmental Psychology, 38*, 1–9. doi: 10.1016/j.jenvp.2013.12.005

Ulrich, R. S. (1983). Aesthetic and affective response to natural environment. *Human Behavior & Environment: Advances in Theory & Research, 6*, 85–125. doi: 10.1007/978-1-4613-3539-9_4

Ulrich, R. S. (1984). View through a window may influence recovery from surgery. *Science, 224*, 420–421. doi: 10.1126/science.6143402

Ulrich, R. S. (1993). Biophilia, biophobia and natural landscapes. In S. R. Kellert, & E. O. Wilson (Eds.), *The biophilia hypothesis* (pp. 73–137). Washington, DC: Island Press.

Ulrich, R. S., Simons, R. F., Losito, B. D., Fiorito, E., Miles, M. A., & Zelson, M. (1991). Stress recovery during exposure to natural and urban environments. *Journal of Environmental Psychology, 11*, 201–230. doi: 10.1016/S0272-4944(05)80184-7

Union of Concerned Scientists. (1992). *World scientists' warning to humanity.* Statement available from the Union of Concerned Scientists, 26 Church St., Cambridge, MA 02238.

Union of Concerned Scientists. (2012). *Industrial agriculture.* Retrieved from http://www.ucsusa.org/food_and_agriculture/our-failing-food-system/industrial-agriculture/

United Nations. (2014). *International decade of action: Water for life 2005–2015.* Retrieved from http://www.un.org/waterforlifedecade/food_security.shtml

U.S. Department of Agriculture. (2013). *Irrigation and water use.* Retrieved from http://www.ers.usda.gov/topics/farm-practices-management/irrigation-water-use.aspx#.U9kstBAg-So

U.S. Department of Agriculture. (2014). *Farmers markets and local food marketing.* Retrieved from http://www.ams.usda.gov/AMSv1.0/farmersmarkets

U.S. Energy Information Administration (EIA). (2013). *Energy in brief: What are the major sources and users of energy in the United States?* Retrieved from http://www.eia.gov/energy_in_ brief/article/major_energy_sources_and_users.cfm

U.S. Environmental Protection Agency (EPA). (2006). *Life cycle assessment: Principles and practice.* Retrieved from http://www.epa.gov/nrmrl/std/lca/lca.html

U.S. Environmental Protection Agency (EPA). (2009). *Environmental justice.* Retrieved from http://www.epa.gov/oecaerth/environmentaljustice/

U.S. Environmental Protection Agency (EPA). (2012a). *Dealing in toxins on the wrong side of the tracks: Lessons from a hazardous waste controversy in Phoenix.* Retrieved from http://www.epa.gov/wastes/conserve/tools/stewardship/products/packaging.htm

U.S. Environmental Protection Agency (EPA). (2012b). *Nonpoint source pollution: The nature's largest water quality problem.* Retrieved from http://water.epa.gov/polwaste/nps/outreach /point1.cfm

U.S. Environmental Protection Agency (EPA). (2012c). *Statistics on the management of used and end-of-life electronics.* Retrieved from http://www.epa.gov/osw/conserve/materials/ecycling/manage.htm

U.S. Environmental Protection Agency (EPA). (2013a). *Earth Day and EPA history: The first Earth Day April 1970.* Retrieved from http://www.epa.gov/earthday/history.htm

U.S. Environmental Protection Agency (EPA). (2013b). *Food waste basics.* Retrieved from http://www.epa.gov/foodrecovery/

U.S. Environmental Protection Agency (EPA). (2013c). *Landfill methane outreach program.* Retrieved from http://www.epa.gov/lmop/basic-info/#a02

U.S. Environmental Protection Agency (EPA). (2014a). *Enhancing sustainable communities with green infrastructure.* Retrieved from http://www2.epa.gov/sites/production/files/2014-10/documents/green-infrastructure.pdf

U.S. Environmental Protection Agency (EPA). (2014b). *Household hazardous waste.* Retrieved from http://www.epa.gov/waste/conserve/materials/hhw.htm

U.S. Environmental Protection Agency (EPA). (2014c, July 9). *Mercury: Basic information.* Retrieved from http://www.epa.gov/hg/about.htm

U.S. Environmental Protection Agency (EPA). (2014d). *Municipal solid waste generation, recycling, and disposal in the United States: Facts and figures for 2012.* Retrieved from http://www.epa.gov/waste/nonhaz/municipal/pubs/2012_msw_fs.pdf

U.S. Environmental Protection Agency (EPA). (2014e). *TCSA Chemical Substance Inventory: Basic information.* Retrieved from http://www.epa.gov/oppt/existingchemicals/pubs/ tscainventory/basic.html

U.S. Environmental Protection Agency (EPA). (2014f). *Textiles*. Retrieved from http://www.epa.gov/wastes/conserve/materials/textiles.htm

USGA (2014). *Mercury in Stream Ecosystems*. Retrieved from http://water.usgs.gov/nawqa/mercury/majorfindings.html

U.S. Nuclear Regulatory Commission. (2007). Radioactive waste. *Backgrounder*. U.S. NRC. Retrieved from http://www.nrc.gov/reading-rm/doc-collections/fact-sheets/radwaste.pdf

University of Minnesota. (2010, May 25). Rising levels of dioxins from common soap ingredient in Mississippi River, study finds. *ScienceDaily*. Retrieved from www.sciencedaily.com/releases/2010/05/100518113236.htm

University of Montana. (2014). Bob Marshall wilderness. *Wilderness.net*. Retrieved from http://www.wilderness.net/NWPS/wildView?WID=64

Vanasselt, W. (2001). *No end to paperwork*. Washington, DC: World Resources Institute: EarthTrends. (Note: This website that is no longer active; see http://www.wri.org/our-work/project/earthtrends-environmental-information)

van de Kerk, G., & Manuel, A.R. (2009). *Sustainable society index. The encyclopedia of earth*. Retrieved from http://www.eoearth.org/article/Sustainable_Society_Index

van den Berg, A.E., & Custers, M.H.G. (2011). Gardening promotes neuroendocrine and affective restoration from stress. *Journal of Health Psychology, 16*(1), 3–11. doi: 10.1177/1359105310365577

van den Berg, A.E., & Heijne, M.T. (2005). Fear versus fascination: An exploration of emotional responses to natural threats. *Journal of Environmental Psychology, 25,* 261–272. doi: 10.1016/j.jenvp.2005.08.004

van den Berg, A.E., Koole, S.L., & van der Wulp, N.Y. (2003). Environmental preference and restoration: (How) are they related? *Journal of Environmental Psychology, 23,* 135–146. doi: 10.1016/S0272-4944(02)00111-1

van den Berg, A.E., & van den Berg, C.G. (2011). A comparison of children with ADHD in a natural and build setting. *Child: Care, Health, and Development, 37*(3), 430–439. doi: 10.1111/j.1365-2214.2010.01172.x

Vandenberg, L.N., Colborn, T., Hayes, T.B., Heindel, J.J., Jacobs, D.R., Jr., Lee, D.H., . . . Myers, J.P. (2012). Hormones and endocrine-disrupting chemicals: low-dose effects and nonmonotonic dose responses. *Endocrinology Review, 33*(3), 378–455. doi: 10.1210/er.2011-1050

van den Bergh, J.C.J.M., & Rietveld, P. (2004). Reconsidering the limits to world population: Meta-analysis and meta-prediction. *BioScience, 54,* 195–204. doi: 10.1641/0006-3568(2004)054[0195:rtltwp]2.0.co;2

van der Wal, A.J., Schade, H.M., Krabbendam, L., & van Vugt, M. (2013). Do natural landscapes reduce future discounting in humans? *Proceedings of the Royal Society B, 280,* 20132295. doi: 10.1098/rspb.2013.2295

van der Werff, E., Steg, L., & Keizer, K. (2013). The value of environmental self-identity: The relationship between biospheric values, environmental self-identity and environmental preferences, intentions and behaviour. *Journal of Environmental Psychology, 34,* 55–63. doi: 10.1016/j.jenvp.2012.12.006

van Riper, C.J., & Kyle, G.T. (2014). Understanding the internal processes of behavioral engagement in a national park: A latent variable path analysis of the value-belief-norm theory. *Journal of Environmental Psychology, 38*, 288–297. doi: 10.1016/j.jenvp.2014.03.002

Vantomme, D., Geuens, M., De Houwer, J., & De Pelsmacker, P. (2005). Implicit attitudes toward green consumer behaviour. *Psychologica Belgica, 45*(4), 217–239. doi: 10.5334/pb-45-4-2171016/j.jenvp.2014.03.002

van Vugt, M. (2002). Central, individual, or collective control? Social dilemma strategies for natural resource management. *American Behavioral Scientist, 45*(5), 783–800. doi: 10.1177/0002764202045005004

van Vugt, M., Griskevicius, V., & Schultz, P.W. (2014). Naturally green: Harnessing Stone Age psychological biases to foster environmental behavior. *Social Issues and Policy Review, 8*(1), 1–32. doi: 10.1111/sipr.12000

Vassar Historian. (2005). Ellen Swallow Richards. *Vassar Encyclopedia*. Retrieved from http://vcencyclopedia.vassar.edu/alumni/ellen-swallow-richards.html

Veblen, T. (1899). *Theory of the leisure class: An economic study in the evolution of institutions*. New York, NY: Macmillan.

Veerman, J.L., Healy, G.N., Cobiac, L.J., Vos, T., Winkler, E.A.H., Owen, N., & Dunstan, D.W. (2011). Television viewing time and reduced life expectancy: A life table analysis. *British Journal of Sports Medicine, 46*, 927–930. doi: 10.1136/bjsports-2011-085662

Verplanken, B., & Holland, R.W. (2002). Motivated decision making: Effects of activation and self-centrality of values on choices and behavior. *Journal of Personality and Social Psychology, 82*(3), 434–447. doi: 10.1037/0022-3514.82.3.434

Verplanken, B., & Wood, W. (2006). Interventions to break and create consumer habits. *Journal of Public Policy & Marketing, 25*(1), 90–103. doi: 10.1509/jppm.25.1.90

Versar, Inc. (1979). *Polychlorinated biphenyls, 1929–1979: Final report*. Washington, DC: U.S. Environmental Protection Agency.

Vine, D., Buys, L., & Morris, P. (2013). The effectiveness of energy feedback for conservation and peak demand: A literature review. *Open Journal of Energy Efficiency, 2*(1), 7–15. doi: 10.4236/ojee.2013.21002

Vroom, V.H. (1964). *Work and motivation*. Oxford, England: Wiley.

Wacker, M., & Holick, M.F. (2013). Sunlight and vitamin D: A global perspective for health. *DermatoEndocrinology, 5*(1), 51–108. doi: 10.4161/derm.24494

Wackernagel, M., & Rees, W. (1996). *Our ecological footprint: Reducing human impact on the earth*. Gabriola Island, BC: New Society Publishers.

Walsh, F. (2009). Human-animal bonds II: The role of pets in family systems and family therapy. *Family Process, 48*, 481–499. doi: 10.1111/j.1545-5300.2009.01297.x

Walter, P. (2009). Philosophies of adult environmental education. *Adult Education Quarterly, 60*(1), 3–25. doi: 10.1177/0741713609336109

Warner, M. (2013). *Pandora's lunchbox: How processed food took over the American meal*. New York, NY: Scribner.

Wason, P. C. (1960). On the failure to eliminate hypotheses in a conceptual task. *Quarterly Journal of Experimental Psychology, 12,* 129–140. doi: 10.1080/17470216008416717

Watson, I. (2013, May 30). China: The electronic wastebasket of the world. *CNN.* Retrieved from http://www.cnn.com/2013/05/30/world/asia/china-electronic-waste-e-waste/

Weart, S. (2013). *The discovery of global warming.* Retrieved from http://www.aip.org/history/climate/summary.htm

Webb, D., Soutar, G. N., Mazzarol, T., & Saldaris, P. (2013). Self-determination theory and consumer behavioural change: Evidence from a household energy-saving behaviour study. *Journal of Environmental Psychology, 35,* 59–66. doi: 10.1016/j.jenvp.2013.04.003

Weber, C. L., & Matthews, H. S. (2008). Food-miles and the relative climate impacts of food choices in the United States. *Environmental Science & Technology, 42*(10), 3508–3513. doi: 10.1021/es702969f

Weber, E. U. (1997). Perception and expectation of climate change: Precondition for economic and technological adaptation. In M. Bazerman, D. Messick, A. Tenbrunsel, & K. Wade-Benzoni (Eds.), *Psychological and ethical perspectives to environmental and ethical issues in management* (pp. 314–341). San Francisco, CA: Jossey-Bass.

Weber, E. U. (2006). Experience-based and description-based perceptions of long-term risk: Why global warming does not scare us (yet). *Climatic Change, 77,* 103–120. doi: 10.1007/s10584-006-9060-3

Weber, E. U., & Stern, P. C. (2011). Public understanding of climate change in the United States. *American Psychologist, 66,* 315–328. doi:10.1037/a0023253

Wei, F.-Y. F., Wang, Y. K., & Fass, W. (2014). An experimental study of online chatting and notetaking techniques on college students' cognitive learning from a lecture. *Computers in Human Behavior, 34,* 148–156. doi: 10.1016/j.chb.2014.01.019

Weiner, M. D., MacKinnon, T. D., & Greenberg, M. R. (2013). Exploring the gender gap and the impact of residential location on environmental risk tolerance. *Journal of Environmental Psychology, 36,* 190–201. doi: 10.1016/j.jenvp.2013.07.012

Weinhold, B. (2012). More chemicals show epigenetic effects across generations. *Environmental Health Perspectives, 120,* A228.

Weiss, B. (2000). Vulnerability of children and the developing brain to neurotoxic hazards. *Environmental Health Perspectives, 108*(3), 375–381.

Weiss, B. (2002). Sexually dimorphic nonreproductive behaviors as indicators of endocrine disruption. *Environmental Health Perspectives, 110,* 387–391.

Weiss, B. (2007). Can endocrine disruptors influence neuroplasticity in the aging brain? *NeuroToxicology, 28,* 938–950.

Weiss, B. (2011). Endocrine disruptors as a threat to neurological function. *Journal of the Neurological Sciences, 305,* 11–21. doi: 10.1016/j.jns.2011.03.014

Weiss, B., Amler, S., & Amler, R. W. (2004). Pesticides. *Pediatrics, 113*(4), 1030–1036.

Wells, N. M., & Lekies, K. S. (2006). Nature and the life course: Pathways from childhood nature experiences to adult environmentalism. *Children, Youth and Environments, 16,* 1–25. Retrieved from http://www.colorado.edu/journals/cye/16_1/16_1_01_NatureAndLifeCourse.pdf

Welte, T.H.L., & Anastasio, P.A. (2010). To conserve or not to conserve: Is status the question? *Environment and Behavior, 42*(6), 845–863. doi: 10.1177/0013916509348461

Werhan, P.O., & Groff, D.G. (2005). Research update: The wilderness therapy trail. *Parks & Recreation, 40*(11), 24.

Werner, C.M. (2003). Changing homeowners' use of toxic household products: A transactional approach. *Journal of Environmental Psychology, 23,* 33–45. doi: 10.1016/S0272-4944(02)00085-3

Werner, C.M., Cook, S., Colby, J., & Lim, H. (2012). "Lights out" in university classrooms: Brief group discussion can change behavior. *Journal of Environmental Psychology, 32*(4), 418–426. doi: 10.1016/j.jenvp.2012.07.001

Werner, C.M., & Stanley, C.P. (2011). Guided group discussion and the reported use of toxic products: The persuasiveness of hearing others' views. *Journal of Environmental Psychology, 31*(4), 289–300. doi: 10.1016/j.jenvp.2011.08.003

White, E.V., & Gatersleben, B. (2011). Greenery on residential buildings: Does it affect preferences and perceptions of beauty? *Journal of Environmental Psychology, 31*(1), 89–98. doi: 10.1016/j.jenvp.2010.11.002

White, M.P., Pahl, S., Ashbullby, K., Herbert, S., & Depledge, M.H. (2013). Feelings of restoration from recent nature visits. *Journal of Environmental Psychology, 35,* 40–51. doi: 10.1016/j.jenvp.2013.04.002

White, M.P., Smith, A., Humphryes, K., Pahl, S., Snelling, D., & Depledge, M. (2010). Blue space: The importance of water for preference, affect, and restorativeness ratings of natural and built scenes. *Journal of Environmental Psychology, 30,* 482–493. doi: 10.1016/j.jenvp.2010.04.004

White, R. (1998). Psychiatry and eco-psychology. In A. Lundberg (Ed.), *The environment and mental health: A guide for clinicians* (pp. 205–212). Mahwah, NJ: Erlbaum.

Whitehead, T.P., Metayer, C., Ward, M.H., Colt, J.S., Gunier, R.B., Deziel, N.C., . . . Buffler, P.A. (2014). Persistent organic pollutants in dust from older homes: Learning from lead. *American Journal of Public Health, 104*(7), 1320–1326. doi: 10.2105/AJPH.2013.301835

Whitmarsh, L. (2009). Behavioral responses to climate change: Asymmetry of intentions and impacts. *Journal of Environmental Psychology, 29,* 13–23. doi: 10.1016/j.jenvp.2008.05.003

Whitmarsh, L., & O'Neill, S. (2010). Green identity, green living? The role of pro-environmental self-identity in determining consistency across diverse pro-environmental behaviours. *Journal of Environmental Psychology, 30,* 305–314. doi: 10.1016/j.jenvp.2010.01.003

Wiener, J.L., & Mowen, J.C. (1986). Source credibility: On the independent effects of trust and expertise. In R.J. Lutz (Ed.), *Advances in consumer research* (Vol. 13, pp. 306–310). Provo, UT: Association for Consumer Research.

Willard, B., Upward, A., Leung, P., & Park, C. (2013). Towards a gold-standard benchmark for a truly sustainable business: Working draft of science-based KPIs and goals. *The Natural Step, Canada.* Retrieved from http://www.naturalstep.ca/sites/default/files/gold-standard-benchmark-latest-version.pdf

Williams, T.T. (1994). *An unspoken hunger: Stories from the field.* New York, NY: Pantheon Books.

Wilmot, E.G., Edwardson, C.L., Achana, F.A., Davies, M.J., Gorely, T., Gray, L.J., . . . Biddle, S.J.H. (2012). Sedentary time in adults and the association with diabetes, cardiovascular disease and death: Systematic review and meta-analysis. *Diabetologia, 55*(11), 2895–2905. doi: 10.1007/s00125-012-2677-z

Wilson, D.S., & Wilson, E.O. (2007). Rethinking the theoretical foundation of sociobiology. *The Quarterly Review of Biology, 82,* 327–348.

Wilson, E.O. (1984). *Biophilia: The human bond with other species.* Cambridge, MA: Harvard University Press.

Wilson, E.O. (1992). *The diversity of life.* New York, NY: W.W. Norton.

Wilson, E.O. (2006). *The creation: An appeal to save life on earth.* New York, NY: W.W. Norton & Co.

Wilson, S.J., & Lipsey, M.W. (2000). Wilderness challenge programs for delinquent youth: A meta-analysis of outcome evaluations. *Evaluation and Program Planning, 23,* 1–12. doi: 10.1016/S0149-7189(99)00040-3

Wines, M. (2014, January 5). Colorado river drought forces a painful reckoning for states. *New York Times.* Retrieved from http://www.nytimes.com/2014/01/06/us/colorado-river-drought-forces-a-painful-reckoning-for-states.html?_r=0

Winter, D.D. (1996) *Ecological psychology: Healing the split between planet and self.* New York, NY: Psychology Press.

Wiseman, M., & Bogner, F.X. (2003). A higher-order model of ecological values and its relationship to personality. *Personality and Individual Differences, 34,* 783–794. doi: 10.1016/S0191-8869(02)00071-5

Wohl, D.L., Arora, S., & Gladstone, J.R. (2004). Functional redundancy supports biodiversity and ecosystem function in a closed and constant environment. *Ecology, 85*(6), 1534–1540. doi: 10.1890/03-3050

Wolsko, C., & Hoyt, K. (2012). Employing the restorative capacity of nature: Pathways to practicing ecotherapy among mental health professionals. *Ecopsychology, 4*(1), 10–24. doi: 10.1089/eco.2012.0002

Wolsko, C., & Lindberg, K. (2013). Experiencing connection with nature: The matrix of psychological well-being, mindfulness, and outdoor recreation. *Ecopsychology, 5*(2), 80–91. doi:10.1089/eco.2013.0008

Wood, W., Tam, L., & Witt, M.G. (2005). Changing circumstances, disrupting habits. *Journal of Personality and Social Psychology, 88*(6), 918–933. doi: 10.1037/0022-3514.88.6.918

World Economic Forum. (2012, December 14). What if the world's soil runs out? *Time.* Retrieved at http://world.time.com/2012/12/14/what-if-the-worlds-soil-runs-out/

World Nuclear Association. (2012). *Radioactive wastes: Myths and realities.* Retrieved from http://www.world-nuclear.org/info/Nuclear-Fuel-Cycle/Nuclear-Wastes/Radioactive-Wastes—-Myths-and-Realities/

World Nuclear Association. (2013). *Nuclear power in the world today.* Retrieved from http://www.world-nuclear.org/info/Current-and-Future-Generation/Nuclear-Power-in-the-World-Today

World Nuclear Association. (2014). *Nuclear basics: Answering some of the key questions about nuclear energy.* Retrieved from http://www.world-nuclear.org/Nuclear-Basics/

World Wildlife Fund. (2014). *Threats: Soil erosion and degradation.* Retrieved from https://worldwildlife.org/threats/soil-erosion-and-degradation

Worldwatch Institute. (2011, October 11). *Global meat production and consumption continue to rise.* Retrieved from http://www.worldwatch.org/global-meat-production-and-consumption-continue-rise-1

Wright, R. (1995, August 28). The evolution of despair. *Time, 146*(9), 50.

Xiao, C., Dunlap, R. E., & Hong, D. (2013). The nature and bases of environmental concern among Chinese citizens. *Social Science Quarterly, 94*(3), 672–690. doi: 10.1111/j.1540-6237.2012.00934.x

Xiao, C., & Hong, D. (2010). Gender differences in environmental behaviors in China. *Population & Environment: Behavioral & Social Issues, 32*(1), 88–104. doi: 10.1007/s11111-010-0115-z

Xie, L., Kang, H., Xu, Q., Chen, M. J., Liao, Y., Thiyagarajan, M., . . . Nedergaard, M. (2013). Sleep drives metabolite clearance from the adult brain. *Science, 342,* 373–377. doi: 10.1126/science.1241224

Yates, K. F., Sweat, V., Yau, P. L., Turchiano, M. M., & Convit, A. (2012). Impact of metabolic syndrome on cognition and brain: A selected review of the literature. *Arterisclerosis, Thrombosis, and Vascular Biology, 32*(9), 2060–2067. doi: 10.1161/ATVBAHA.112.252759

Yerkes, R. M., & Dodson, J. D. (1908). The relation of strength of stimulus to rapidity of habit-formation. *Journal of Comparative Neurology and Psychology, 18,* 459–482. doi: 10.1002/cne.920180503

Yolton, K., Cornelius, M., Ornoy, A., McGough, J., Makris, S., & Schantz, S. (2014). Exposure to neurotoxicants and the development of attention deficit hyperactivity disorder and its related behaviors in childhood. *Neurotoxicology and Teratology, 44,* 30–45. doi: 10.1016/j.ntt.2014.05.003

Young, H. P. (2009). Innovation diffusion in heterogeneous populations: Contagion, social influence, and social learning. *American Economic Review, 99*(5), 1899–1924. doi: 10.1257/aer.99.5.1899

Young, L. R., & Nestle, M. (2012). Reducing portion sizes to prevent obesity: A call to action. *American Journal of Preventive Medicine, 43*(5), 565–568. doi: 10.1016/j.amepre.2012.07.024

Zakin, S. (1993). *Coyotes and town dogs: Earth First! and the environmental movement.* Tucson, AZ: University of Arizona Press.

Zalasiewicz, J., Williams, M., Smith, A., Barry, T. L., Coe, A. L., Brown, P. R., . . . Stone, P. (2008). Are we now living in the Anthropocene? *GSA Today, 18*(2), 4–8. doi: 10.1130/GSAT01802A.1

Zelenski, J. M., & Nisbet, E. K. (2014). Happiness and feeling connected: The distinct role of nature relatedness. *Environment and Behavior, 46*(1), 3–23. doi: 10.1177/0013916512451901

Zelezny, L.D., Chua, P.P., & Aldrich, C. (2000). Elaborating on gender differences in environmentalism. *Journal of Social Issues, 56*, 443–458. doi: 10.1111/0022-4537.00177

Zhao, E., Tranovich, M.J., & Wright, V.J. (2014). The role of mobility as a protective factor of cognitive functioning in aging adults: a review. *Sports Health, 6*, 63–69. doi: 10.1177/1941738113477832

Zimbardo, P.G., & Boyd, J.N. (1999). Putting time in perspective: A valid, reliable individual-differences metric. *Journal of Personality and Social Psychology, 77*, 1271–1288. doi: 10.1037/0022-3514.77.6.1271

Zuk, M. (2013). *Paleofantasy: What evolution really tells us about sex, diet, and how we live*. New York, NY: W.W. Norton & Company.

Name Index

Note: Italicized page numbers indicate a figure on the corresponding page.

Subject Index

Note: *Italicized* page numbers indicate a table or figure on the corresponding page.